Advanced Array Signal Processing for Target Imaging and Detection (Second Edition)

Advanced Array Signal Processing for Target Imaging and Detection (Second Edition)

Guest Editors

Jiahua Zhu
Xuebo Zhang
Wei Guo
Xiaotao Huang
Hongtu Xie

MDPI

Basel • Beijing • Wuhan • Barcelona • Belgrade • Novi Sad • Cluj • Manchester

Guest Editors

Jiahua Zhu
College of Meteorology and Oceanography
National University of Defense Technology
Changsha
China

Xuebo Zhang
College of Physics and Electronic Engineering
Northwest Normal University
Lanzhou
China

Wei Guo
College of Meteorology and Oceanography
National University of Defense Technology
Changsha
China

Xiaotao Huang
College of Electronic Science and Technology
National University of Defense Technology
Changsha
China

Hongtu Xie
School of Electronics and Communication Engineering
Sun Yat-sen University
Guangzhou
China

Editorial Office
MDPI AG
Grosspeteranlage 5
4052 Basel, Switzerland

This is a reprint of the Special Issue, published open access by the journal *Remote Sensing* (ISSN 2072-4292), freely accessible at: https://www.mdpi.com/journal/remotesensing/special_issues/Q12K62GKEW.

For citation purposes, cite each article independently as indicated on the article page online and as indicated below:

Lastname, A.A.; Lastname, B.B. Article Title. *Journal Name* **Year**, *Volume Number*, Page Range.

ISBN 978-3-7258-4605-4 (Hbk)
ISBN 978-3-7258-4606-1 (PDF)
https://doi.org/10.3390/books978-3-7258-4606-1

Contents

Contents

About the Editors

Jiahua Zhu

Jiahua Zhu (Senior Member, IEEE) received a B.S. degree in Electronic Engineering and a Ph.D. degree in Information and Communication Engineering from the National University of Defense Technology, Changsha, China, in 2012 and 2018, respectively. He is currently an Associate Professor and Master's Supervisor at the College of Meteorology and Oceanography, National University of Defense Technology. From November 2015 to November 2017, he was a Visiting Ph.D. Student with the School of Engineering, RMIT University, and the Department of Electrical and Electronic Engineering, University of Melbourne, Australia. His current research interests include waveform design and target detection for radar and sonar. He received the IEEE ICCCAS Outstanding Young Scholar Award, the SPIE ICSPS Young Researcher Award, three Best/Excellent Paper Awards, and the outstanding Ph.D. degree thesis of the Chinese People's Liberation Army (PLA) award, 2020. He is a Remote Sensing Technical Committee Co-Chair of IEEE Oceanic Engineering Society, a Program Co-Chair of IEEE ICSIP and IEEE ICCCAS, an Editorial Board Member of *IEEE Signal Processing Letters* and *IET Signal Processing*, a Topical Advisory Panel Member/Lead Guest Editor of *Remote Sensing*, and the Guest Editor of the *IEEE Journal of Selected Topics in Applied Earth Observations and Remote Sensing*.

Xuebo Zhang

Xuebo Zhang (Senior Member, IEEE) received a B.Eng. degree in Electronic Engineering and a Ph.D. degree in Underwater Acoustic Engineering from the Naval University of Engineering, Wuhan, China, in 2009 and 2014, respectively. He worked at the Underwater Antagonizing Laboratory, Zhanjiang, China, from 2014 to 2020, followed by a position with Northwest Normal University, Lanzhou, China. Dr. Zhang currently holds the role of Technology Committee Co-Chair for the IEEE Oceanic Engineering Society. He has served in various capacities, such as Session Chair, Invited Speaker, and Special Session Chair, for several conferences. He now serves as the Guest Editor for the Special Issues of the *IEEE Journal of Selected Topics in Applied Earth Observations and Remote Sensing* and *IET Electronics Letters*. His contributions have earned him recognition, including the titles of "Outstanding Reviewer" and "Outstanding Editorial Board Member" for a prestigious journal, along with the Science and Technology Progress First Prize and other honors.

Wei Guo

Wei Guo received her B.E. and M.E. degrees in Underwater Acoustic Engineering from Harbin Engineering University, Harbin, China, in 2013 and 2015, respectively. In 2020, she received her doctorate in Underwater Acoustic Engineering from Harbin Engineering University. In the same year, she continued her research on underwater acoustics as a research associate at the National University of Defense Technology. Her major area of focus is underwater acoustic signal processing, including target detection and identification.

Xiaotao Huang

Xiaotao Huang (Senior Member, IEEE) received his B.S. and Ph.D. degrees in Information and Communications Engineering from the National University of Defense Technology, Changsha, China, in 1990 and 1999, respectively. He is currently a Professor at the National University of Defense Technology. His research interests include radar theory, signal processing, and radio frequency signal suppression.

Hongtu Xie

Hongtu Xie (Member, IEEE) received his B.S. degree in Electronic Information Engineering from Hunan University, Changsha, China, his M.S. degree in Electronic Science and Technology, and his Ph.D. degree in Information and Communication Engineering from the National University of Defense Technology, Changsha, in 2008, 2010 and 2015, respectively. He is currently an Associate Professor and an M.S. Supervisor at the School of Electronics and Communication Engineering, Sun Yat-sen University, Guangzhou, China, and the Judge's Experts of the National Natural Science Foundation of China, the Guangdong Basic and Applied Basic Research Foundation, and the Shenzhen Science and Technology Innovation Committee. In his research fields, he is the principal investigator and an important participant in more than 20 projects in receipt of grants and funding from different governmental sources, including mainly the National Natural Science Foundation of China, the Guangdong Basic and Applied Basic Research Foundation, and the Shenzhen Science and Technology Innovation Committee. He has authored or co-authored more than 80 refereed journals and conference papers and has granted 9 and filed 20 national invention patents in China. His research interests include high-resolution synthetic aperture radar imaging and target detection and recognition. Dr. Xie received the outstanding Ph.D. degree thesis of the Chinese People's Liberation Army award in 2018 and third prize in the Teaching Competition for the Young Teacher of the School of Electronics and Communication Engineering in 2020. He is the Promotion Chair of the 2024 IEEE International Conference on Software Systems and Information Processing and a Guest Editor for *Remote Sensing*.

Preface

In recent decades, considerable progress has been made in the theory and methodology of array signal processing for airborne, ground, marine, and underwater target detection and imaging. However, further development faces increasing challenges regarding improving target illumination performance due to the influences of clutter, interference, and noise. It is valuable to attain a comprehensive understanding of current array signal processing theory and approaches for detecting various targets in the air, on the land, in the sea, and underwater, thus solving future problems exerted by the new application requirements.

This Reprint is the second edition of "Advanced Array Signal Processing for Target Imaging and Detection", and contains an updated selection of 14 papers within the research field of radar/sonar array signal processing and analysis, which are of interest to readers working in the areas of waveform design, detection, and imaging, as well as in target recognition, etc. We hope that this Reprint delivers exciting discoveries from the frontier of this field and inspires further investigation and discussion.

Jiahua Zhu, Xuebo Zhang, Wei Guo, Xiaotao Huang, and Hongtu Xie
Guest Editors

remote sensing

MDPI

Article

Low-Sidelobe Imaging Method Utilizing Improved Spatially Variant Apodization for Forward-Looking Sonar

Lu Yan [1,*], Juan Yang [1], Feng Xu [1] and Shengchun Piao [2]

[1] Institute of Acoustics, Chinese Academy of Sciences, Beijing 100190, China; yangjuan@mail.ioa.ac.cn (J.Y.); xf@mail.ioa.ac.cn (F.X.)

[2] College of Underwater Acoustic Engineering, Harbin Engineering University, Harbin 150001, China; piaoshengchun@hrbeu.edu.cn

* Correspondence: yanlu@mail.ioa.ac.cn

Abstract: For two-dimensional forward-looking sonar imaging, high sidelobes significantly degrade the quality of sonar images. The cosine window function weighting method is often applied to suppress the sidelobe levels in the angular and range dimensions, at the expense of the main lobe resolutions. Therefore, an improved spatially variant apodization imaging method for forward-looking sonar is proposed, to reduce sidelobes without degrading the main lobe resolution in angular-range dimensions. The proposed method is a nonlinear postprocessing operation in which the raw complex-valued sonar image produced by a conventional beamformer and matched filter is weighted by a spatially variant coefficient. To enhance the robustness of the spatially variant apodization approach, the array magnitude and phase errors are calibrated to prevent the occurrence of beam sidelobe increase prior to beamforming operations. The analyzed results of numerical simulations and a lake experiment demonstrate that the proposed method can greatly reduce the sidelobes to approximately −40 dB, while the main lobe width remains unchanged. Moreover, this method has an extremely simple computational process.

Keywords: forward-looking imaging sonar; spatially variant apodization; low-sidelobe; main lobe resolution; image quality

1. Introduction

Imaging sonars can directly present underwater scene information, providing strong support for underwater target detection and recognition [1–3]. Imaging sonars can be divided by their detection coverage and system construction type into: forward-looking [4], side-looking [5] and down-looking solutions [6]. In this paper, two-dimensional forward-looking sonar (FLS) is discussed, and a forward-looking sonar imaging method is proposed to reduce sidelobes without sacrificing main lobe resolution.

Generally, for FLS, the angular dimension is obtained via conventional beamforming for a uniform linear array, and the range dimension is obtained via match filtering for a linear frequency modulated signal [7,8]. For the output envelopes of conventional beamforming and match filtering, the main-to-side ratio of both is approximately −13 dB [9,10]. Typically, in practical engineering, the ratio needs to be approximately −40 dB to acquire high-quality sonar images. In practice, high sidelobe levels from a strong target echo can weaken or even completely mask the main lobe of a smaller target echo, so additional sidelobe suppression for FLS images is essential [11,12].

Conventional amplitude weighting cosine window functions are commonly applied to reduce the sidelobe levels with angular and range dimensions for FLSs, which may lead to the degradation of the angular and range resolutions. Compared with conventional beamforming, sonar images are obtained with lower sidelobe levels and narrower main lobe widths by utilizing adaptive beamforming (ADBF) [13–16]. Nevertheless, the performance of the ADBF methods degrades in the presence of array model mismatch due to imperfect array channels and inaccurately estimated data covariance matrices; hence, several robust adaptive beamforming algorithms have been investigated. The calculation of the adaptive weighed vector in robust adaptive beamforming is equivalent to the second-order cone programming problems, which has high complexity; furthermore, the prior knowledge of the programming problem is difficult to obtain due to complicated underwater propagation [17]. Moreover, for forward-looking imaging sonar, these existing adaptive beamforming approaches are not applicable to the output in the range dimension. Thus, robust adaptive beamforming cannot be easily used to obtain lower sidelobe levels for forward-looking sonar imaging.

Compressive sensing (CS) has been proposed for use in arrays to improve the resolution of the spatial spectrum of underwater source locations in passive sonar. Compressed beamforming has also proven to be an effective approach for improving sonar image quality without increasing the array size [18–21]. Additionally, a few transducers can be implemented in compressed beamforming to obtain a similar image quality to that produced by a larger-aperture array. Nevertheless, the performance of CS in underwater acoustics imaging will be limited, because prior knowledge cannot be properly matched to the reverberation-rich scenes of underwater environments.

Deconvolution algorithms have been proposed to improve the angular resolution in spatial spectrum estimation by processing beamformer output data. These methods were subsequently extended to high-frequency sonar imaging to achieve higher resolution in both the angular and range dimensions simultaneously [7,8,12,22–27]. However, the deconvolution results require several iterations, and the performance is related to the number of iterations. Thus, the application of this method in engineering is hindered by its high computational complexity.

The spatially variant apodization (SVA) method was proposed and applied for synthetic aperture radar (SAR) imaging. In this method, the properties of the cosine-on-pedestal weighing functions are utilized to achieve range dimension sidelobe suppression without broadening the main lobe [28–31]. Furthermore, it is regarded as a postprocessing method with a lower computational cost in the image domain. Moreover, it can preserve the raw sonar image. However, the principle of SAR imaging is different from that of forward-looking sonar imaging, and the frequency domain characteristics of windows are utilized for the SVA method in the field of SAR imaging. For forward-looking sonar, the multibeams are obtained by beamforming at the sonar system receiving end, and the sidelobes need to be reduced in the beam domain. An image is formed per frame in forward-looking imaging sonar, while an SAR image is generated by platform continuous motion fusion. So, the raw image data properties and image coordinate system of the forward-looking imaging sonar are different from those of SAR. Hence, the SVA method used in SAR cannot be directly applied to imaging sonar sidelobe reduction. Nevertheless, the use of the SVA method has rarely been reported in the forward-looking imaging sonar field, and the original SVA method is less robust to existing array amplitude phase errors. In this paper, we analyze the peculiarities of the beamforming output and matched filter output using cosine window weighting for forward-looking imaging sonar, combined with the SVA method to reduce sidelobes in angular-range dimensions. Additionally, array amplitude phase errors are calibrated to enhance the robustness of the original SVA method.

To obtain lower sidelobe levels without degrading the main lobe resolution in angular-range dimensions, an improved spatially variant apodization forward-looking sonar imaging method is proposed in this article. The proposed method is a nonlinear postprocessing operation in which the raw complex-valued sonar image data produced by a conventional beamformer and matched filter are weighted by a spatially variant coefficient. To enhance the robustness of the spatially variant apodization approach, the array magnitude and phase errors are calibrated to avoid beam sidelobe heightening prior to beamforming operations. The analyzed results of the numerical simulations and a lake experiment demonstrate that the proposed method can greatly reduce the sidelobes to approximately −40 dB while the main lobe width remains unchanged. Moreover, this method uses an extremely simple computational procedure.

The rest of the paper is organized as follows. In Section 2, the conventional forward-looking sonar imaging procedure is described. The improved forward-looking sonar imaging method, which utilizes spatially variant apodization combined with array magnitude and phase error calibration, is described in Section 3. Analyses of the simulation and lake experiment results are presented to demonstrate the effectiveness of the proposed approach in Section 4. In Section 5, the conclusions are provided.

2. Conventional Variant Apodization for Forward-Looking Sonar Imaging

In this section, the signal model of forward-looking imaging sonar is first described. Then, conventional variant apodization methods are introduced.

2.1. Signal Model

We assume that forward-looking imaging sonar is composed of a single transmitter and a receiving uniform linear array (ULA) with N elements at a half-wavelength spacing. For simplicity, the narrowband far-field underwater targets are assumed to be ideal point scatterers. The transmission loss and the medium absorption loss associated with free-space propagation are neglected, and the underwater reverberations and noise are ignored.

For a target, the discrete baseband echo received by the nth receiver can be expressed as follows:

$$x_n(t) = \sigma_0 s_0(t) \exp(-\mathrm{j}2\pi f_0 \tau_n) \tag{1}$$

where σ_0 denotes the complex backscattering coefficient of the target and $s_0(t)$ denotes the envelope of the transmitted waveforms. τ_n is the propagation delay time between the scatterer and the nth receiver, and f_0 denotes the carrier frequency. The transmitted linear modulated frequency (LFM) waveforms are expressed as follows:

$$s_0(t) = \mathrm{rect}\left(\frac{t}{T}\right) \exp\left[\mathrm{j}2\pi\left(-\frac{B}{2}t + \frac{1}{2}\frac{B}{T}t^2\right)\right] \exp(\mathrm{j}2\pi f_0 t) \tag{2}$$

where $\mathrm{rect}(\cdot)$ denotes the rectangular window function, and B and T represent the bandwidth and the chirp duration, respectively.

The time–domain conventional beamforming (CBF) output is given by:

$$\begin{aligned} y_q(t) &= \sum_{n=1}^{N} w_n^* x_n(t) \\ &= \sigma_0 s_0(t) \sum_{n=1}^{N} w_n^* \exp(-\mathrm{j}2\pi f_0 \tau_n) \end{aligned} \tag{3}$$

where w_n denotes the complex weight, $w_n = \exp[-\mathrm{j}2\pi f_0 n d \sin\theta_0 / c] / \sqrt{N}$, d represents the spacing between adjacent receiving elements, $(\cdot)^*$ denotes the conjugate, θ_0 denotes the beam pointing, and c is the sound speed underwater. Then, the beamforming complex output can be expressed as follows:

$$y_q(t) = \sigma_0 s_0(t) \frac{\mathrm{sinc}\left[\frac{\pi N d (u_\theta - u_{\theta 0})}{\lambda}\right]}{\mathrm{sinc}\left[\frac{\pi d (u_\theta - u_{\theta 0})}{\lambda}\right]} = \sigma_0 s_0(t) BP(u_\theta) \qquad -1 \le u_\theta \le 1 \tag{4}$$

where $u_\theta = \sin\theta$, θ represents the beam steering angle, $\mathrm{sinc}(x) = \frac{\sin x}{x}$, λ is the wavelength, and $BP(u_\theta)$ denotes the array beam pattern expressing the PSF in the angular dimension of sonar images. Equation (4) indicates that the angular dimension of the sonar image is only related to the angle [9].

In the range dimension of the sonar image, assuming that the angular dimension and range dimension are independent of each other, the matched filter output envelope of the qth beam can be expressed as follows:

$$z_{Bq}(t) = \sigma_0 T \frac{\sin \pi B(1-\frac{|t|}{T})t}{\pi B t} \mathrm{rect}(\frac{t}{2T}) = \sigma_0 \boldsymbol{R}(t) \tag{5}$$

When t is less than or equal to T, $\boldsymbol{R}(t)$ is approximated as

$$\boldsymbol{R}(t) = T \sin\mathrm{c}(\pi B t) \mathrm{rect}(\frac{t}{2T}) \tag{6}$$

where $\boldsymbol{R}(t)$ denotes the PSF in the range dimension of the sonar images. Equation (6) indicates that the envelope of the matched filter output approximately follows the sinc function. Therefore, the sidelobe levels are high in two dimensions of the sonar images.

Sidelobe reduction is accomplished via amplitude weighting applied angular dimension echoes or range dimension echoes, and the cosine-on-pedestal weighting functions are expressed as follows:

$$\begin{aligned} w(n) &= 1 + 2w\cos(\tfrac{2\pi n}{N-1}) \\ &= 1 + we^{(\mathrm{j}\frac{2\pi n}{N-1})} + we^{(-\mathrm{j}\frac{2\pi n}{N-1})}, \quad 0 \le w \le 1/2, n = -(N-1)/2, \cdots, 0, \cdots, (N-1)/2 \end{aligned} \tag{7}$$

Similarly, any unweighted sonar image angular-range dimension sidelobe can be suppressed using one of the families of cosine-on-pedestal weighting functions. However, lower sidelobes have been achieved at the expense of the main lobe width by cosine-on-pedestal weighting functions.

2.2. Dual Apodization and Multiapodization

To suppress sidelobe levels while preserving the main lobe resolution, nonlinear apodization operators are proposed, i.e., dual and multiapodization methods. For simplicity of analysis, since there is only scatter in the echoes, the beam pattern is used to calculate the azimuth beamforming output.

Taking the angular dimension beam pattern of imaging sonar as an example, the dual apodization (DA) method calculates the beam pattern of the receiving array by using two window functions, first using rectangular window processing and second using cosine or Chebyshev window processing. Then, at each spatial sampling point, the minimum value of the two processing responses is selected as the final response output value. The DA calculation process is expressed as follows:

$$\widetilde{\boldsymbol{B}}_{DA}(u_{\theta i}) = min[|\boldsymbol{B}_{w1}(u_{\theta i})|, |\boldsymbol{B}_{w2}(u_{\theta i})|] \qquad i = 1, 2, 3, \cdots, M_{u_\theta} \tag{8}$$

where $\boldsymbol{B}_{w1}(u_{\theta i})$ represents the beamforming complex output obtained using uniform weighting, and $\boldsymbol{B}_{w2}(u_{\theta i})$ represents the beamforming complex output obtained using Hanning window weighting.

DA can be extended to tri-apodization (TA) by window function weighting, and the output result of TA can be obtained by repeating the above steps. Then, the TA calculation process can be expressed as follows:

$$\widetilde{\boldsymbol{B}}_{TA}(u_{\theta i}) = min[|\boldsymbol{B}_{w1}(u_{\theta i})|, |\boldsymbol{B}_{w2}(u_{\theta i})|, |\boldsymbol{B}_{w3}(u_{\theta i})|] \qquad i = 1, 2, 3, \cdots, M_{u_\theta} \tag{9}$$

where $\boldsymbol{B}_{w3}(u_{\theta i})$ represents the beamforming complex output obtained using the Hamming window weighting calculation. We consider a ULA consisting of 48 array elements, where the spacing between adjacent array elements is a half wavelength. The beam patterns obtained by using uniform weighting, Hanning window weighting, Hamming window weighting, DA and TA are shown in Figure 1. In Equation (7), Hanning weighting is a cosine window function when $w = 0.5$, and Hamming weighting is a special case of cosine-on-pedestal window function when $w = 0.43$. Evidently, the DA and TA have narrow main lobe widths of the sinc function and small sidelobes of the cosine-on-pedestal window function. However, with the larger number of weighting functions, the ideal sidelobe levels are still not reached, and these methods only make use of the amplitude information of the beam pattern, ignoring complex information [29].

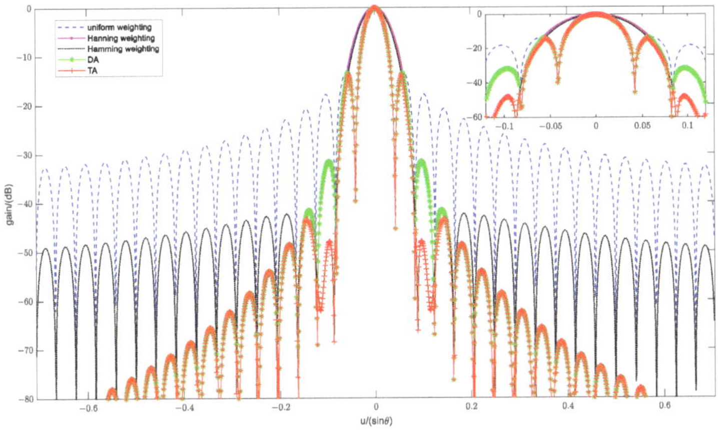

Figure 1. Comparison of the beam patterns with different weighing methods.

The complex DA (CDA) method makes full use of the amplitude and sign information of the real part and the imaginary part of the beamforming output. In the main lobe region, the sign of the uniform weighted real part (imaginary part) is the same as that of the cosine window weighted real part (imaginary part). In the sidelobe region, the sign of the uniform weighted real part (imaginary part) is opposite to that of the cosine window weighted real part (imaginary part); then, the processing method of CDA is expressed as:

$$I_{cda}(u_{\theta i}) = \begin{cases} 0, & I(u_{\theta i})I_w(u_{\theta i}) < 0 \\ \min(|I(u_{\theta i})|, |I_w(u_{\theta i})|), & \text{others} \end{cases} \tag{10}$$

where $I(u_{\theta i})$ represents the real part of the uniform weighting beamforming output and $I_w(u_{\theta i})$ represents the real part of the cosine base window function weighting beamforming output. Moreover, the imaginary part can be determined using Equation (10). The simulation conditions are the same as those in Figure 1, and the beam pattern response with different weighting processes is shown in Figure 2.

As observed from Figure 2, compared with other weighting methods, CDA obtains significantly lower sidelobes below approximately −40 dB without degrading the main lobe resolution (the position of the first sidelobe obtained by CDA labeled using the diamond shape), and has the advantages of simplicity and easy implementation.

Similarly, the matched filter output in the range dimension also reduces the sidelobes by using CDA at each sampling time. Considering the LFM signal bandwidth of 8 kHz and pulse width of 10 ms, there is a single target point, and simulation results with different weighting methods are shown in Figure 3. It is clear that the matched filter output has lower sidelobe levels of less than −40 dB without decreasing the main lobe resolution.

CDA has a dramatic positive impact on the visual appearance of sonar images in two dimensions but at the expense of an increased computational burden. Additionally, the envelope of the PSF deviates from the theoretical value; thus, such methods are less robust with respect to existing array amplitude phase errors.

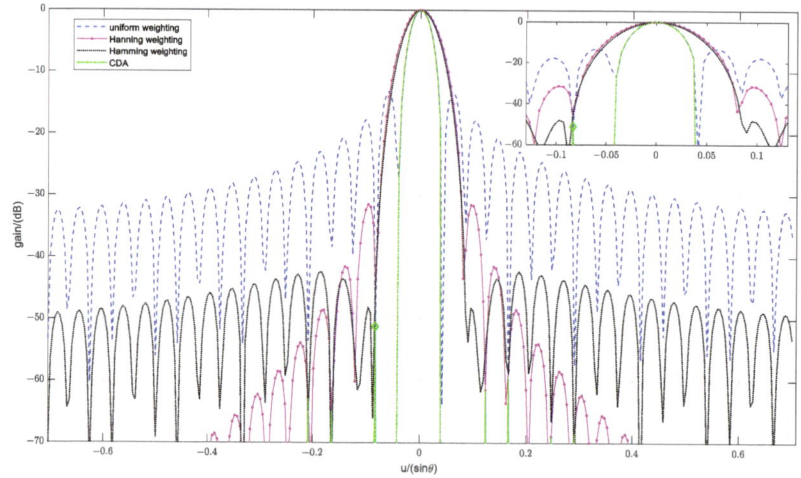

Figure 2. Comparison of the beam patterns with different weighing methods.

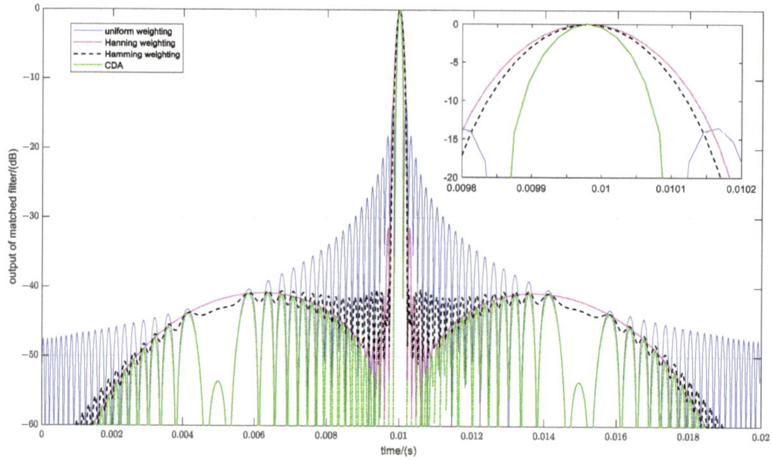

Figure 3. Output of the matched filter with different weighing methods.

3. Improved Spatially Variant Apodization Forward-Looking Sonar Imaging Method

Sidelobe improvements can be achieved by using cosine window weighting, which decreases the main lobe resolution. DA and multiapodization can be performed to obtain a narrow main lobe; these methods can also simultaneously achieve improved sidelobes if larger numbers of weighting windows are used. Furthermore, the CDA method achieves lower sidelobes of approximately −40 dB, while also preserving the main lobe resolution. However, this method has a higher computational complexity and is less robust. Thus, the improved SVA forward-looking sonar imaging algorithm is proposed to reduce the sidelobes without degrading the main lobe resolution in angular-range dimensions.

The proposed method is a nonlinear postprocessing operation in which the raw complex-valued sonar image produced by a conventional beamformer and matched filter is weighted by a spatially variant coefficient. To enhance the robustness of the spatially variant apodization approach, the array magnitude and phase errors are calibrated to avoid beam sidelobe heightening prior to beamforming operation. The analyzed results of numerical simulations and a lake experiment demonstrate that the proposed method can greatly reduce the sidelobes to approximately −40 dB and simultaneously maintain the main lobe width. Furthermore, this method is extremely simple computationally.

3.1. SVA for Two-Dimensional Forward-Looking Sonar Imaging

To preserve the main lobe width and suppress the sidelobes, the SVA exploits the sinc function characteristics and the special properties of raised-cosine weighting functions, allowing each spatial-time sampling location in a sonar image to use its amplitude weighting function.

By combining the expression of the cosine-on-pedestal weighting window functions (7) with the beamforming output (3), the beamforming output with cosine-on-pedestal weighting can be expressed as follows:

$$
\begin{aligned}
\boldsymbol{B}_s(u_\theta) &= \frac{1}{N}\sum_{n=-(N-1)/2}^{(N-1)/2}\left[1+we^{\mathrm{j}\frac{2\pi n}{N-1}}+we^{-\mathrm{j}\frac{2\pi n}{N-1}}\right]^{*} e^{\mathrm{j}nd\frac{2\pi}{\lambda}u_\theta} \\
&= \frac{1}{N}\sum_{n=-(N-1)/2}^{(N-1)/2} e^{\mathrm{j}nd\frac{2\pi}{\lambda}u_\theta} + \frac{1}{N}w\sum_{n=-(N-1)/2}^{(N-1)/2} \mathrm{e}^{\mathrm{j}2n\pi\left(\frac{du_\theta}{\lambda}+\frac{1}{N-1}\right)} + \frac{1}{N}w\sum_{n=-(N-1)/2}^{(N-1)/2} \mathrm{e}^{\mathrm{j}2n\pi\left(\frac{du_\theta}{\lambda}-\frac{1}{N-1}\right)} \\
&= \frac{\sin\mathrm{c}\left(\frac{Ndu_\theta}{\lambda}\right)}{\sin\mathrm{c}\left(\frac{du_\theta}{\lambda}\right)} + w\frac{\sin\mathrm{c}\left(\frac{Ndu_\theta}{\lambda}+\frac{N}{N-1}\right)}{\sin\mathrm{c}\left(\frac{du_\theta}{\lambda}+\frac{1}{N-1}\right)} + w\frac{\sin\mathrm{c}\left(\frac{Ndu_\theta}{\lambda}-\frac{N}{N-1}\right)}{\sin\mathrm{c}\left(\frac{du_\theta}{\lambda}-\frac{1}{N-1}\right)} \\
&= \frac{\sin\mathrm{c}\left(\frac{Ndu_\theta}{\lambda}\right)}{\sin\mathrm{c}\left(\frac{du_\theta}{\lambda}\right)} + w\frac{\sin\mathrm{c}\left[\frac{Nd}{\lambda}\left(u_\theta+\frac{\lambda}{d(N-1)}\right)\right]}{\sin\mathrm{c}\left[\frac{d}{\lambda}\left(u_\theta+\frac{\lambda}{d(N-1)}\right)\right]} + w\frac{\sin\mathrm{c}\left[\frac{Nd}{\lambda}\left(u_\theta-\frac{\lambda}{d(N-1)}\right)\right]}{\sin\mathrm{c}\left[\frac{d}{\lambda}\left(u_\theta-\frac{\lambda}{d(N-1)}\right)\right]}
\end{aligned} \tag{11}
$$

Using Equation (4), Equation (11) can be expressed as follows:

$$
\boldsymbol{B}_s(u_\theta) = \boldsymbol{B}(u_\theta) + w\boldsymbol{B}\left(u_\theta+\frac{\lambda}{d(N-1)}\right) + w\boldsymbol{B}\left(u_\theta-\frac{\lambda}{d(N-1)}\right) \tag{12}
$$

According to Equation (12), the beamforming output weighted by cosine-on-pedestal functions is equivalent to the superposition of the conventional beamforming output uniform weighting and weighted shifted conventional beamforming output uniform weighting, in which the beam domain data are complex data composed of real and imaginary parts, i.e., $\boldsymbol{B}_s(u_\theta) = \boldsymbol{I}_s(u_\theta) + \mathrm{j}\boldsymbol{Q}_s(u_\theta)$.

To effectively suppress the beam sidelobes, the SVA algorithm attempts to estimate the optimal solution w for each special location. The real and imaginary parts of the beam domain data cannot be used to obtain the minimum value when $\boldsymbol{B}_s(u_\theta)$ is simultaneously and directly calculated. However, the real and imaginary parts can be calculated separately. Then, the minimum value of the real part can be solved under the constraint of the $|\boldsymbol{I}_s(u_\theta)|^2$ minimum of the real part, and the minimum value of the imaginary part can be solved in the same manner. Then, the minimum value of the beam domain data can be obtained.

In the angular dimension, sample points in the beam domain are used as variables to represent the beam domain data.

$$
\boldsymbol{B}_s(m) = \boldsymbol{B}(m) + w(m)\boldsymbol{B}(m+k_a) + w(m)\boldsymbol{B}(m-k_a) \tag{13}
$$

where k_a denotes the shift sampling number and is calculated with $k_a = \left\lfloor \frac{\lambda}{d_B(N-1)d} \right\rfloor$. In practical applications, $k_a \le 10$. Here, d_B is the u_θ spacing of adjacent beams, $\lfloor\cdot\rfloor$ denotes the downward integer operation, and $\boldsymbol{B}_s(m) = \boldsymbol{I}_s(m)+\mathrm{j}\boldsymbol{Q}_s(m)$, $\boldsymbol{B}(m) = \boldsymbol{I}(m)+\mathrm{j}\boldsymbol{Q}(m)$.

The SVA algorithm uses the amplitude weighting method to obtain high-quality images, and its aim is to minimize the sidelobe while preserving the main lobe. For the real part of the beam domain data, the minimum $|\boldsymbol{I}_s(m)|^2$, subject to $0 \le w_I(m) \le 1/2$, is applied in the SVA algorithm to estimate the optimal solution $w_I(m)$, which is expressed as follows:

$$\begin{cases} \min\limits_{w_I(m)} \left\{ |I_s(m)|^2 \right\} \\ \text{s.t. } 0 \le w_I(m) \le 1/2 \end{cases} \tag{14}$$

$I(m)$ is assumed to be a real-valued component, and $w_I(m)$ can also be obtained by directly solving for $I_s(m) = 0$. Then, solving for $w_I(m)$ minimizes $|I_s(m)|^2$, and $w_I(m)$ can be determined with the unconstrained $w_I(m)$ as follows:

$$w_I(m) = \frac{-I(m)}{I(m+1) + I(m-1)} \tag{15}$$

By inserting Equation (15) into Equation (13) and constraining $w_I(m)$ in (15) to $0 \le w_I(m) \le 1/2$, and combining the properties of the CDA method, the output beam domain of the sonar image by the SVA can be written as follows:

$$I_s(m) = \begin{cases} I(m), & w_I(m) < 0 \\ 0, & 0 \le w_I \le 1/2 \\ I(m)+(1/2)[I(m+k_a) + I(m-k_a)], & w_I(m) > 1/2 \end{cases} \tag{16}$$

Equation (16) is also independently applied to the imaginary part, and $w_Q(m)$ and $Q_s(m)$ are obtained. When $w_I(m) > 1/2$, $I_s^2(m)$ is expressed as:

$$I_s^2(m) = I^2(m)\left[1 - \frac{1}{2w_I(m)}\right]^2 \le I^2(m) \tag{17}$$

Therefore, the beam domain SVA algorithm applies the optimal weight at the sampling point of each beam on the condition of minimizing $|I_s(m)|^2$ and $|Q_s(m)|^2$ independently.

$$B_s(m) = \begin{cases} B(m), & w_I(m) < 0 \\ 0, & 0 \le w_I \le 1/2 \\ B(m) + 1/2[B(m+k_a) + B(m-k_a)], & w_I(m) > 1/2 \end{cases} \tag{18}$$

Thus, the main lobe of the beam can be preserved, and the sidelobe of the beam can be suppressed. To further reduce the computational burden, the variables $I_a(m)$ are assumed to be:

$$I_a(m) = (1/2)[I(m+k_a) + I(m-k_a)] \tag{19}$$

The weight $w_I(m)$ and the SVA output $I_s(m)$, normalized by $I_a(m)$, are expressed as:

$$\begin{aligned} & w_I(m) = \begin{cases} 0, & I_a(m)I(m) \ge 0 \\ -I(m)/I_a(m)/2, & I_a(m)I(m) < 0 \;\&\; |I(m)| < |I_a(m)| \\ 1/2, & \text{else} \end{cases} && \text{(a)} \\ & I_s(m)/I_a(m) = \begin{cases} I(m)/I_a(m), & I_a(m)I(m) \ge 0 \\ 0, & I_a(m)I(m) < 0 \;\&\; |I(m)| < |I_a(m)| \\ 1 + I(m)/I_a(m), & \text{else} \end{cases} && \text{(b)} \end{aligned} \tag{20}$$

According to Equation (20), assuming that $I(m)$ is within a main lobe, $I(m)$ and $I_a(m)$ have the same signs, $I_a(m)I(m) \ge 0$ and $w_I(m) = 0$, i.e., the beam main lobes are preserved. When $I(m)$ is within an area of pure sidelobes, $I(m)$ and $I_a(m)$ have opposite signs, $|I(m)| < |I_a(m)|$, and $0 \le w_I(m) \le 1/2$, i.e., the beam sidelobes are completely suppressed. Assuming that $I(m)$ is in an area of a beam main lobe superimposed with beam sidelobes, $I(m)$ and $I_a(m)$ have opposite signs and $|I(m)| > |I_a(m)|$, $w_I(m) = 1/2$, i.e., the beam domain data are suppressed somewhat in an attempt to reduce the impact of the beam sidelobes.

Combined with Equations (16) and (20), the beam domain SVA algorithm output can also be expressed as follows:

$$I_s(m) = \begin{cases} I(m), & I_a(m)I(m) \geq 0 \\ 0, & I_a(m)I(m) < 0 \;\&\; |I(m)| < |I_a(m)| \\ I(m) + I_a(m), & \text{else} \end{cases} \tag{21}$$

Moreover, the real and imaginary parts of the SVA algorithm are equivalent to those of the CDA algorithm. The beam domain data are weighted $w_I(m)$ and $w_Q(m)$ by the point amplitude to realize the apodization process, and a high-quality acoustic image is obtained. Hence, after processing by the SVA method, the beam main lobe is preserved with a beam width equal to that of conventional beamforming $\theta_{BW} = 50.7\lambda/(Nd)$, and the sidelobe levels are approximately equal to zero and lower than -40 dB. In practical engineering, for forward-looking sonar imaging, high-quality acoustic images can be obtained with the sidelobe levels lower than -40 dB.

From Equation (21), the sidelobes of the beamforming results are set to exactly zero by the SVA, but the main lobes are preserved. Therefore, the sidelobes are suppressed without degrading the main lobe width in the angular dimension using the SVA method.

Since the output envelope matching filter in the range dimension is also in the form of a sinc function, lower sidelobes can be achieved, while the main lobe resolution can be preserved by using the SVA algorithm.

The effects of frequency–domain cosine window weighting and time–domain cosine window weighting for distance sidelobe reduction are essentially the same. Furthermore, using the property of the Fourier transform, the matching template of the frequency–domain window functions can be represented as follows:

$$H'(f) = w(f)s^*(f) \tag{22}$$

where $w(f)$ is the cosine base window function with respect to f, and $s(f)$ is the frequency domain of the transmitted pulse signal. Thus, the matched filter output using the frequency–domain window weighting can be expressed as follows:

$$\begin{aligned} R'(f) &= s(f)H'(f) = s(f)[s^*(f)w(f)] \\ &= s(f)s^*(f)\left[1 + we^{(j\frac{2\pi f}{B})} + we^{(-j\frac{2\pi f}{B})}\right] \\ &= R(f) + R(f)we^{(j\frac{2\pi f}{B})} + R(f)we^{(-j\frac{2\pi f}{B})} \end{aligned} \tag{23}$$

Equation (23) is transferred to the time domain as follows:

$$R'(t) = R(t) + R(t + 1/B) + R(t - 1/B) \tag{24}$$

According to Equation (24), the matched filter output weighted by cosine-on-pedestal functions is equivalent to the superposition of the conventional matched filter output uniform weighting and weighted shifted conventional matched filter output uniform weighting. Then, the sonar image range dimension discrete output by the SVA can be expressed as follows:

$$R_s(p) = R(p) + wR(p + k_r) + wR(p - k_r) \tag{25}$$

where p represents the samples in the range dimensions, k_r denotes the shift sampling number, $k_r = \left\lfloor \frac{1}{T_s B} \right\rfloor$, and T_s is the sampling rate time domain of the echoes. According to the matched filter theory, the relationship between the main lobe width and bandwidth determined by the conventional method can be expressed as $\Delta r_{ML} = \frac{c}{2B}$. Δr_{ML} is the main lobe width, and c is the underwater acoustic velocity. Then, the main lobe width of the

SVA method in the range dimension is equal to that of the conventional method, and the and the sidelobe levels are approximately equal to zero, lower than −40 dB.

In the SVA method, the raw complex-valued sonar image data are first obtained via uniform weighing. Subsequently, for the real parts of the angular dimension, the variables k_a and $\boldsymbol{I}_a(m)$ are calculated. Then, the real parts of the SVA algorithm are obtained via Equation (21), and the imaginary parts of the angular dimension, are processed in the same manner. Similarly, the range dimension sonar image data are performed and the reduced sidelobe levels data are obtained.

For the computational burden of conventional multibeamforming, it can be estimated that, per sample, the complex multiplication number is $N_B N$ (N_B is the number of beams), and the addition number is $N_B(N-1)$. The time consumption of complex multiplication is far greater than that of the addition of a digital signal processor or computer, and N_B is far greater than N, so the computational complexity of the conventional multibeamforming method is $O[N_B]$. For the angular dimension of the SVA method, compared with the conventional beamforming method, the increase in the multiplication number is $2N_B$, and the increase in the addition number is approximately $2N_B$. Hence, the computational complexity of the SVA method in the angular dimension is also $O[N_B]$.

For the matched filtering process, for each element, the computational complexity of the FFT is between $O[N_T]$ and $O[{N_T}^2]$ (N_T is the number of time samples). For the SVA method in the range dimension, compared with the conventional method, the increased multiplication number is $2N_T$, and the increased addition number is approximately $2N_T$. Hence, the computational complexity of the SVA method in the range dimension is also between $O[N_T]$ and $O[{N_T}^2]$.

According to the above derivation, we propose a two-step SVA algorithm in which sonar imaging is first applied to the angular dimension and then to the range dimension. This process achieves sonar imaging with angular-range sidelobe suppression, while preserving the angular-range resolution.

3.2. Amplitude and Phase Error Calibration

In practical engineering, the magnitude and phase errors of sonar arrays are always present due to their nonideal status. In this case, sidelobe heightening occurs, resulting in the deviation in the output of the envelope of the angular-range dimensions from the theoretical values. Thus, the SVA algorithm is less robust, and the array amplitude and phase inconsistencies should be calibrated prior to the use of the SVA algorithm.

The array phase difference ψ_n represents the measured phase difference, which is the sum of the additional phase shift φ_n caused by array channel nonuniformity and the geometric phase difference ϕ_n caused by the sound path difference, i.e., $\psi_n = \phi_n + \varphi_n$.

In practice, the geometric phase difference ϕ_n can be calculated based on the sonar array parameters. Assuming that a sound source is located at the (r_s, θ_s) near field, the additional phase shift φ_n can be calibrated by minimizing an appropriate cost function as follows:

$$\mathrm{Y} = \min_{r_s,\theta_s}\left[\sum_{n=1}^{N}\left[\hat{\psi}_n - \frac{2\pi}{\lambda}(r_n - r_s)\right]^2\right] \tag{26}$$

where r_n is the distance between the sound source and the $_n$th receiver. Equation (26) is estimated by adopting least squares of the phase difference, and when the value of Y is minimal, the location of the source can be precisely estimated. Within the preset position range of the sound source, the minimum value has unique convergence, that is:

$$\begin{cases} \frac{\partial \mathrm{Y}}{\partial r} = 0 \\ \frac{\partial \mathrm{Y}}{\partial \theta} = 0 \end{cases} \tag{27}$$

Then, the additional phase shift of the $_n$th receiver φ_n can be obtained as follows:

$$\hat{\varphi}_n = \hat{\psi}_n - \frac{2\pi}{\lambda}\left\{\left[\hat{r}_s^2 + d_n^2 - 2\hat{r}_s d_n \sin\hat{\theta}_s\right]^{1/2} - \hat{r}_s\right\} \tag{28}$$

where $\hat{\theta}_s$ and $\hat{r}_s$ are the angle and range of the sound source estimated by Equation (26), respectively. To improve the measurement accuracy of the phase difference, phase inconsistencies are usually measured and calculated at multiple angles of the sound source.

Multiple samples of the sound source direct wave signal received by each element are averaged, and the amplitude of each element is estimated as follows:

$$A_n = \frac{1}{K}\sum_{i=1}^{K}|x_n(t_i)| \tag{29}$$

where K denotes the number of samples. The amplitude of the reference matrix element is subtracted to obtain the amplitude inconsistency.

According to the estimated phase inconsistency $\hat{\varphi}_n$ and the amplitude inconsistency $\Delta\hat{A}_n$ between array elements, the array amplitude and phase inconsistency calibrated matrix can be expressed as follows:

$$\hat{\Gamma} = \text{diag}\left[\frac{1}{(1+\Delta\hat{A}_1)}e^{-j\Delta\hat{\varphi}_1}, \frac{1}{(1+\Delta\hat{A}_2)}e^{-j\Delta\hat{\varphi}_2}, \frac{1}{(1+\Delta\hat{A}_n)}e^{-j\Delta\hat{\varphi}_n}, \ldots, \frac{1}{(1+\Delta\hat{A}_N)}e^{-j\Delta\hat{\varphi}_N}\right] \tag{30}$$

where $\hat{\Gamma}^{-1}\hat{\Gamma} = \boldsymbol{I}_N$ compensates for the beamforming weighting vector and $\boldsymbol{I}_N$ denotes unit matrix. The calibrated beamforming results obtained using $\hat{\Gamma}$ are close to the theoretical sinc function values, enhancing the robustness of the SVA algorithm. The amplitude and phase errors are compensated before sonar imaging, and its time consumption has little effect on real-time imaging.

The amplitude and phase errors in the range dimension are ignored here due to the complicated underwater propagation environment, and further research will be conducted in the future. In the proposed method, the amplitude and phase inconsistency are first calibrated. Then, a two-step SVA is applied to the azimuth-range dimensions. Compared with the conventional imaging method, the proposed method also includes addition, multiplication, and comparison operations, and there is no iterative loop operation. The computational cost of the proposed method is similar to that of conventional imaging methods, which is beneficial for real-time imaging applied in engineering. Therefore, the sidelobe levels are suppressed without sacrificing the main lobe resolution by the improved SVA algorithm for forward-looking sonar imaging, which is extremely simple computationally and more robust.

A flowchart outlining the proposed algorithm is depicted in Figure 4.

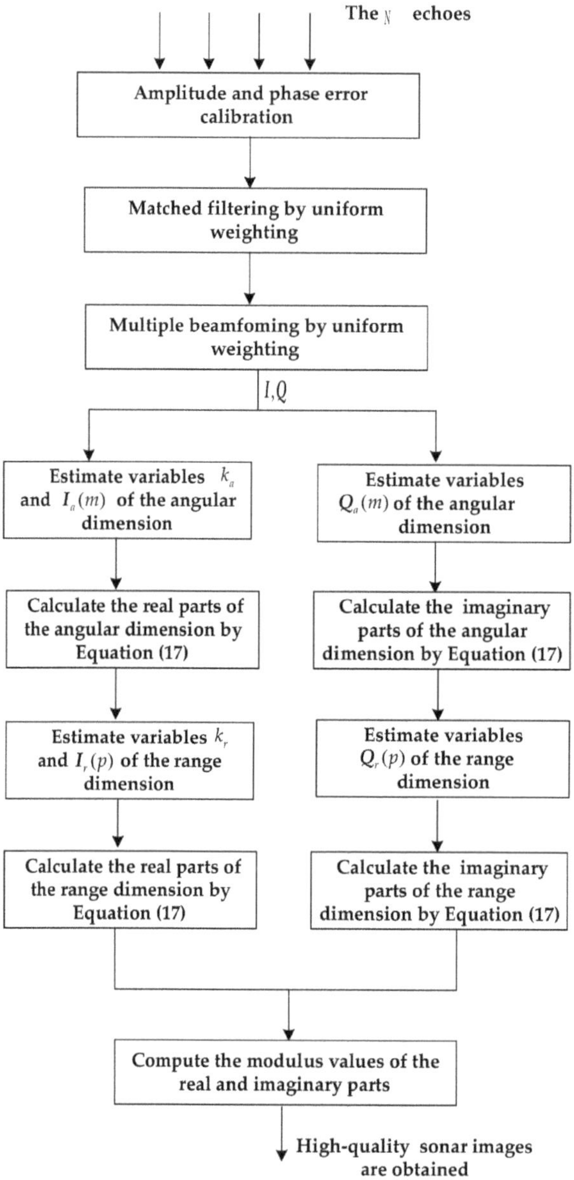

Figure 4. Flowchart outlining the proposed algorithm, where the variables $\boldsymbol{I}_r(p)$ of the range dimension are assumed to be $\boldsymbol{I}_r(p) = (1/2)[\boldsymbol{I}_{ru}(p+k_r) + \boldsymbol{I}_{ru}(p-k_r)]$, and $\boldsymbol{I}_{ru}(p)$ represents the real part of the uniform weighting matched filter output.

4. Simulation and Experimental Results

In this section, we mainly analyze sonar single-ping imaging, where the conventional method, CDA method and proposed improved SVA method are employed, and a comparative study is conducted via simulations and a lake experiment.

4.1. Simulation Results

4.1.1. Analysis of the Resolutions and Sidelobes

A ULA spacing half wavelength with elements from 40 to 200 is considered and the carrier frequency of the imaging sonar is 80 kHz. The dependences of the beam main lobe widths and peak sidelobe levels on the number of receivers are investigated by uniform weighting, Hamming weighting, CDA and SVA. The corresponding results are given in Figures 5 and 6, respectively.

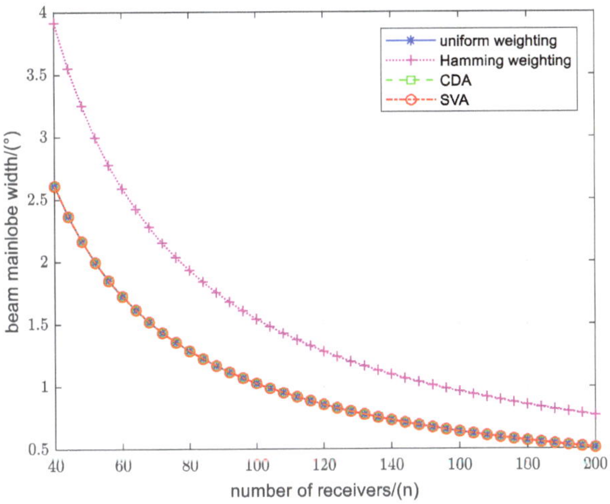

Figure 5. Beam width versus the number of receivers.

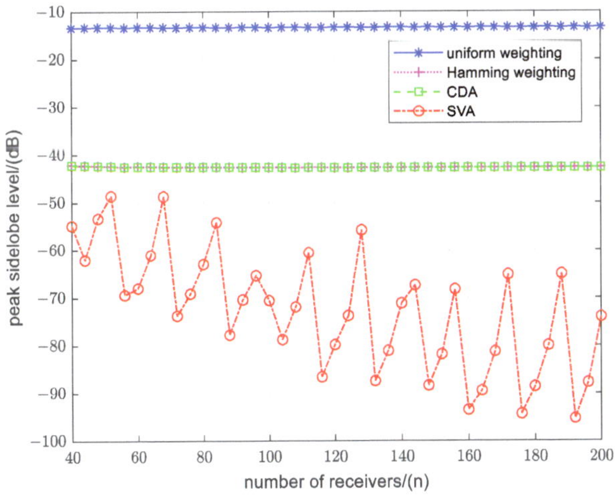

Figure 6. Peak sidelobe level versus the number of receivers.

It is clear that the beam main lobe widths of the CDA and SVA are both consistent with those of the uniform weighting method with 40 to 200 receivers. Moreover, the beam peak sidelobe levels (PSLs) of the CDA and SVA are both lower at −40 dB compared to the PSL of the uniform weighting method, which is approximately −13 dB. However, the sidelobe levels are reduced with decreasing main lobe resolution via Hamming weighting.

Since the carrier frequency of the imaging sonar is 80 kHz, the transmitted signal is a linear frequency modulation signal. Additionally, the pulse width is 40 ms, and the bandwidth changes from 5 kHz to 20 kHz. The outputs of matched filtering are obtained by using uniform weighting, Hamming window weighting, CDA processing, and SVA processing methods. The simulation results of the main lobe width and the peak sidelobe level are shown in Figures 7 and 8, respectively.

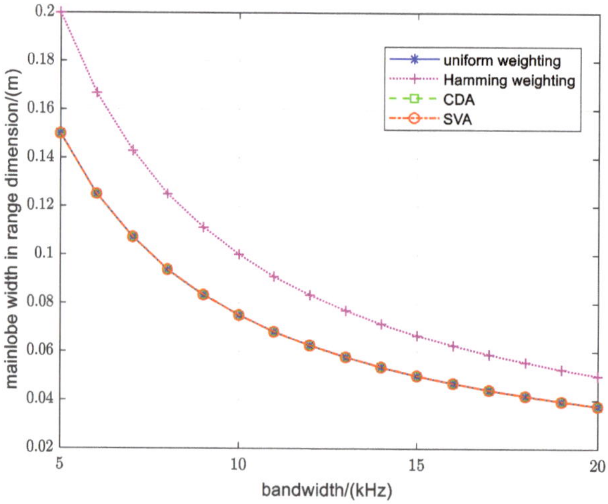

Figure 7. Range dimension of the main lobe width versus bandwidth.

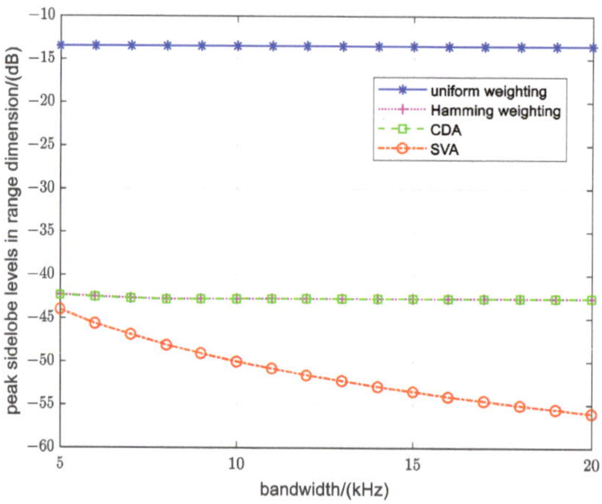

Figure 8. Range dimension peak sidelobe level versus bandwidth.

Figure 7 shows that the main lobe width of the target point in the range dimension is related to the bandwidth of the transmitted signal, where the main lobe width obtained by the SVA method is consistent with that of the conventional method and is also consistent with the theoretical value, e.g., when the bandwidth is 10 kHz, the main lobe width is 0.075 m. Therefore, the relationship between the main lobe width and bandwidth for different methods is nonlinear. As shown in Figure 8, in the bandwidth range of 8 kHz to 20 kHz, the main lobe and sidelobe ratios obtained by using the conventional method are approximately 13 dB, while the main lobe and sidelobe ratios obtained by the SVA method are approximately 40 dB. Furthermore, the range dimension SVA method can also achieve lower sidelobe levels while maintaining the same main lobe width.

4.1.2. Analysis of the Array Gain

The improvement of the output signal-to-noise ratio (SNR) between the SVA method and CBF beamforming is estimated analytically by adopting the array directivity index, which is defined as the array gain in isotropic noise according to:

$$\mathrm{AG_{sva}} = 10\log_{10}\left(\frac{2}{\int_{-1}^{1} B_{p\_sva}^{2}(u_\theta)du_\theta}\right) \tag{31}$$

where $B_{p\_sva}^{2}(u_\theta)$ denotes the beam power of the SVA algorithm. The array gain versus the number of receivers is simulated using uniform weighting, Hamming window weighting, CDA and SVA. The carrier frequency of the imaging sonar is 80 kHz in the simulation, and the corresponding results are shown in Figure 9. The simulation results clearly demonstrate that the AG of the uniform weighting method yields a $10\log_{10}(N)$ theoretical curve, while the AG of the Hamming window weighting method decreases. Nevertheless, compared with that of the uniform weighting method, the AG of the SVA method slightly increases by approximately 0.5 dB.

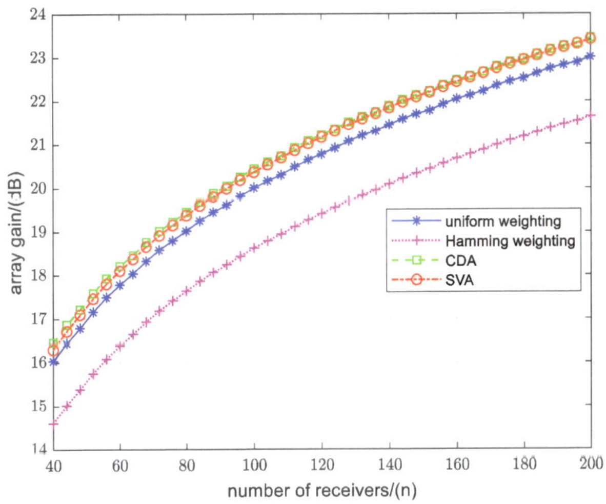

Figure 9. Comparison of the array gain versus the number of receivers.

4.1.3. Analysis of the Quality of Sonar Images

The FLS array is composed of a transmitter and an 80-receiver ULA, and the carrier frequency of the imaging sonar is 80 kHz. The transmitted pulse is LFM with a bandwidth of 8 kHz and a pulse length of 40 ms, where the SNR is assumed to be 40 dB. The two-dimensional uniform weighting separately method, two-dimensional cosine window weighting separately method, and one-dimensional two-step SVA are processed, and sonar images are obtained. Additionally, the angular dimension slices and range dimension slices of the sonar images are obtained.

The 2-D sonar images and the target slices are shown in Figure 10. Figure 10a shows the sonar image when the angular and range dimensions are weighted by the uniform method separately. Notably, the sidelobe levels in the two dimensions around the targets are higher, causing the target images to be blurry. Figure 10b presents the image result obtained by the angular and range dimensions Hamming function weighting method, where the sidelobe levels are suppressed at the expense of strongly decreased resolution. Figure 10c shows a sonar image obtained by the proposed method, in which the sidelobe levels around the targets are reduced considerably to −40 dB with no changes the resolution in two dimensions, resulting in enhanced targets. Furthermore, Figure 10d,e show the angular slice and range slice of the sonar images at the target, where the sidelobe levels in two dimensions are −40 dB while the main lobe width remains unchanged.

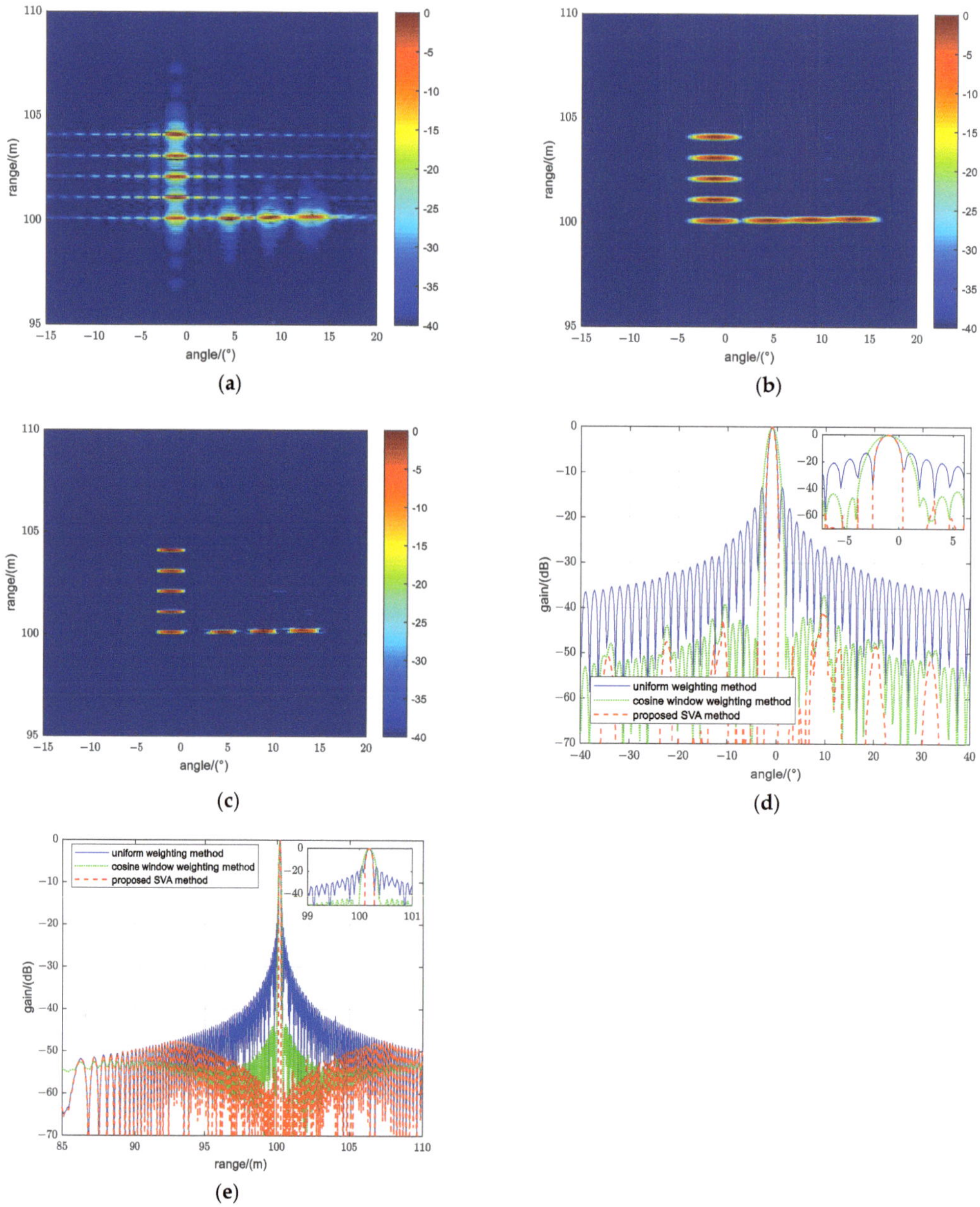

Figure 10. Comparison of the 2D acoustic images obtained using different processing methods: (**a**) simulation image obtained by the uniform weighting method; (**b**) simulation image obtained by the Hamming window weighting method; (**c**) simulation image obtained by the proposed method; (**d**) results in the angular dimension corresponding to a target at 104 m; (**e**) results in the range dimension corresponding to a target at 12°.

4.1.4. Analysis of the Computational Burden

To analyze the computational burden of the proposed method, the computational times of the conventional uniform weighting method, cosine window weighting method and one-dimensional two-step SVA method are given in Figure 11, and the matrix size of the raw image is 256 × 1667. A PC with an Intel (R) Core (TM) i7-10510U CPU @ 1.8 GHz was used, and the computation times were obtained by the MATLAB functions CLOCK and ETIME.

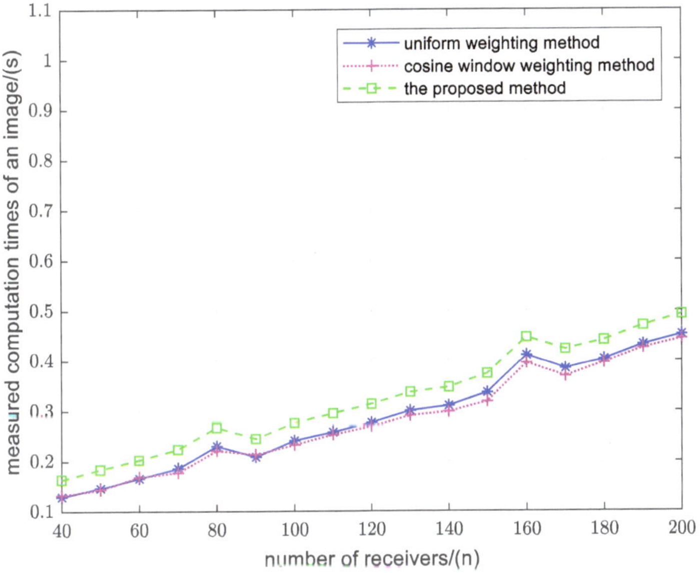

Figure 11. Comparison of the computing time for forming a raw image.

In Figure 11, the results are the average values of 50 repeated runs. For the sonar array number of 100, the computation times for both conventional window weighting methods were approximately 0.4 s. The one-dimensional SVA sequential processing method involves postprocessing of the raw image data with extremely simple computations, and the calculation amount is slightly greater than that of the conventional methods; therefore, the proposed SVA method is particularly suitable for real-time imaging.

4.1.5. Analysis of the Robustness

In this simulation, the amplitude and phase errors in the angular dimension of the system and the environmental errors are considered. Assuming that a receiving ULA of 80 elements is a spaced at half-wavelength, the array amplitude and phase errors are Gaussian distributed with variances equal to (1 dB, 5°) and (2 dB, 10°), respectively. The output beamforming is shown in Figure 12.

As shown in Figure 12, for both the window weighting methods and the SVA method, the performance decreases with beam sidelobe levels higher than −30 dB in the presence of array amplitude and phase errors. In contrast, the proposed improved SVA method combining the amplitude and phase error calibration method with the SVA method is not sensitive to the amplitude and phase errors, and the main side lobe ratio of the beam pattern is greater than 40 dB, indicating its high robustness.

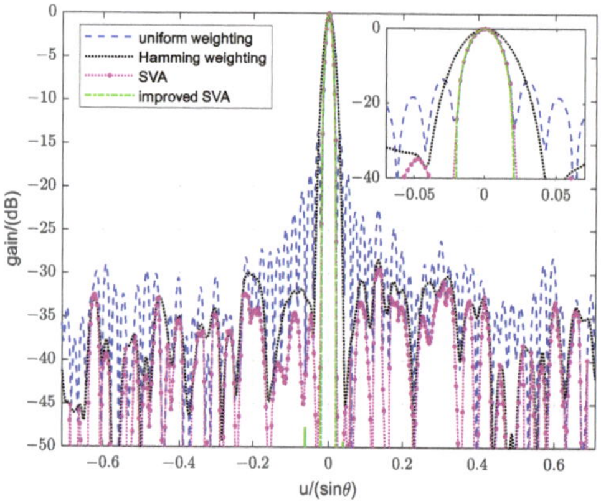

(a)

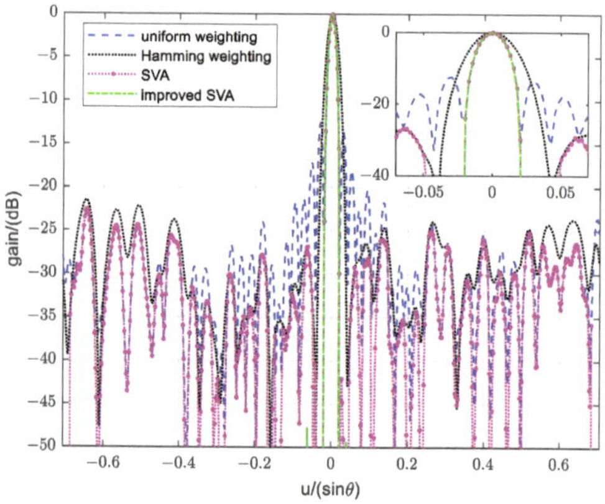

(b)

Figure 12. Comparison of the beam patterns under different amplitude and phase errors: (**a**) simulation results under amplitude and phase errors (1 dB, 5°); (**b**) simulation results under amplitude and phase errors (2 dB, 10°).

4.1.6. Analysis of the Algorithm Performance with Respect to the SNR

Considering that a receiving ULA of 100 elements is a spaced at half-wavelength, the carrier frequency of the imaging sonar is 80 kHz. The beam main lobe widths and peak sidelobe levels versus the input SNR are investigated by uniform weighting, Hamming weighting, CDA and SVA. The corresponding results are given in Figures 13 and 14, respectively.

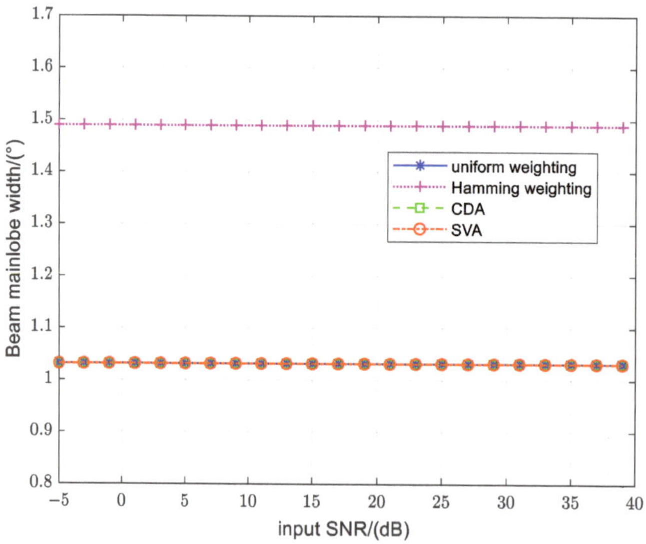

Figure 13. Beam main lobe width versus input SNR.

According to the results presented in Figures 13 and 14, the beam main lobe widths of the proposed SVA method are consistent with those of the uniform weighting method over a wide input SNR range, while those of the Hamming weighting method are wider. When the input SNR is less than 0 dB, the peak sidelobe levels of the proposed SVA method are essentially the same as those of the uniform weighting method. With the increase in the input SNR, the peak sidelobe levels of the proposed SVA method are lower than those of the uniform weighting method. CDA shows the same performance as the proposed SVA method, but its computational burden is greater.

As shown from the above simulations and analysis, compared to the uniform weighting method, cosine window weighting method and CDA method, the proposed method can reduce the sidelobes without degrading the main lobe width for forward imaging sonar in a wide input SNR and slightly increases the array gain, leading to a satisfactory improvement in the image quality. Additionally, the proposed method has good robustness and a low computational cost.

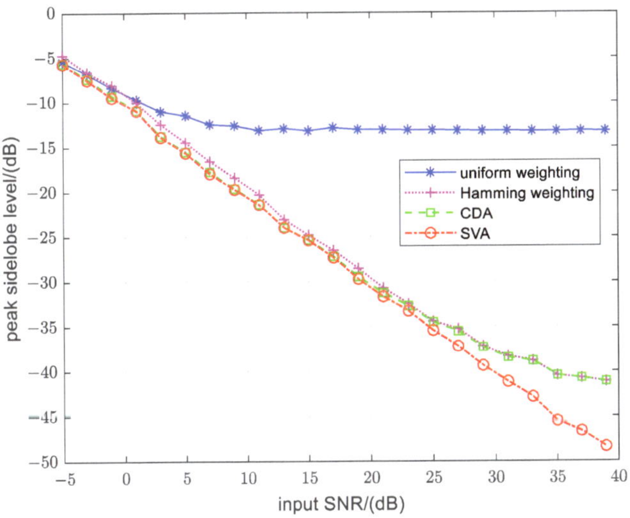

Figure 14. Peak beam sidelobe levels versus the input SNR.

4.2. Lake Experimental Data Processing Results

A lake experiment for forward-looking sonar imaging was conducted to evaluate the proposed improved SVA method. The water depth in the experimental area was approximately 65 m. In this section, first, the array amplitude and phase errors are analyzed and calibrated. Then, improved sonar images with low sidelobes and an unchanged resolution are obtained by processing the experimental data using the improved SVA method. The sonar system, which is composed of a transmitter and an 80-receiver ULA, is suspended 5 m underwater by a connecting rod from an anchored ship. The sonar system is applied to detect underwater targets and lake-bottom mountains.

4.2.1. Analysis of Calibration and Point Scatterer Imaging Results

The underwater sound source is placed underwater at the same depth as the sonar array with a horizontal distance of approximately 111 m, and a view of the lake calibration experiment is shown in Figure 15. The sound source emission signal is a single-frequency short pulse when the direct wave is selected for analysis and processing.

Based on the position of the sound source, the least squares estimation method is used to estimate the phase errors and amplitude errors. The underwater sound source is assumed to be the ideal point scatterer target. Then, the imaging performances of the uniform weighting method [7,9], Chebyshev weighting method [9], and improved SVA method before and after amplitude and phase error calibration are compared, and the experimental results are shown in Figure 16.

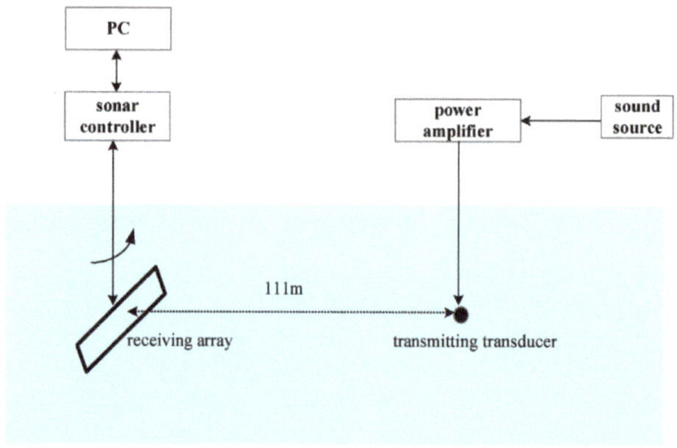

Figure 15. View of the lake calibration experiment.

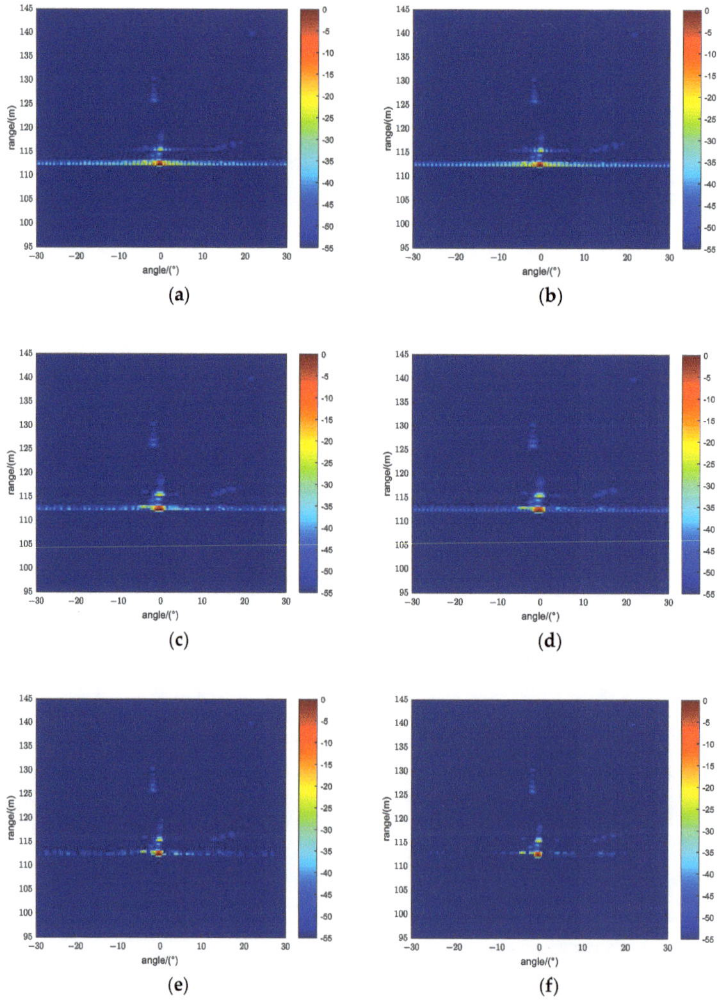

Figure 16. *Cont.*

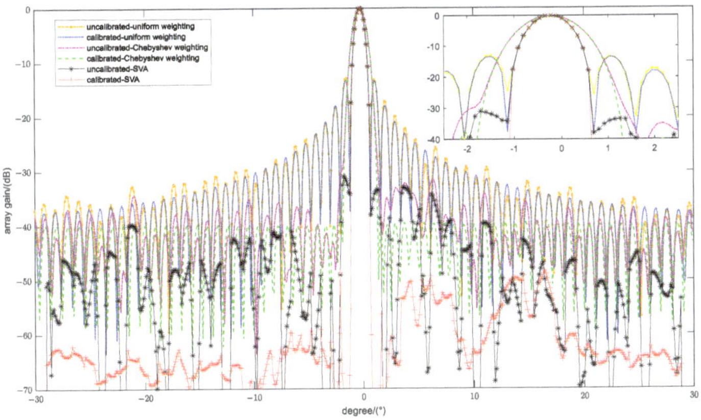

(g)

Figure 16. Comparison of the imaging performance between the uncalibrated and the calibrated arrays using three methods:(**a**) FLS image obtained by the uncalibrated uniform weighting method; (**b**) FLS image obtained by the calibrated uniform weighting method; (**c**) FLS image obtained by the uncalibrated Chebyshev window weighting method; (**d**) FLS image obtained by the calibrated Chebyshev window weighting method; (**e**) FLS image obtained by the uncalibrated SVA method; (**f**) FLS image obtained by the calibrated SVA method; (**g**) the beam response near underwater sound source corresponding to a target at 111.7 m.

Figure 16 shows the imaging results of the underwater sound source target between the uncalibrated and the calibrated arrays using three methods for lake experiments. A comparison of the sonar image obtained by the conventional uncalibrated uniform weighting method presented in Figure 16a, and the sonar image generated by the uncalibrated Chebyshev window weighting method presented in Figure 16b shows that the beam sidelobe levels are reduced when increasing the beam width, while the beam sidelobe levels obtained by the proposed method are suppressed with the beam width, approximately equal to that of the image in Figure 16a. Figure 16b,g show that the calibrated beam pattern envelope is closer to the sinc function and that the beam sidelobe levels are lower when uniform weighting is used. Moreover, the sidelobe levels are lower (−40 dB) after calibration but have greater main lobe width. For SVA filtering, the beam sidelobe levels after calibration are all below −45 dB, and the azimuth resolution remains unchanged. Hence, the proposed improved SVA method can successfully obtain images of the point scatterer target; simultaneously, it has high robustness.

4.2.2. Analysis of Imaging Results for the Whole Target Scene

The transmitted signal of the FLS is an LFM pulse with a carrier frequency of 80 kHz, and the receiving array mainly receives the reflected echoes of targets and mountains in the water. The layout and scene of the lake experiment are shown in Figure 17. After calibrating the amplitude and phase errors of the array echo signal, the conventional imaging method [7,9], the azimuth-range Hamming window weighting imaging method [9], and the azimuth-range two-step SVA imaging method are used. The imaging results are shown in Figure 18.

Figure 18 shows the sonar images of the underwater mountains and underwater targets processed by the three methods. A comparison of the sonar image obtained by the conventional method presented in Figure 18a and the sonar image generated by the proposed method presented in Figure 18c shows that the two-dimensional resolution of the image in Figure 18c is approximately equal to that of the image in Figure 18a, and the azimuth-range dimension sidelobe levels are effectively suppressed, with sidelobe levels approximately equal to those in Figure 18b. Furthermore, Figure 18d shows the azimuthal section of the underwater target at a distance of 337 m, and it is observed that the beam main lobe width remains unchanged at approximately 0.9°. Due to the influence of underwater reverberation and noise, the beam sidelobe levels near the target point are approximately −30 dB. Figure 18e shows the range dimension of the underwater target at an azimuth of 26°. At the same time, the proposed method partly reduced the sidelobe levels near the target point. However, with the influence of the high complexity of underwater acoustic channel propagation and the frequency fluctuation characteristic of transmitting-receiving transducers, the echo LFM waveforms deviate from the theoretical waveforms, so the range dimension sidelobe suppression effect of the SVA method is not evident in practice. Further research on the impact of underwater acoustic channel propagation on the emission waveform can be carried out to improve the algorithm. The above analysis shows that two-dimensional lower sidelobe level sonar images with unchanged azimuth-range resolution, obtained by using an improved imaging method, improved sonar imaging performance, which is beneficial for the subsequent target detection and recognition.

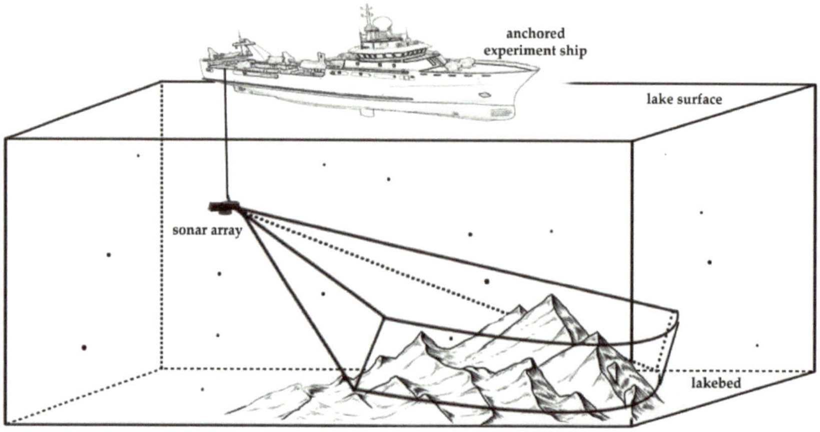

Figure 17. Lake experiment diagram.

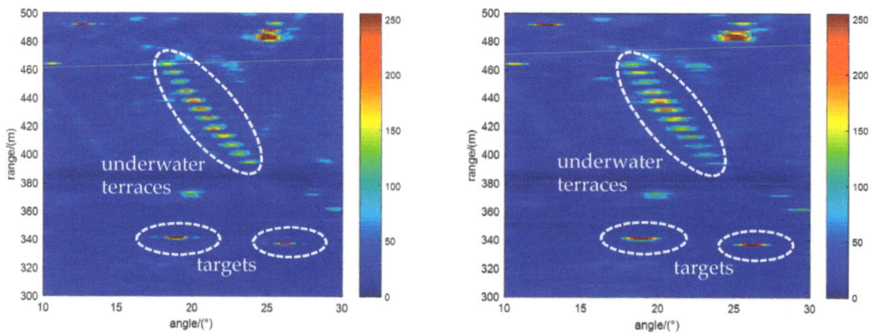

(a) (b)

Figure 18. *Cont.*

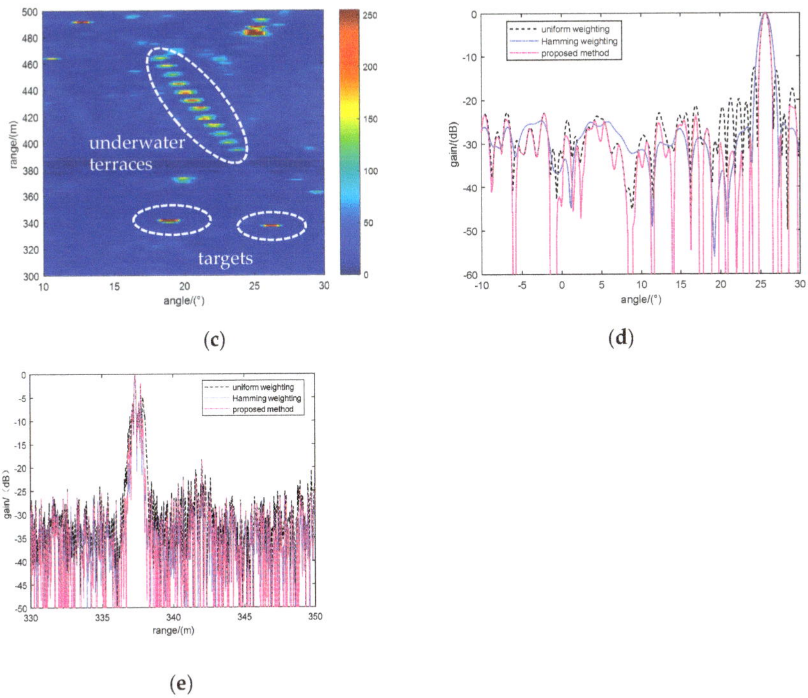

Figure 18. Two-dimensional images of sonar with different processing methods for sidelobe suppression: (**a**) FLS image obtained by the uniform weighting method; (**b**) FLS image obtained by the Hamming window weighting method; (**c**) FLS image obtained by the proposed method; (**d**) results in the angular dimension corresponding to a target at 337 m; (**e**) results in the range dimension corresponding to a target at 26°.

5. Conclusions

Generally, the cosine window function weighting method is often used to reduce sidelobe levels for forward-looking sonar systems. This method has a lower computational cost and robustness, but it degrades the main lobe resolution. To solve this problem, we first derived the two-step SVA method for forward-looking sonar and improved its robustness to suppress sidelobes without degrading the main lobe resolution in angular-range dimensions. Through theoretical analyses, numerical simulations and lake experiment results, we have shown that the improved spatially variant apodization algorithm can adequately improve image quality by greatly reducing sidelobes while keeping the main lobe width unchanged. This process is extremely simple computationally. Due to the complex underwater environment, the range dimension sidelobe suppression performance of the SVA method decreases slightly in practice, Therefore, the propagation and variation characteristics of echoes in the range dimension will be studied in further research.

Author Contributions: Conceptualization, J.Y. and F.X.; methodology, L.Y. and F.X.; software, L.Y.; supervision, S.P. validation, L.Y. and S.P.; formal analysis, S.P.; investigation, J.Y.; data curation, F.X.; writing—original draft preparation, L.Y.; writing—review and editing, J.Y., F.X. and S.P.; funding acquisition, L.Y. All authors have read and agreed to the published version of the manuscript.

Funding: This work was funded by the IACAS Frontier Exploration Project under grant QYTS202107 and the National Key Research and Development Program of China under grant 2022YFC2806101.

Data Availability Statement: The data presented in this study are available upon request from the corresponding author. The data are not publicly available due to privacy or ethical restrictions.

Conflicts of Interest: The authors declare no conflicts of interest.

References

1. Murino, V.; Trucco, A. Three-dimensional image generation and processing in underwater acoustic vision. *Proc. IEEE* **2000**, *88*, 1903–1946. [CrossRef]
2. Wang, S.; Chi, C.; Wang, P.; Zhao, S.; Huang, H.; Liu, X.; Liu, J. Feature-enhanced beamforming for underwater 3-D acoustic imaging. *IEEE J. Ocean. Eng.* **2023**, *48*, 401–415. [CrossRef]
3. Petillot, Y.; Ruiz, T.I.; Lane, D.M. Underwater vehicle obstacle avoidance and path planning using a multi-beam forward looking sonar. *IEEE J. Ocean. Eng.* **2001**, *26*, 240–251. [CrossRef]
4. Zhou, T.; Si, J.; Wang, L.; Xu, C.; Yu, X. Automatic detection of underwater small targets using forward-looking sonar images. *IEEE Trans. Geosci. Remote Sens.* **2022**, *60*, 4207912. [CrossRef]
5. Kiang, C.W.; Kiang, J.F. Imaging on underwater moving targets with multistatic synthetic aperture sonar. *IEEE Trans. Geosci. Remote Sens.* **2022**, *60*, 4211218. [CrossRef]
6. Thomas, C.W. A CFAR detection approach for identifying gas bubble seeps with multibeam echo sounders. *IEEE J. Ocean. Eng.* **2021**, *46*, 1346–1355.
7. Liu, X.; Fan, J.; Sun, C.; Yang, Y.; Zhuo, J. High-resolution and low-sidelobe forward-look sonar imaging using deconvolution. *Appl. Acoust.* **2021**, *178*, 107986. [CrossRef]
8. Wang, P.; Chi, C.; Liu, J.; Huang, H. Improving performance of three-dimensional imaging sonars through deconvolution. *Appl. Acoust.* **2021**, *175*, 107812. [CrossRef]
9. Van Trees, H.L. *Optimum Array Processing: Part 4 of Detection, Estimation, and Modulation Theory;* John Wiley & Sons Inc.: New York, NY, USA, 2002; pp. 17–79.
10. Fredric, J.H. On the use of windows for harmonic analysis with the discrete Fourier transform. *Proc. IEEE* **1978**, *66*, 51–83.
11. Yan, L.; Piao, S.C.; Xu, F. Orthogonal waveform separation in multiple-input and multiple-output imaging sonar with fractional Fourier filtering. *IET Radar Sonar Navig.* **2021**, *15*, 471–484. [CrossRef]
12. Ma, C.; Sun, D.; Mei, J.; Teng, T. Spatiotemporal two-dimensional deconvolution beam imaging technology. *Appl. Acoust.* **2021**, *183*, 108310. [CrossRef]
13. Stoica, P.; Wang, Z.; Li, J. Robust Capon beamforming. *IEEE Signal Process. Lett.* **2003**, *10*, 172–175. [CrossRef]
14. Lønmo, T.I.B.; Austeng, A.; Hansen, R.E. Improving swath sonar water column imagery and bathymetry with adaptive beamforming. *IEEE J. Ocean. Eng.* **2020**, *45*, 1552–1563. [CrossRef]
15. Buskenes, J.I.; Hansen, R.E.; Austeng, A. Low-complexity adaptive sonar imaging. *IEEE J. Ocean. Eng.* **2017**, *42*, 87–96. [CrossRef]
16. Blomberg, A.E.A. Adaptive Beamforming for Active Sonar Imaging. Ph.D. Thesis, University of Oslo, Oslo, Norway, 2011.
17. Liu, J.; Gershman, A.B.; Luo, Z.Q.; Wong, K.M. Adaptive beamforming with sidelobe control: A second-order cone programming approach. *IEEE Signal Process. Lett.* **2015**, *10*, 331–334.
18. Gerstoft, P.; Xenaki, A.; Mecklenbräuker, C.F. Multiple and single snapshot compressive beamforming. *J. Acoust. Soc. Amer.* **2015**, *138*, 2003–2014. [CrossRef]
19. David, G.; Robert, J.L.; Zhang, B.; Laine, A.F. Time domain compressive beam forming of ultrasound signals. *J. Acoust. Soc. Amer.* **2015**, *137*, 2773–2784. [CrossRef] [PubMed]
20. Guo, Q.; Yang, S.; Zhou, T.; Wang, Z.; Cui, H.L. Underwater acoustic imaging via online bayesian compressive beamforming. *IEEE Geosci. Remote Sens. Lett.* **2022**, *19*, 4024505. [CrossRef]
21. Guo, Q.; Xin, Z.; Zhou, T.; Yin, J.; Cui, H.L. Adaptive compressive beamforming based on bi-sparse dictionary learning. *IEEE Trans. Instrum. Meas.* **2022**, *71*, 6501011. [CrossRef]
22. Yang, T.C. Deconvolved conventional beamforming for a horizontal line array. *IEEE J. Ocean. Eng.* **2018**, *43*, 160–172. [CrossRef]
23. Yang, T.C. Deconvolution of decomposed conventional beamforming. *J. Acoust. Soc. Amer.* **2020**, *148*, 195–201. [CrossRef] [PubMed]
24. Guo, W.; Piao, S.; Yang, T.C.; Guo, J.; Iqbal, K. High-resolution power spectral estimation method using deconvolution. *IEEE J. Ocean. Eng.* **2020**, *45*, 489–499. [CrossRef]
25. Mei, J.; Pei, Y.; Zakharov, Y.; Sun, D.; Ma, C. Improved underwater acoustic imaging with non-uniform spatial resampling RL deconvolution. *IET Radar Sonar Navig.* **2020**, *14*, 1697–1707. [CrossRef]
26. Sun, D.; Ma, C.; Mei, J.; Shi, W. Improving the resolution of underwater acoustic image measurement by deconvolution. *Appl. Acoust.* **2020**, *165*, 107292. [CrossRef]
27. Liu, X.; Yang, Y.; Sun, C.; Zhuo, J.; Wang, Y.; Zhai, Q. High-angular- and range-resolution imaging using MIMO sonar with two-dimensional deconvolution. *IET Radar Sonar Navig.* **2023**, *17*, 991–1001. [CrossRef]
28. Stankwitz, H.C.; Dallaire, R.J.; Fienup, J.R. Spatially variant apodization for sidelobe control in SAR imagery. In Proceedings of the IEEE National Radar Conference, Atlanta, GA, USA, 29–31 March 1994.
29. Stankwitz, H.C.; Dallaire, R.J.; Fienup, J.R. Nonlinear apodization for sidelobe control in SAR imagery. *IEEE Trans. Aero. Electron. Syst.* **1995**, *31*, 267–279. [CrossRef]

30. Zhu, R.Q.; Zhou, J.X.; Fu, Q. Spatially variant apodization for sidelobe suppression in near range radar imagery. *IET Radar Sonar Navig.* **2022**, *16*, 986–999. [CrossRef]
31. Ding, Z.; Zhu, K.; Wang, Y.; Li, L.; Liu, M.; Zeng, T. Spatially variant sidelobe suppression for linear array MIMO SAR 3-D imaging. *IEEE Trans. Geosci. Rem. Sens.* **2022**, *60*, 5220915. [CrossRef]

remote sensing

MDPI

Article

A Steering-Vector-Based Matrix Information Geometry Method for Space–Time Adaptive Detection in Heterogeneous Environments

Runming Zou, Yongqiang Cheng *, Hao Wu, Zheng Yang, Xiaoqiang Hua and Hanjie Wu

College of Electronic Science and Technology, National University of Defense Technology, Changsha 410073, China; zourunming18@nudt.edu.cn (R.Z.); wuhao13c@nudt.edu.cn (H.W.); yangzheng18@nudt.edu.cn (Z.Y.); huaxiaoqiang12@nudt.edu.cn (X.H.); wuhanjie21@nudt.edu.cn (H.W.)

* Correspondence: yqcheng@nudt.edu.cn

Abstract: Plagued by heterogeneous clutter, it is a serious challenge for airborne radars to detect low-altitude, weak targets. To overcome this problem, a novel matrix information geometry detector for airborne multi-channel radar is proposed in this paper. The proposed detector applies the given steering vector and array structure information to the matrix information geometry detection method so that it can be used for space–time adaptive detection. While improving the detection performance, the matrix information geometry detector's original anti-clutter advantage is enhanced as well. The simulation experiment results indicate that the proposed detector has advantages in several of the properties related to space–time adaptive detection, while its computational complexity does not increase significantly. Moreover, experiment results based on the measured data verify the superior performance of the proposed method. Sea-detecting data-sharing-program data, mountaintop data, and phased-array radar data are employed to verify the performance advantage of the proposed method in heterogeneous clutter and the ability for weak target detection.

Keywords: information geometry; space–time adaptive detection; heterogeneous clutter; airborne radar

1. Introduction

Due to the movement of a radar platform, airborne radar clutter appears to be significantly broadened in the frequency domain, which is characterized by two-dimensional coupling in space–time. In such a case, it is difficult for the traditional clutter suppression and target detection methods to achieve satisfactory detection performance. In order to solve this problem, space–time adaptive processing (STAP) has been developed and is widely used in airborne radar clutter suppression [1]. By transmitting and receiving multiple pulses from an array radar, STAP filters the clutter in both the spatial and temporal dimensions. Unfortunately, although STAP can effectively improve the airborne radar's clutter suppression ability when the homogeneous training samples are sufficient [2,3], it suffers from heterogeneous clutter in a complex environment, and in such a case, its clutter suppression effectiveness greatly affects the performance of target detection. In recent years, the space–time adaptive detection (STAD) [4,5] method has been studied by many scholars. Different from the existing filter-then-CFAR detection methods [6], STAD is a target detection method that jointly processes the received signal in both the spatial and temporal domains and directly outputs the detection statistics. Compared to STAP-based airborne radar target detection methods, STAD has the following advantages [7]: (1) It usually has CFARness and does not require specialized constant false alarm rate (CFAR) techniques. (2) It has better detection performance, especially under low signal-to-clutter ratio (SCR) conditions.

It can be seen that STAD is a radar target detection technique for multi-channel radar. Multi-channel adaptive signal detection was first investigated by Kelly in 1986 [8]. In [8], Kelly proposed Kelly's generalized likelihood ratio test (KGLRT) to detect a rank-one signal in a homogeneous environment. A rank-one signal refers to a target with a known steering vector but an unknown amplitude [6]. Based on the generalized likelihood ratio test (GLRT) criterion, Chen and Robey proposed the adaptive matched filter (AMF) independently [9,10]. According to the Rao detector and Wald detector, De Maio proposed the corresponding detectors in [11,12] and proved that the AMF is equivalent to the Wald detector. The probability of detection (Pd) of the Rao detector obtained by De Maio (DMRao) [11] is lower than those of the KGRLT and AMF detectors, but it is superior to them in terms of rejecting mismatched signals.

It is worth highlighting that the aforementioned algorithms are considered in homogeneous environments [13–18]; however, since airborne radar usually works in a downward-looking state [19], it faces a severely heterogeneous clutter environment. Heterogeneous clutter refers to clutter for which its characteristics are inconsistent with that of the cell under test (CUT), such as clutter with a different power [20] or probability density function (PDF). When detecting the target, heterogeneous clutter interferes with the detector's estimation of the clutter, thus affecting the correctness of the detection results. The irregular terrain and sea surface make the environment heterogeneous, and the available training samples are drastically reduced, leading to performance degradation of existing algorithms in heterogeneous clutter environments [21]. To solve this problem, a class of clutter covariance matrix (CCM) estimation methods and detection algorithms [22,23] are investigated. Many studies have improved the estimation method of the CCM to enhance the robustness of detectors in heterogeneous environments [24–28], but these methods usually need iterative operations and are accompanied by heavy computational burden. In addition, based on the generalization of the Euclidean center, some geometric-center-based CCM estimation methods are presented in [29–31]. These methods have shown distinct superiority in radar signal processing, but their algorithm complexity is too complicated as well.

Recently, related studies on radar target detection methods based on information geometry received a lot of attention for their distinct advantages in hetero-clutter environments [32–44]. Specifically, the matrix information geometry (MIG) detector, which was pioneered by Barbaresco [45] and developed by Cheng et al. [46], shows a novel detection scheme for target detection. Thanks to the excellent discriminative ability of the geodesic distance on the matrix manifold, the MIG detector has outstanding performance in hetero-clutter environments [47]. Originally, the Riemann distance was adopted by the MIG detector [45,48], but its computational burden is too heavy. To overcome this problem, many other measurements have been introduced into the MIG detector [49]. Typically, the Kullback–Leibler divergence (KLD) is widely used [50]. While the MIG detector has made great progress and proved its advantages in hetero-clutter environments [51–53], which matters much in STAD, it is not applied in STAD because

- The given steering vector of the target is abandoned by existing MIG detectors, causing underutilized information in STAD. In addition, the MIG detector fails to suppress the clutter in the spatial and temporal domains; therefore, the performance of the MIG detector can be further improved for airborne radar.
- The various existing MIG detectors are designed for single-channel radar. For multi-channel radar, there is a strong correlation between channels, which has not been considered before. This problem is especially accentuated when the dimensionality of the data matrix increases. At present, MIG detectors can only work with dimensionality reduction algorithms, which put a price on performance loss.

To apply MIG methods on STAD, a novel MIG detector is proposed in this paper. Inspired by the existing STAD methods and the STAP method, we employ the space–time steering vectors to describe the target information during each temporal Doppler scan and the correlation between channels. Based on the GRLT, we deduce the detection formula of the proposed method and analyze its performance. The superiority of the proposed method is demonstrated by experimental results. Particularly, the principal contributions of this paper are listed as follows.

- We provide a novel MIG detector for STAD. The proposed steering-vector-based matrix information (S-MIG) detector introduces the steering vector into target detection, overcoming the shortcomings of the MIG detector and making it feasible to apply information geometry methods to STAD.
- We perform a detailed analysis and validation of the performance of the S-MIG detector, especially the important property that matters in target detection. The mathematical derivation and experimental results illustrate that the proposed S-MIG detector has competitive performance, while its computational burden is comparable to existing methods.
- We describe a novel detection scheme for detectors based on matrix information geometry. In cases for which the target's steering vector is given, a detection scheme that is a hyperbola on the Hermitian positive definite (HPD) manifold is proposed, and many detection methods based on various metrics can be extended from it.

The remainder of this paper is organized as follows. Section 2 reviews the signal model of space–time coupling clutter for STAD and introduces the preliminaries of matrix information geometry. In Section 3, the proposed steering-vector-based MIG detector is deduced, and its geometry explanation is given. In addition, several important properties of the detector are analyzed in Section 3 as well. Section 4 presents experimental results based on both simulated and real data. Finally, Section 5 provides a brief conclusion of this paper.

The following notations are adopted in this paper: the math italic x, lowercase italic $\mathbf{x}$, and uppercase bold $\mathbf{A}$ denote scalars, vectors, and matrices, respectively; $\overline{x}$ is the conjugate of x. Symbols $(\cdot)^{-1}$ and $(\cdot)^{\mathrm{H}}$ denote the inverse and the conjugate transpose of matrices, respectively. $E(\cdot)$ denotes the statistical expectation, and $\otimes$ denotes the Kronecker product. $\mathbf{A} \succ 0$ represents that $\mathbf{A}$ is a positive definite matrix. Constant matrix $\mathbf{I}$ indicates the identity matrix.

2. Problem Formulation and Preliminaries of MIG

2.1. Signal Model and Detection Model

Consider a side-looking airborne radar that is a uniform linear array (ULA) with N elements. Assuming that each antenna receives P coherent pulses, the degree of freedom (DoF) M of this system is $M = NP$. As shown in Figure 1, we appoint the parameters of the radar as follows. The airborne platform travels with a velocity v, and the azimuth angle and the elevation angle between the radar and clutter are θ and φ, respectively.

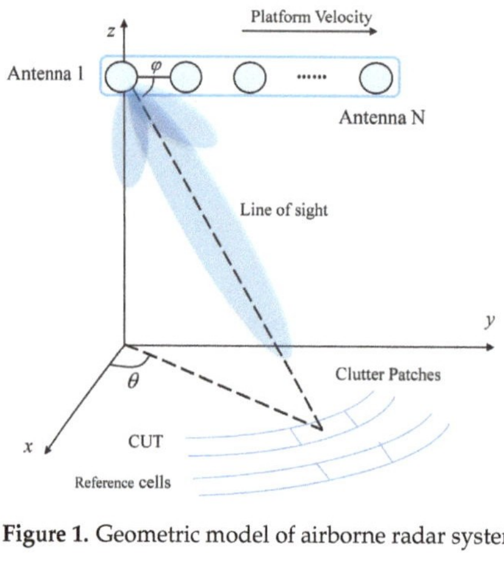

Figure 1. Geometric model of airborne radar system.

Denote $\mathbf{x}_k$ as the received signal of the ULA in the kth range cell, which is composed of the target echo $\mathbf{x}_\mathrm{t}$ (if the target exists), clutter signal $\mathbf{x}_\mathrm{c}$, and noise $\mathbf{n}$. According to the Wald clutter model of airborne radar [54], the clutter received by antennas is the summation of the echo of each clutter patch: namely,

$$\begin{aligned}\mathbf{x}_k &= \alpha_\mathrm{t}\mathbf{s}_\mathrm{t} + \mathbf{x}_{ck} + \mathbf{n} \\ &= \alpha_\mathrm{t}\mathbf{s}_\mathrm{t} + \sum_{i=1}^{N_\mathrm{c}} \alpha_i \mathbf{s}_\mathrm{i} + \mathbf{n}\end{aligned} \tag{1}$$

where N_c is the number of clutter patches in the kth range cell; α_t and $\mathbf{s}_\mathrm{t}$ are the reflection coefficient and the steering vector, respectively, of the target; analogously, α_i and $\mathbf{s}_\mathrm{i}$ are the reflection coefficient and the steering vector of the ith clutter patch in kth range cell, respectively.

The steering vector is the Kronecker product of the spatial steering vector $\mathbf{s}_\mathrm{s}(f_\mathrm{s})$ and the temporal steering vector $\mathbf{s}_d(f_\mathrm{d})$, which are determined by the position of the tangential and radial Doppler. The space–time steering vector is written as

$$\mathbf{s} = \mathbf{s}_\mathrm{t}(f_\mathrm{d}) \otimes \mathbf{s}_\mathrm{s}(f_\mathrm{s}) \tag{2}$$

where

$$\begin{aligned}\mathbf{s}_\mathrm{t}(f_\mathrm{d}) &= [1, e^{\mathrm{j}2\pi f_\mathrm{d}}, \cdots, e^{\mathrm{j}(P-1)2\pi f_\mathrm{d}}]^T \\ \mathbf{s}_\mathrm{s}(f_\mathrm{s}) &= [1, e^{\mathrm{j}2\pi f_\mathrm{s}}, \cdots, e^{\mathrm{j}(N-1)2\pi f_\mathrm{s}}]^T\end{aligned} \tag{3}$$

and the temporal Doppler $f_\mathrm{d} = 2vT/\lambda \cos\theta \cos\varphi$ and the spatial Doppler $f_\mathrm{s} = d/\lambda \cos\theta$ $\cos\varphi$; λ is the wavelength, and d is the distance between adjacent array elements.

In radar target detection, the received signals in different range cells can be divided into two sorts: the data in the CUT (also referred to as the primary data), which may include the target echo and the reference range cells located on both sides of the CUT, and the data in the reference range cells, which are called secondary data. Based on this classification, the problem of radar target detection can be expressed as a binary hypothesis problem:

$$\begin{aligned}\mathcal{H}_0 &: \begin{cases}\mathbf{x}_\mathrm{CUT} = \mathbf{x}_\mathrm{c} + \mathbf{n} \\ \mathbf{x}_k = \mathbf{x}_{ck} + \mathbf{n} \quad k = 1, \ldots, K\end{cases} \\ \mathcal{H}_1 &: \begin{cases}\mathbf{x}_\mathrm{CUT} = \alpha_\mathrm{t}\mathbf{s}_\mathrm{t} + \mathbf{x}_\mathrm{c} + \mathbf{n} \\ \mathbf{x}_k = \mathbf{x}_{ck} + \mathbf{n} \quad k = 1, \ldots, K\end{cases}\end{aligned} \tag{4}$$

where $\mathbf{x}_\mathrm{CUT}$ is the data in the CUT, and $\mathbf{x}_k, k = 1, \cdots, K$ represent the data in the reference range cells. K is the number of reference cells, and $\mathbf{x}_\mathrm{c}$ is the clutter in the CUT and is usually unknown. The clutter in the CUT and the reference range cells are supposed to be independent and identically distributed (IID) so that the secondary data $\mathbf{x}_k, k = 1, \cdots, K$ can be used to estimate $\mathbf{x}_\mathrm{c}$.

To make use of the correlation between different array elements and pulses, the clutter covariance matrix of the CUT is often employed to improve the detection performance, which is shown as

$$\mathbf{R}_k = \mathrm{E}[(\mathbf{x}_k\mathbf{x}_k^\mathrm{H})] = \begin{bmatrix} x_0 & \overline{x}_1 & \cdots & \overline{x}_{M-1} \\ x_1 & x_0 & \cdots & \overline{x}_{M-2} \\ \vdots & \ddots & \ddots & \vdots \\ x_{M-1} & x_{M-2} & \cdots & x_0 \end{bmatrix} \tag{5}$$

where $x_k = \sum_{i=k+1}^{M} x_i x_{i+k}, 0 \leq k \leq M-1$.

The problem of radar target detection in the form of the covariance matrix is expressed as

$$\begin{aligned}\mathcal{H}_0 &: \begin{cases}\mathbf{R}_{\mathrm{CUT}} = \mathbf{R}_{\mathrm{c}} + \mathbf{R}_{\mathrm{n}} \\ \mathbf{R}_k = \mathbf{R}_{\mathrm{c}k} + \mathbf{R}_{\mathrm{n}} \quad k = 1, \ldots, K\end{cases} \\ \mathcal{H}_1 &: \begin{cases}\mathbf{R}_{\mathrm{CUT}} = \mathbf{R}_{\mathrm{t}} + \mathbf{R}_{\mathrm{c}} + \mathbf{R}_{\mathrm{n}} \\ \mathbf{R}_k = \mathbf{R}_{\mathrm{c}k} + \mathbf{R}_{\mathrm{n}} \quad k = 1, \ldots, K\end{cases}\end{aligned} \tag{6}$$

where $\mathbf{R}_{\mathrm{CUT}}$, $\mathbf{R}_{\mathrm{c}}$, $\mathbf{R}_{\mathrm{n}}$, $\mathbf{R}_k$, $\mathbf{R}_{\mathrm{c}k}$, and $\mathbf{R}_{\mathrm{t}}$ are the covariance matrices constructed by $\mathbf{x}_{\mathrm{CUT}}$, $\mathbf{x}_{\mathrm{c}}$, $\mathbf{n}$, $\mathbf{x}_k$, $\mathbf{x}_{\mathrm{c}k}$, and $\alpha_{\mathrm{t}}\mathbf{s}_{\mathrm{t}}$, respectively.

2.2. Preliminaries of Matrix Information Geometry

According to Formula (5), the covariance matrices of the secondary data are Hermitian and Toeplitz. In addition, these matrices are positive definite matrices as well. Therefore, the matrices utilized in radar signal processing are located on the HPD manifold, which is defined as follows:

$$\mathcal{M}_{H_{++}} = \left\{\mathbf{R} \in \mathcal{T}(M, \mathbb{C}) | \mathbf{R}^{\mathrm{H}} = \mathbf{R}, \mathbf{R} \succ \mathbf{0}\right\} \tag{7}$$

where $\mathcal{M}_{H_{++}}$ denotes the HPD manifold, and $\mathcal{T}(M, \mathbb{C})$ is the set of $M \times M$ dimensional complex Toeplitz matrices.

The HPD manifold is a set constructed by all HPD matrices and has a certain spatial structure, which is distinct from the Euclidean space. On the HPD manifold, the different matrices of secondary data respond to different points, and their relationship in statistic characters can be expressed by the geometric measurement on the HPD manifold. Because the manifold is a curved space, the geodesic distance between two matrices $\mathbf{P}, \mathbf{Q} \in \mathcal{M}$ on the manifold $\mathcal{M}$ is called the Fisher information distance, which is defined as

$$\mathcal{D}(\mathbf{P}, \mathbf{Q}) = \inf\{\mathcal{L}(\zeta) | \zeta : [a, b] \to \mathcal{M}, \zeta(a) = \mathbf{P}, \zeta(b) = \mathbf{Q}\} \tag{8}$$

where $\mathcal{L}(\zeta)$ is the length of ζ, which is a sufficiently smooth curve on $\mathcal{T}(M, \mathbb{C})$. Specifically, the Fisher information distance on the HPD manifold is called the Riemann distance (RD), which is

$$\mathcal{D}_{\mathrm{Rie}}(\mathbf{P}, \mathbf{Q}) = \left\|\log(\mathbf{P}^{-1}\mathbf{Q})\right\|_{\mathrm{F}} \tag{9}$$

where $\log(\cdot)$ is the logarithm of the matrix, and $||\cdot||_{\mathrm{F}}$ is the Frobenius norm of the matrix. It is worth indicating that, while the Riemann distance is the true geodesic distance on the HPD manifold, many other measurements, such as geometric divergence, are widely used in practice. Some of these measurements do not strictly satisfy the definition of distance on the manifold, but they are easily calculated and have preferable detection performance in some cases. The most commonly used KL divergence on the HPD manifold is defined as

$$\mathcal{D}_{\mathrm{KL}}(\mathbf{P}, \mathbf{Q}) = \mathrm{tr}(\mathbf{P}^{-1}\mathbf{Q}) - \log(|\mathbf{P}^{-1}\mathbf{Q}|) - M \tag{10}$$

Based on various geometric measurements on the matrix manifold, different MIG detectors are studied, but their detection formulations can be summarized as

$$\mathcal{D}(\mathbf{R}_{\mathrm{CUT}}, \hat{\mathbf{R}}) \underset{\mathcal{H}_0}{\overset{\mathcal{H}_1}{\gtrless}} \gamma \tag{11}$$

where $\hat{\mathbf{R}}$ is the estimated value of the clutter covariance matrix by the secondary data, and γ is the detection threshold. $\mathcal{D}(\cdot)$ represents the geometric distance employed by the MIG detector.

3. Steering-Vector-Based Matrix Information Geometry Detector and Its Performance Analysis

In this section, the S-MIG detector for STAD is proposed from the likelihood ratio test, and its geometric explanation on the HPD manifold is given. In addition, as a target detection method, the performance of the S-MIG detector is analyzed in theory.

3.1. Steering-Vector-Based Matrix Information Geometry Detector

In statistical signal processing, the likelihood ratio test is a basic form of hypothesis testing. Denote $x_1, x_2, \cdots, x_M$ as the M independent and identically distributed observations of Q, and $\mathbf{q}$ is the probability density function. Assume that there are two hypotheses for $\mathbf{q}$, denoted as $\mathbf{p}_0$ and $\mathbf{p}_1$, respectively. The likelihood ratio of this hypothesis test problem is

$$L = \prod_{i=1}^{M} \frac{\mathbf{p}_1(x_i)}{\mathbf{p}_0(x_i)}, \tag{12}$$

and the normalized log-likelihood ratio is

$$l = \frac{1}{M} \ln L = \frac{1}{M} \sum_{i=1}^{M} \ln \frac{\mathbf{p}_1(x_i)}{\mathbf{p}_0(x_i)}. \tag{13}$$

Actually, l is the arithmetic average of log-likelihood ratio l_i. According to the strong law of large numbers, l convergences in probability to the expectation of $l_i, i = 1, \cdots, M$: namely,

$$\begin{aligned} l = E[l_i] &= \int \mathbf{q}(x) \ln \frac{\mathbf{p}_1(x)}{\mathbf{p}_0(x)} dx \\ &= \int \mathbf{q}(x) \ln \left[\frac{\mathbf{p}_1(x)}{\mathbf{p}_0(x)} \frac{\mathbf{q}(x)}{\mathbf{q}(x)} \right] dx \\ &= \int \mathbf{q}(x) \left[\ln \frac{\mathbf{q}(x)}{\mathbf{p}_0(x)} - \ln \frac{\mathbf{q}(x)}{\mathbf{p}_1(x)} \right] dx \\ &= \int \mathbf{q}(x) \ln \frac{\mathbf{q}(x)}{\mathbf{p}_0(x)} dx - \int \mathbf{q}(x) \ln \frac{\mathbf{q}(x)}{\mathbf{p}_1(x)} dx. \end{aligned} \tag{14}$$

Denote $\mathcal{D}_{\mathrm{KL}}(\mathbf{q}||\mathbf{p}_0)$ and $\mathcal{D}_{\mathrm{KL}}(\mathbf{q}||\mathbf{p}_1)$ as the KLD from $\mathbf{q}_0$ to $\mathbf{p}$ and the KLD from $\mathbf{q}_1$ to $\mathbf{p}$, respectively; the formula mentioned above can be written as

$$l = \mathcal{D}_{\mathrm{KL}}(\mathbf{q}||\mathbf{p}_0) - \mathcal{D}_{\mathrm{KL}}(\mathbf{q}||\mathbf{p}_1) \tag{15}$$

Formula (15) reveals the equivalence between the likelihood ratio and the Kullback–Leibler divergence; therefore, the likelihood ratio test also has its KL divergence form, expressed as

$$\mathcal{D}_{\mathrm{KL}}(\mathbf{q}||\mathbf{q}_0) - \mathcal{D}_{\mathrm{KL}}(\mathbf{q}||\mathbf{q}_1) \underset{\mathcal{H}_0}{\overset{\mathcal{H}_1}{\gtrless}} \gamma \tag{16}$$

Compare Formulas (13) and (16): the KLD can be interpreted as a measurement of probability. The larger $\mathcal{D}_{\mathrm{KL}}(\mathbf{q}||\mathbf{p}_0)$ is, the lower probability that hypothesis $\mathbf{p}_0$ has of being true.

For data in matrix form, the detection formula for radar target detection is

$$\mathcal{D}_{\mathrm{KL}}(\mathbf{R}_{\mathrm{CUT}}, \mathbf{R}_{\mathrm{c}}) - \mathcal{D}_{\mathrm{KL}}(\mathbf{R}_{\mathrm{CUT}}, \mathbf{R}_{\mathrm{c}} + \mathbf{R}_{\mathrm{t}}) \underset{\mathcal{H}_0}{\overset{\mathcal{H}_1}{\gtrless}} \gamma \tag{17}$$

Noted that there is no a priori knowledge of the parameter α_t: the matrix $\mathbf{R}_t$ in Formula (17) is unknown. In this case, the generalized likelihood ratio test (GLRT) is a suitable means for circumventing such a priori uncertainty. The GLRT is

$$\max_{\alpha_t} l \underset{\mathcal{H}_0}{\overset{\mathcal{H}_1}{\gtrless}} \gamma \tag{18}$$

which is equal to

$$\max_{\alpha_t}(\mathcal{D}_{\text{KL}}(\mathbf{R}_{\text{CUT}}, \mathbf{R}_c) - \mathcal{D}_{\text{KL}}(\mathbf{R}_{\text{CUT}}, \mathbf{R}_c + \mathbf{R}_t)) \underset{\mathcal{H}_0}{\overset{\mathcal{H}_1}{\gtrless}} \gamma \tag{19}$$

The gradient of the left-hand side (LHS) of Formula (19) is

$$\nabla_{\mathbf{R}_t} LHS = -\mathbf{R}_{\text{CUT}}{}^{-1} + \frac{1}{\mathbf{R}_{\text{CUT}}(\mathbf{R}_c + \mathbf{R}_t)} \cdot \mathbf{R}_{\text{CUT}} \tag{20}$$

From the equation $\nabla_{\mathbf{R}_t} LHS = 0$, the LHS of Formula (19) can be easily shown to be minimum at:

$$\mathbf{R}_{\text{CUT}} = \mathbf{R}_c + \mathbf{R}_t \tag{21}$$

The least-square solution of this contradictory equation is

$$\alpha_t \alpha_t^{\text{H}} = \frac{< \mathbf{R}_{\text{CUT}} - \mathbf{R}_c, \mathbf{s}_t \mathbf{s}_t^{\text{H}} >}{< \mathbf{s}_t \mathbf{s}_t^{\text{H}}, \mathbf{s}_t \mathbf{s}_t^{\text{H}} >} \tag{22}$$

where $< \cdot >$ is the inner product of the matrix. Taking Formulas (22) and (10) into (17), we get the output of the S-MIG detector

$$\text{tr}(\mathbf{R}_{\text{CUT}}^{-1}\mathbf{R}_c) - \text{tr}(\mathbf{R}_{\text{CUT}}^{-1}(\mathbf{R}_c + \mathbf{R}_t)) - \log(|\mathbf{R}_{\text{CUT}}^{-1}\mathbf{R}_c|) + \log(|\mathbf{R}_{\text{CUT}}^{-1}(\mathbf{R}_c + \mathbf{R}_t)|) \tag{23}$$

in which

$$\mathbf{R}_t = \frac{< \mathbf{R}_{\text{CUT}} - \mathbf{R}_c, \mathbf{s}_t \mathbf{s}_t^{\text{H}} >}{< \mathbf{s}_t \mathbf{s}_t^{\text{H}}, \mathbf{s}_t \mathbf{s}_t^{\text{H}} >} \mathbf{s}_t \mathbf{s}_t^{\text{H}} \tag{24}$$

where $\text{tr}(\cdot)$ and $|\cdot|$ denote the trace and determinant of the matrix, respectively.

3.2. Geometric Explanation of the S-MIG Detector

Consider the situation that the PDF of the hypothesis that the target exists is unknown; detection Formula (16) is simplified to

$$\mathcal{D}(\mathbf{q} \| \mathbf{p}_0) \underset{\mathcal{H}_0}{\overset{\mathcal{H}_1}{\gtrless}} \gamma, \tag{25}$$

which is consistent with the detection formula of the MIG detector (11). Therefore, the MIG detector can be regarded as a centroid detector on the manifold. The decision boundary of the MIG detector on the manifold is a circle, as shown in Figure 2. If the geometric distance between $\mathbf{R}_{\text{CUT}}$ and $\hat{\mathbf{R}}$ is greater or equal to the threshold γ, the target is considered to exist in the CUT. Therefore, the decision domain of hypothesis $\mathcal{H}_0$ is a solid circle, and others are the decision domain of hypothesis $\mathcal{H}_1$.

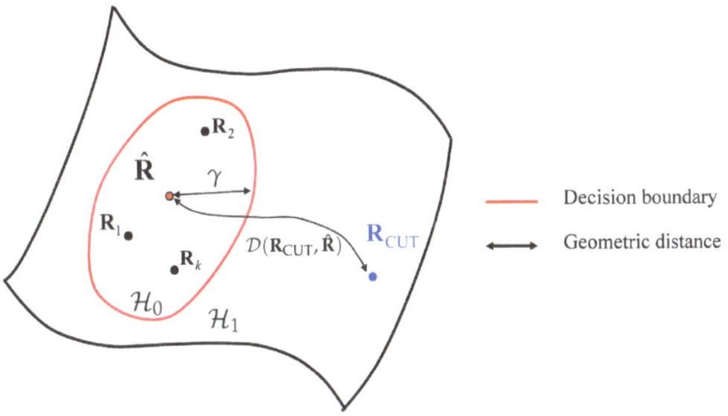

Figure 2. Geometric explanation of the MIG detector.

Different from the case that the target information is unknown, for the S-MIG detector, the target steering vector $\mathbf{s}_\mathrm{t}$ is known and α_t can be estimated; thus, the PDF and covariance matrix of hypothesis $\mathcal{H}_1$ is utilizable. Benefiting from the target's steering vector, a rectification term is added to the MIG detector: from Formula (11) to Formula (17). On the manifold, the decision boundary of the S-MIG detector is a hyperbola defined by $\hat{\mathbf{R}}$, $\hat{\mathbf{R}} + \mathbf{R}_\mathrm{t}$, and γ. The decision domain of the S-MIG detector is shown in Figure 3.

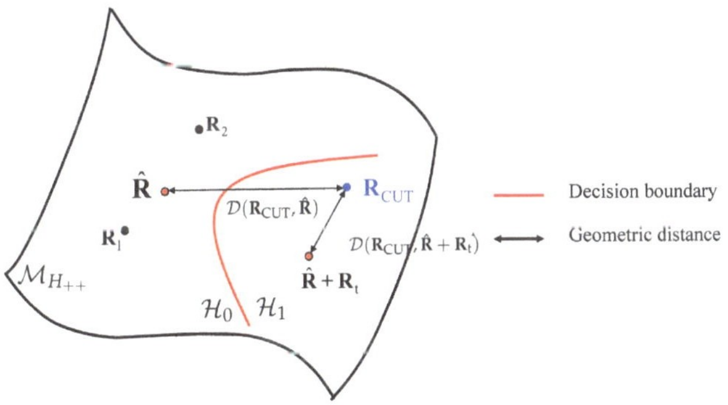

Figure 3. Geometric explanation of the S-MIG detector.

The geometric distance is an effective method to measure the similarities and differences between two matrices. The statistical meaning of the geometric distance can be interpreted as: the smaller the geometric distance is, the more similar the matrices are and the closer their statistical properties are. From this perspective, the MIG detector can be regarded as a detector that compares whether the CUT and the estimation of the clutter are "similar" enough—if they are not, the target exists, regardless of whether it is the target of interest or not. Likewise, the S-MIG detector is a detector that measures which hypothesis's likelihood is greater, focusing on the target of interest and the clutter. The S-MIG detector considers which hypothesis is more similar instead of caring about resemblance to hypothesis $\mathcal{H}_0$ as the MIG detector does. Therefore, the robustness of the S-MIG detector is better than that of the MIG detector. Much heterogeneous clutter will lead to false alarms in the MIG detector, but the wider decision domain of the S-MIG detector leads to more heterogeneous clutter remaining in the decision domain of hypothesis $\mathcal{H}_0$.

Based on the scheme analyzed in this subsection and shown in Figure 3, a variety of S-MIG detectors can be designed by using different geometric metrics. As for the selection of geometric metrics, the optimal metric is different in different clutter environments. In this paper, we focus on the KLD-based S-MIG detector.

3.3. Performance Analysis

This subsection is devoted to analyzing the detection performance of the S-MIG detector in theory, including the computation complexity and robustness. In this subsection, AMF [8], adaptive normalized matched filter (ANMF) [55], and MIG detectors are selected as comparison methods. The detection formulas of the AMF detector and the ANMF detector, respectively, are

$$\Lambda_{\mathrm{AMF}} = \frac{|\mathbf{s}_t^H \mathbf{R}_c^{-1} \mathbf{x}_{\mathrm{CUT}}|^2}{\mathbf{s}_t^H \mathbf{R}_c^{-1} \mathbf{s}_t} \tag{26}$$

$$\Lambda_{\mathrm{ANMF}} = \frac{|\mathbf{s}_t^H \mathbf{R}_c^{-1} \mathbf{x}_{\mathrm{CUT}}|^2}{(\mathbf{s}_t^H \mathbf{R}_c^{-1} \mathbf{s}_t)(\mathbf{x}_{\mathrm{CUT}}^H \mathbf{R}_c^{-1} \mathbf{x}_{\mathrm{CUT}})} \tag{27}$$

3.3.1. Computation Complexity Analysis

Adaptive radar target detectors usually include three steps: the construction of covariance matrices, the estimation of the clutter covariance matrix, and the calculation of the test statistic.

In step I of each detection operation, the primary data and K secondary data should be modeled as HPD matrices by Formula (5). In Formula (5), each element in the covariance matrix is calculated by $x_k = \sum_{i=k+1}^{M} x_i x_{i+k}, 0 \leq k \leq M-1$; therefore, the computation complexity of each covariance matrix is $\mathcal{O}(M^2)$. Because the covariance matrix of $\mathbf{x}_{\mathrm{CUT}}$ is not required in AMF or ANMF detectors, their computation complexity is $\mathcal{O}(KM^2)$, while the calculation complexity of MIG and S-MIG detectors is $\mathcal{O}((K+1)M^2)$.

In step II, the computation complexity of different CCM estimate methods is various. For the MIG detector, many information-geometry-based estimation methods have outstanding performance but heavy computation burden. To evaluate various detection methods, the algebraic mean of the secondary data is employed as the estimation result of the CCM in this paper, which is shown below.

$$\hat{\mathbf{R}} = \frac{1}{K} \sum_{i=1}^{K} \mathbf{R}_i \tag{28}$$

The computation complexity of the algebraic mean is $\mathcal{O}(KM)$.

In step III, note that the matrix inversion operation is contained in these four detectors, which occupies the main computing cost; thus, the computation complexity of (10), (26), and (27) is $\mathcal{O}(M^3)$. For the S-MIG detector, an extra operation (22) is added compared to the MIG detector, for which the computation complexity is $\mathcal{O}(M)$, and the computation complexity of Formula (23) is $\mathcal{O}(M^3)$.

In summary, the computation complexities of four detectors are listed in Table 1. In most cases, the DoF M is far greater than the number of reference cells K without dimension reduction; thus, the summary of the computation complexities of detectors is $\mathcal{O}(M^3)$.

Table 1. The computation complexities of different detectors.

Detector	Step I	Step II	Step III	Summary
AMF	$\mathcal{O}(KM^2)$	$\mathcal{O}(KM)$	$\mathcal{O}(M^3)$	$\mathcal{O}(M^3)$
ANMF	$\mathcal{O}(KM^2)$	$\mathcal{O}(KM)$	$\mathcal{O}(M^3)$	$\mathcal{O}(M^3)$
MIG	$\mathcal{O}((K+1)M^2)$	$\mathcal{O}(KM)$	$\mathcal{O}(M^3)$	$\mathcal{O}(M^3)$
S-MIG	$\mathcal{O}((K+1)M^2)$	$\mathcal{O}(KM)$	$\mathcal{O}(M^3)$	$\mathcal{O}(M^3)$

According to Table 1, the computation costs of the proposed S-MIG detector and other detectors are on the same scale, indicating that the complexity of the S-MIG detector is acceptable.

3.3.2. Robustness Analysis

For the reason that the order of magnitudes of different algorithms are conflicting and their metrics have different statistical implications, it is improper to assess the robustness of detectors in heterogeneous clutter by the normalized deviation of the detection statistic. In this paper, an influence function (IF) is proposed to evaluate the robustness of space–time adaptive detectors in heterogeneous environments, which is the unitized deviation of the detection statistic defined below.

The influence function of space–time adaptive detectors is

$$IF(m,n) = E\left[\frac{|\Lambda(\mathbf{R}_{\mathrm{CUT}}, \hat{\mathbf{R}}) - \Lambda(\mathbf{R}_{\mathrm{CUT}}, \mathbf{R}_c)|}{mean(metrics(\mathbf{R}_{\mathrm{CUT}}, \mathbf{R}_c), metrics(\mathbf{R}_{\mathrm{CUT}}, \mathbf{R}_J))}\right] \tag{29}$$

where $\mathbf{R}_J$ is the covariance matrix of heterogeneous clutter, and $mean(x,y)$ denotes the average of x and y; m and $n = K - m$ are the number of the range cells of homogeneous clutter and heterogeneous clutter, respectively; $metrics(\cdot,\cdot)$ represents the metrics of detectors. For the AMF and ANMF detectors, $metrics(\mathbf{R}_{\mathrm{CUT}},\cdot) = \Lambda(\mathbf{x}_{\mathrm{CUT}},\cdot)$, and for the MIG detector and S-MIG detector, $metrics(\mathbf{R}_{\mathrm{CUT}},\cdot) = \mathcal{D}_{\mathrm{KL}}(\mathbf{R}_{\mathrm{CUT}},\cdot)$. Based on the analysis of measured data, heterogeneous clutter may be localized or distributed [56], as shown in Figure 4. In the former case, the homogeneous clutter is disturbed by discrete mountains or man-made buildings, but the other terrain is consistent with homogeneous clutter. The covariance matrix of hetero-clutter can be expressed by

$$\mathbf{R}_J = \mathbf{R}_c + \sum \alpha_J \alpha_J^H \mathbf{s}_J \mathbf{s}_J^H \tag{30}$$

where α_J and $\mathbf{s}_J$ are the reflection coefficient and steering vector, respectively, of discrete interference. Usually, the steering vector $\mathbf{s}_J$ is considered separate from the target.

In the second case, where the hetero-clutter is distributed, the hetero-clutter has entirely different statistical properties and a covariance matrix from the homo-clutter. Distributed hetero-clutter occurs when two or more kinds of terrain are illuminated by airborne radar, such as a sea–land interface scene. In this case, we have

$$\mathbf{R}_J \neq \mathbf{R}_c. \tag{31}$$

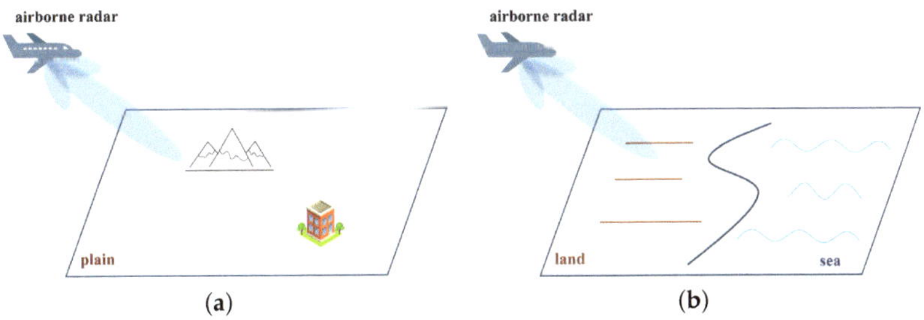

Figure 4. Schematic diagrams of hetero-clutter. (**a**) localized clutter. (**b**) distributed clutter.

Theorem 1. *In heterogeneous clutter environments, the robustness of the S-MIG detector is superior to that of the MIG detector: namely,*

$$IF_{\mathrm{S-MIG}}(m,n) \leq IF_{\mathrm{MIG}}(m,n) \tag{32}$$

Proof. See Appendix A. □

The theorem introduced above indicates that in heterogeneous environments, the S-MIG detector will not be interfered with more seriously than the MIG detector. Considering that airborne radar usually works in heterogeneous environments, this property is of great significance.

4. Experimental Results and Analysis

In this section, the performance of the proposed S-MIG is compared with that of other STAD methods, such as the AMF and ANMF detectors; in addition, target detection methods based on STAP and CFAR are selected as comparison methods as well. It should be pointed out that the loaded sample covariance matrix (LSCM) algorithm is designated as the exclusive CCM estimate method; therefore, the detection performances of alternative methods are different because of their detection abilities rather than something else. The experiments in this section are divided into two parts: experiments based on simulated data to investigate the properties of the S-MIG detector and experiments based on measured data to validate the superiority of the proposed method.

4.1. Experiments Based on Simulated Data

Performance results are tested on simulated data in this subsection. In this subsection, we focus on three important properties of multi-channel detectors: direction sensitivity, robustness, and computational efficiency. Direction sensitivity represents the ability to suppress interference in other directions, and robustness reflects the reliability of detectors in heterogeneous environments. In addition, computational efficiency is vital because the application feasibility of the proposed method depends on it.

4.1.1. Directional Sensitivity

To assess the direction sensitivity of detectors, 10^3 independent trails are repeated by the standard Monte Carlo counting technique. In the simulation, the clutter shows the same covariance matrix Σ generated by

$$\Sigma = \Sigma_0 + \mathbf{I} \tag{33}$$

where

$$\Sigma_0(i,k) = \sigma_c^2 \rho^{|i-k|} e^{j2\pi f_{dc}(i-k)}, \quad i,k = 1, \cdots, M \tag{34}$$

Here, ρ is the one-lag correlation coefficient, σ_c denotes the clutter-to-noise power ratio, and f_{dc} is the clutter Doppler frequency. In this part, we set $\sigma_c^2 = 20$ dB, $\rho = 0.6$, and $f_{dc} = 0.2$. The number of array elements N is 16, and the number of coherent pulses P is 10. These parameters are used in the following simulated experiments unless otherwise specified.

A simulated target with normalized spatial Doppler 0.55 and normalized temporal Doppler 0.3 is added to the clutter. The signal-to-clutter ratio (SCR) in this simulation is 5 dB. Figure 5 shows the normalized output of different directions (normalized spatial Doppler), which is obtained by detecting the target with different steering vectors. According to Figure 5, the detectors achieve the best performance when the steering vector is consistent with the target, and performance decreases if the directions are mismatched. The directional sensitivities of the AMF and ANMF detectors are similar but are worse than that of the S-MIG detector. The S-MIG detector still has better detection ability in the case of a small amount of direction mismatch, but the ability drops rapidly. The result indicates that for the S-MIG detector, a spot of target direction mismatch is tolerable, but interference from other directions will be efficiently suppressed, especially when the normalized spatial Doppler mismatch exceeds 0.3.

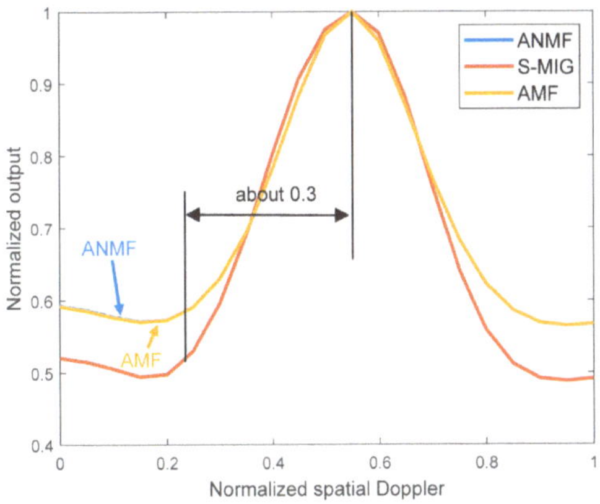

Figure 5. Normalized output in different directions.

4.1.2. Robustness

The robustness of detectors is studied by analyzing their influence functions in heterogeneous environments with different numbers of reference cells. In this experiment, the number of hetero-clutter range cells is set to 24. Considering the localized hetero-clutter and distributed hetero-clutter introduced in Section 3.3.2, two experiments are performed. The homogeneous clutter used in the experiments is given by Formula (33), and the clutter follows a K distribution. In case I, two interferences are added into homo-clutter to imitate discrete buildings. The covariance matrix of hetero-clutter is shown as Formula (30). In case II, the parameters in Formula (33) are entirely changed to establish a completely different covariance matrix. The parameters for the simulation experiments are listed in Table 2.

Table 2. Parameters for simulation.

Parameters	Clutter	Hetero-Clutter in Case I	Hetero-Clutter in Case II
N	4	4	4
P	6	6	6
n	-	24	24
ρ	0.6	0.6	0.1
σ^2	20 dB	20 dB	5 dB
f_{dc}	0.2	0.2	0.8
shape parameter v	0.4	0.4	0.4
scale parameter b	0.6	0.6	0.6
number of interference	0	2	0
(f_{d1}, f_{s1})	-	(0.2, 0.2)	-
(f_{d2}, f_{s2})	-	(0.2, 0.9)	-
(f_{dt}, f_{st})	(0.55, 0.55)	-	-

Figure 6 describes the IF performances of case I and case II. As seen in Figure 6a, in localized clutter, the IF curve of the S-MIG detector is superior to that of the MIG detector, which means that the detection result of the S-MIG detector will be less affected in a localized heterogeneous environment. The ANMF detector achieves the best performance and is followed by the AMF detector. However, in the distributed heterogeneous clutter environment, the AMF detector is inferior to other detectors, and the IF curves of the MIG and S-MIG detector are almost identical. Different from the ANMF detector, which has similar robustness in both cases, the MIG detector suffers from localized clutter much more than from distributed clutter. Compared to the MIG detector, the robustness of the proposed S-MIG detector is effectively improved.

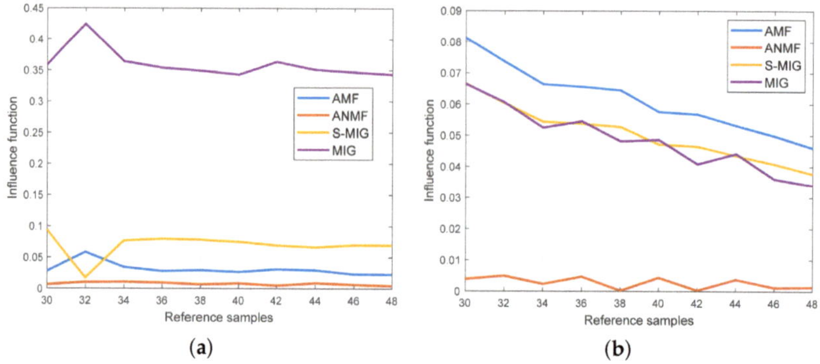

Figure 6. Influence function curves for different numbers of reference cells. (**a**) localized clutter. (**b**) distributed clutter.

4.1.3. Computational Efficiency

In this numerical simulation experiment, the average times for 500 runs of the proposed detector and the competitive methods are tested. To assess the run-times of various detectors, all these algorithms are performed based on the same estimate method. In addition, a CFAR detector is added after STAP. The experiment results are presented in Figure 7, which shows that there is no significant increase in computation time, and the practical run-times of the proposed method and the other methods are roughly equal. Under different DoFs, the run-time of the S-MIG detector is approximately 5 times that of the AMF detector and less than 2 times that of the MIG detector, which is close to that of the STAP-CFAR detector.

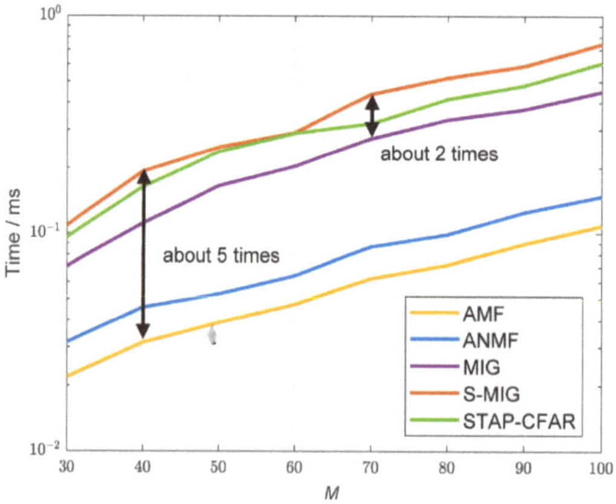

Figure 7. The average run-times for detectors.

4.2. Experiments Based on Measured Data

Experiments based on measured data are performed to investigate the weak target detection ability in heterogeneous environments. The experiment based on measured data includes three parts. The first part is based on the sea-detecting data-sharing program (SDDSP) [57,58], and the second is based on the mountaintop data set to verify the robustness of the S-MIG detector, as shown in Section 4.2.2. The third one is an experiment based on the PHA radar data set published in [59] to investigate the weak target detection ability, as shown in Section 4.2.3.

4.2.1. Real Data of the Sea-Detecting Data-Sharing Program

In order to investigate the suppression ability and robustness of different detectors, a real scene containing ground and sea clutter is considered. In this experiment, data cube 20191025165159_01_*scanning* is utilized, in which range cells from 200 to 500 are land–sea interface clutter, and range cells from 1 to 200 and from 500 to 1320 are land clutter and sea clutter, respectively, as presented in Figure 8. The measured data are expanded into multi-channel data by a phase shift. Three simulated targets are added in range cells 200, 400, and 800. Their SCR is -15 dB, and their normalized Doppler is 0.1. Both the clutter and target are located on the normal line of the antennae array. The number of reference samples is 24. Figure 9 presents the clutter suppression and target detection results of detectors with different types of clutter. As we can see, the proposed method achieves the best detection performance in three types of clutter environments. The STAP method can detect the target, but there are a lot of false alarms around it. The AMF detector and the MIG detector fail to detect the target in three clutter environments. The experimental results illustrate that the proposed method can suppress the different types of clutter and detect weak targets in the heterogeneous environment.

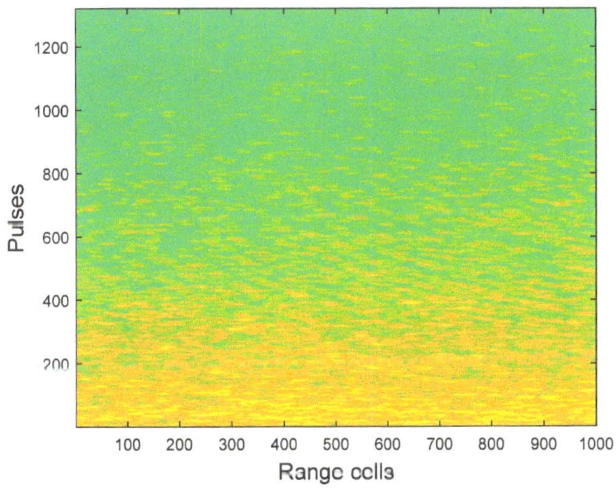

Figure 8. Measured data of SDDSP.

4.2.2. Real Data of the Mountaintop Program

The mountaintop data set was collected to support STAP studies at first and is now widely used in research on clutter suppression and target detection for airborne radars. The radar surveillance technology experimental radar (RSTER) is central to the mountaintop program. The antenna for the system is a 5 m wide by 10 m high horizontally polarized array made up of 14 column elements. Behind each element are an independent phase shifter, transmitter, and receiver. The RSTER is deployed on North Oscura Peak (NOP) on the White Sands Missile Range (WSMR). NOP is at the northeast corner of the WSMR and is 8000 feet above sea level and approximately 3500 feet above the desert floor [60–62], as shown in Figure 10. The site offers line-of-sight to a variety of terrain types including desert, bare and wooded hills, mountains, lava flows, and small suburban areas.

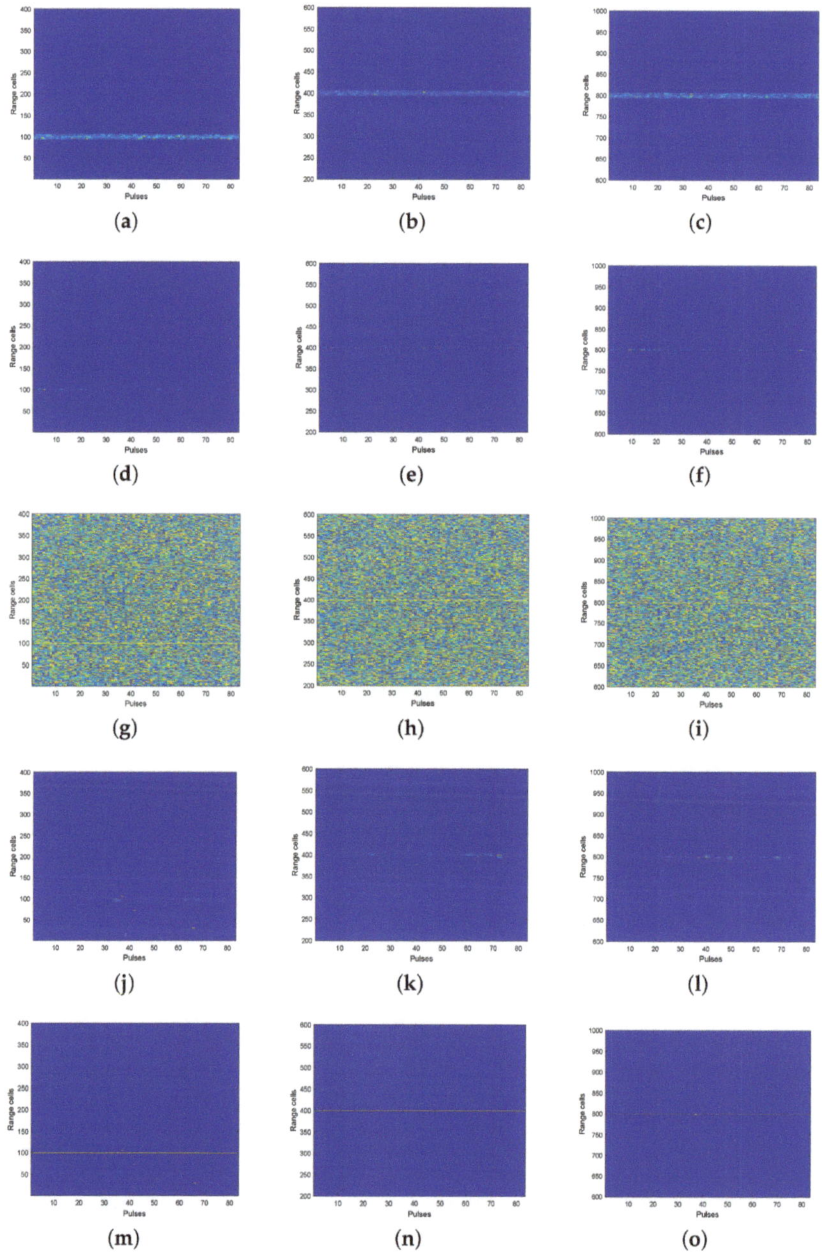

Figure 9. The clutter suppression and target detection performances of different methods in various clutter environments. (**a**) STAP, land clutter. (**b**) STAP, land-sea clutter. (**c**) STAP, sea clutter. (**d**) AMF, land clutter. (**e**) AMF, land-sea clutter. (**f**) AMF, sea clutter. (**g**) ANMF, land clutter. (**h**) ANMF, land-sea clutter. (**i**) ANMF, sea clutter. (**j**) MIG, land clutter. (**k**) MIG, land-sea clutter. (**l**) MIG, sea clutter. (**m**) S-MIG, land clutter. (**n**) S-MIG, land-sea clutter. (**o**) S-MIG, sea clutter.

Figure 10. The RESTER system (**left**) and its antenna (**right**) [63].

The data used in this paper are data file *stap3001v1.mat*, which contains 2 coherent processing intervals (cpi) and 404 range cells. A remotely sited broadband jammer provides the only signal in the data. In addition to the signal observed in the jammer's direction, some terrain-scattered jamming energy may be observed in the data. The jamming signal is pseudo-random noise with a bandwidth of 600 kHz, which is broadband relative to the radar's bandwidth, so it appears as broadband barrage jamming. The angle between the jammer and the radar direction is 42° (normalized spatial Doppler -0.31). Figure 11 presents the Fourier spectrum of the received signal. According to Figure 11, the power of the jammer signal is about 50 dB greater than that of the clutter.

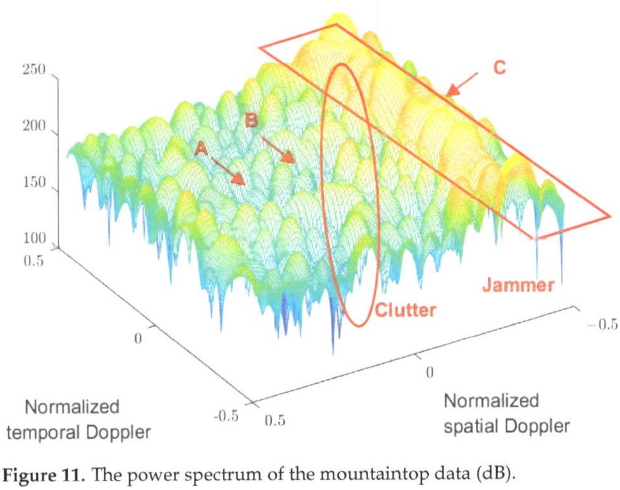

Figure 11. The power spectrum of the mountaintop data (dB).

Three simulated targets with different Doppler are added into the clutter, and their detection performances under different signal-to-clutter-plus-interference ratios (SCIR) is analyzed. The Doppler of simulated targets and other parameters in this experiment are listed in Table 3. Their spatial Doppler were set up differently to test the effectiveness of various detectors on targets in different directions, where target A imitates an ordinary target, target B is a simulated slowly moving target whose Doppler is close to that of the clutter, and target C is a target near the jammer. As shown in Figure 12, the spectrum of clutter in different range cells is different, and the clutter is heterogeneous as well, which is challenging for detectors. First, the experimental results are shown in the form of output curves that focus on the average, minimum, and maximum outputs of detectors in hypotheses $\mathcal{H}_1$ and $\mathcal{H}_0$.

Table 3. Parameters for simulation.

Parameter	Symbol	Value
Doppler for target A	(f_{dA}, f_{sA})	(0.25, −0.16)
Doppler for target B	(f_{dB}, f_{sB})	(0.01, −0.16)
Doppler for target C	(f_{dC}, f_{sC})	(−0.41, −0.16)
DoF	M	224
number of reference cells	K	28
number of guard cells	-	1
CPI sequence number	cpi	1

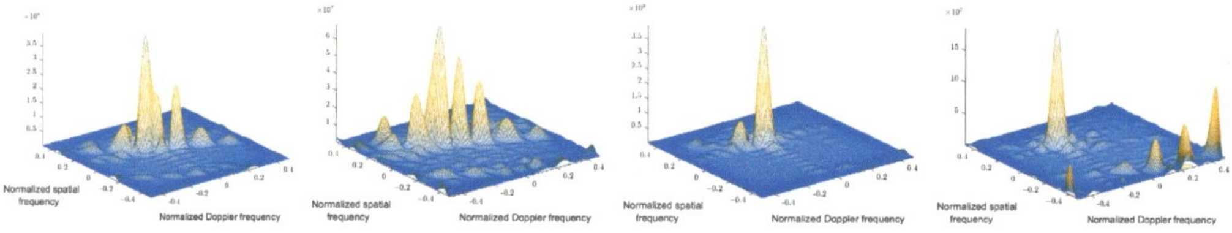

Figure 12. The spectrum of the clutter in different range cells. (**a**) range cell 20. (**b**) range cell 27. (**c**) range cell 35. (**d**) range cell 46.

The output curves are presented in Figures 13–15. In these figures, the curves represent the average outputs for both hypotheses, and the shaded regions around the curves indicate the distribution intervals of the output statistics, whose upper bound and lower bound are the maximum and minimum, respectively. In this paper, we define the absolutely detected point (ADP) to assess the performance of detectors. The ADP refers to the SCIR that has its minimum output in hypothesis $\mathcal{H}_1$ greater than its maximum output in hypothesis $\mathcal{H}_0$, as marked in Figure 13. As we can see, the ADP of the proposed S-MIG detector is approximately 5 dB lower than those of the other detectors, meaning that a lower SCIR is needed for the proposed method to detect the target without a missed alarm. For target A, the S-MIG detector achieves the best performance and is followed by the AMF detector. As the target's Doppler approaches the clutter, the ADPs of detectors increase in Figure 14, but in such a case, the S-MIG detector still maintains advantageous performance. Furthermore, another unfavorable case is considered. The direction of simulated target C is close to the jammer, with a spatial Doppler difference of 0.1; here, the performance of the S-MIG detector is superior to those of the others as well. Particularly, the ADPs are listed in Table 4. The result proves that in heterogeneous environments, the proposed method can detect the target in a lower SCNR, which represents superior detection performance for weak targets.

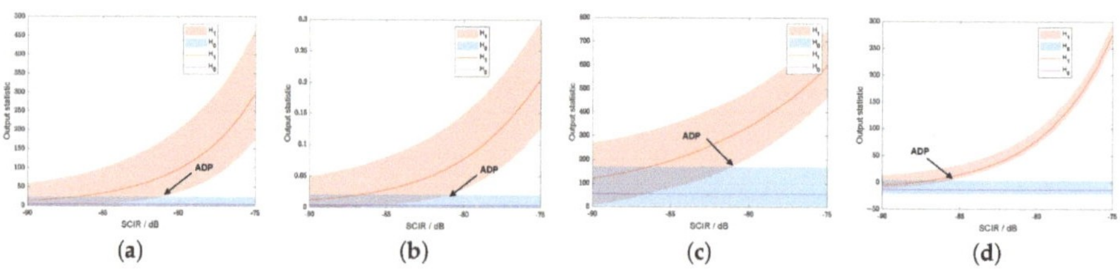

Figure 13. Output statistic curves for different SCIRs for target A. (**a**) AMF. (**b**) ANMF. (**c**) STAP. (**d**) S-MIG.

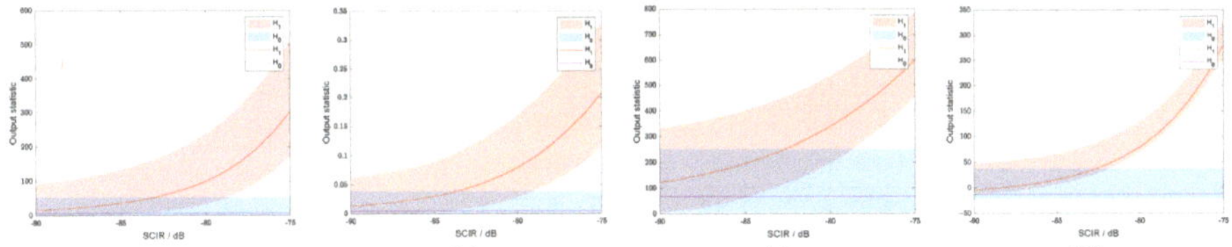

(a) (b) (c) (d)

Figure 14. Output statistic curves for different SCIRs for target B. (**a**) AMF. (**b**) ANMF. (**c**) STAP. (**d**) S-MIG.

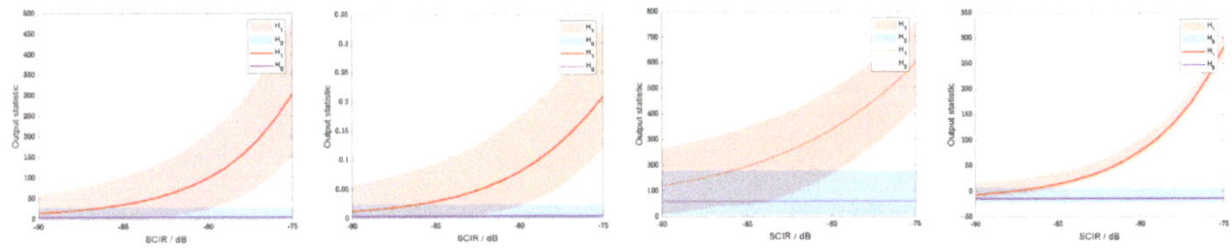

(a) (b) (c) (d)

Figure 15. Output statistic curves for different SCIRs for target C. (**a**) AMF. (**b**) ANMF. (**c**) STAP. (**d**) S-MIG.

Table 4. ADPs for different detectors.

	AMF	ANMF	STAP	S-MIG
Target A	−81.06 dB	−81.05 dB	−81.08 dB	−85.63 dB
Target B	−78.69 dB	−79.03 dB	−78.62 dB	−81.76 dB
Target C	−79.65 dB	−79.66 dB	−79.63 dB	−84.88 dB

Furthermore, the probability-of-detection curves for the various detectors are presented in Figure 16. In this experiment, the simulated target is located in the direction of 90°, and its normalized Doppler is 0.15. To perform a Monte Carlo experiment, the received signals of different coherent processing intervals (CPI) are employed to calculate the Pd values of various detectors.

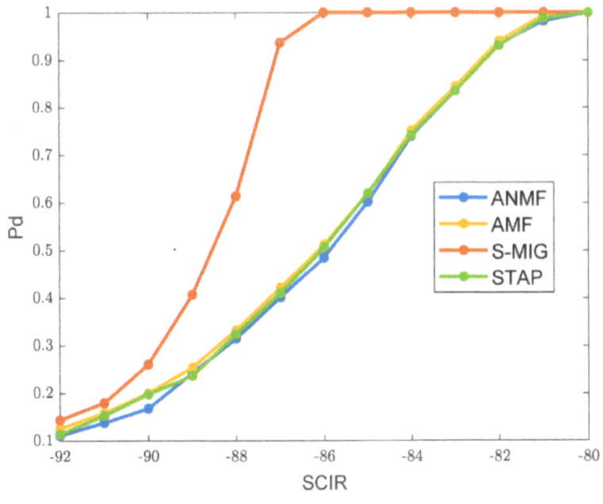

Figure 16. The Pd curves of the various detectors.

4.2.3. Real Data of the PHA Radar Database

In order to develop detection and tracking techniques, a series of low-altitude targets and radar clutter measurement trials were conducted by a high-resolution phased-array

(PHA) radar in 2023 at Dongying, Shandong, China. The PHA radar is a Ku-band radar with a synthetic bandwidth of 1 GHz and detects targets from 300 to 2000 m. The PHA data set includes radar data for clutter measurement trails, migrating insect and bird observations, and drone observations [59].

In this experiment, the measured radar data of a single drone moving at a low speed is employed to verify the effectiveness of the proposed method, as shown in Figure 17. Here, the cooperative target is a DJI M300 drone that is moving horizontally, and the radar works in tracking mode. For the reason that there is no highlight clutter in the horizontal directions and the target is located at a low elevation angle (11.68°), spatial adaptive detection is performed in the pitch dimension to suppress the ground clutter.

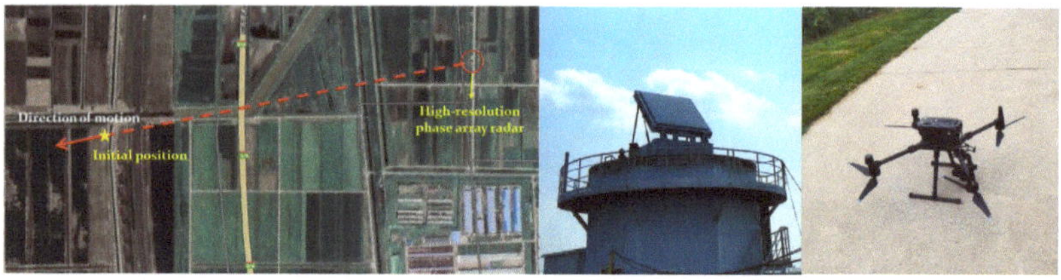

Figure 17. The test scene, radar, and cooperative target [59].

Figures 18 and 19 present the detection results for different methods, including STAD and STAP. In this experiment, the DoF is set to 12, 24 reference cells are valuable, and 4 range cells adjacent to the CUT are designated as guard cells. As we can see, the performances of filter-then-detect methods (Figure 18a–d) are unsatisfactory. Meanwhile, the ANMF detector, which detects the target by phase information, fails to highlight the target. Without the utilization of the given steering vector, the target can be detected by the MIG detector, but it is accompanied by many false alarms caused by clutter (shown in Figure 18g). One can see that after being improved by the target steering vector, the clutter is effectively suppressed in Figure 18h. Both the S-MIG detector and AMF detector have the best performance. In detail, the output signal-to-clutter-plus-noise ratios (SCNRs) are listed in Table 5. The statistics in Table 5 are calculated by the average of the selected range cells in this experiment (excluding two range cells around the target). Moreover, the performance advantage of STAD is clearly presented in Figure 19. The figure demonstrates that the S-MIG detector and AMF detector achieve the best SCNR, followed by the MIG detector and STAP. The S-MIG detector has an SCNR gain of approximately 15 dB compared to the MIG detector and approximately 3 dB compared to the AMF detector.

Table 5. Output SCNRs of different methods.

Method	SCNR	Method	SCNR
STAP without CFAR	16.626 dB	AMF	28.831 dB
CA-CFAR after STAP	12.559 dB	ANMF	7.751 dB
OS-CFAR after STAP	12.255 dB	MIG	17.977 dB
GO-CFAR after STAP	12.888 dB	S-MIG	32.060 dB

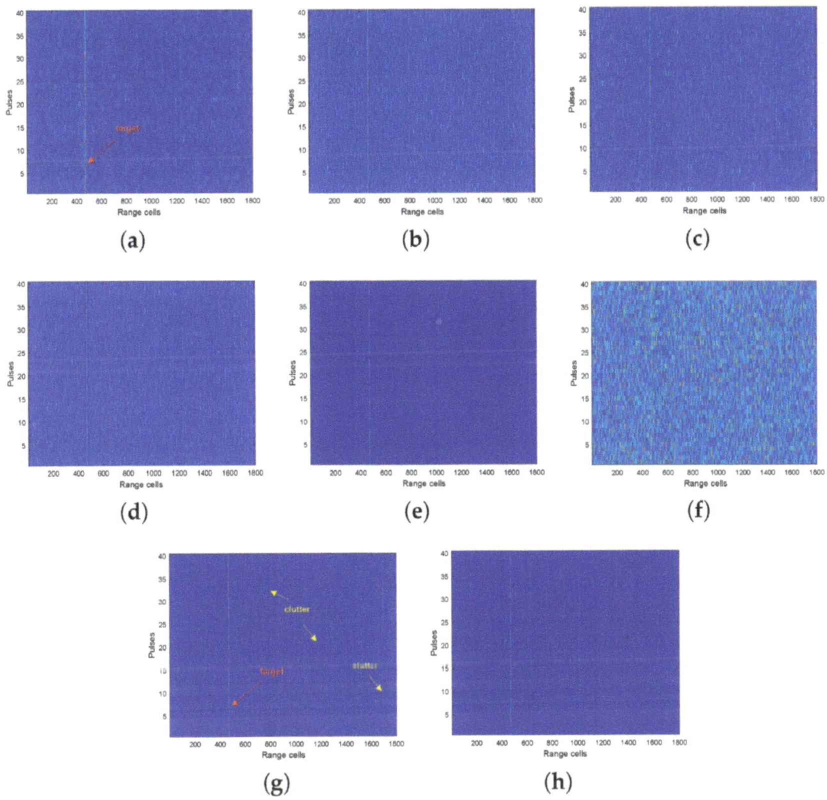

Figure 18. The range-pulse maps for different methods. (**a**) STAP without CFAR. (**b**) CA-CFAR after STAP. (**c**) OS-CFAR after STAP. (**d**) GO-CFAR after STAP. (**e**) AMF. (**f**) ANMF. (**g**) MIG. (**h**) S-MIG.

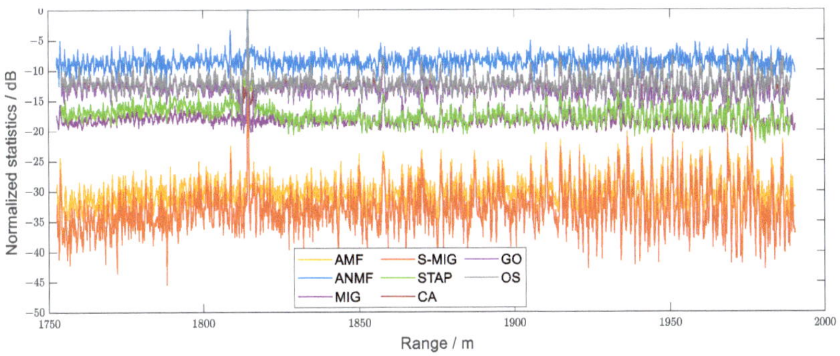

Figure 19. Normalized output statistics curves.

Then, to investigate the detection performance for weak moving targets in heterogeneous clutter, the clutter data from 1830 m to 1990 m are selected as the experimental environment, and a simulated target is added at 1885 m. To imitate a low-altitude drone hidden in ground clutter, the velocity of the target is set to 50 km/h, and the elevation angle is 30°. Figure 20a illustrates the amplitude of ground clutter in the range of 1800 to 2000 m. The ground clutter shows strong heterogeneity, as we can see, and the target is added in a position where the clutter power on both sides is significantly stronger than the CUT. In general, the number of homogeneous samples should be no less than two times the DoF to obtain a satisfying result [2]. However, considering that the clutter in this experiment is heterogeneous, 16 neighboring reference cells, which is less than twice the DoF, are valuable in each detection to reduce the negative influence of hetero-clutter. The detection performances for detectors against different SCRs in this scene are presented in Figure 20b. In this experiment, the false alarm rate is 10^{-5}, and 10^6 trials are performed to determine the detection threshold γ. To calculate the Pd, 10^5 Monte Carlo experiments are performed. Compared to competitive methods, the Pd curve of the proposed method increases faster and has the best performance. According to the Pd curves, the AMF detector and ANMF detector perform better for a lower SCR, but STAP becomes better as the SCR increases. The detection performance of the S-MIG detector is an improvement of about 4 dB and 5 dB over the ANMF detector and the AMF detector, respectively. Moreover, the SCR is improved by about 6.5 dB over the MIG detector when Pd = 0.5. In addition, the Pd curves of the persymmetric adaptive normalized matched filter (P-ANMF), the recursive P-ANMF (RP-ANMF), the fixed point adaptive normalized matched filter (FP-ANMF), and the persymmetric FP-ANMF (PFP-ANMF) are presented in Figure 20b as well. These four methods improved the CCM estimation algorithm of the ANMF detector to detect the weak target better. As shown in Figure 20b, the performances of the ANMF detector based on different CCM estimation algorithms are different, but the ANMF detector is still inferior to the S-MIG detector, which utilizes the ordinary LSCM estimation method. Compared to the RP-ANMF detector, the S-MIG detector has a SCR gain of 2 dB.

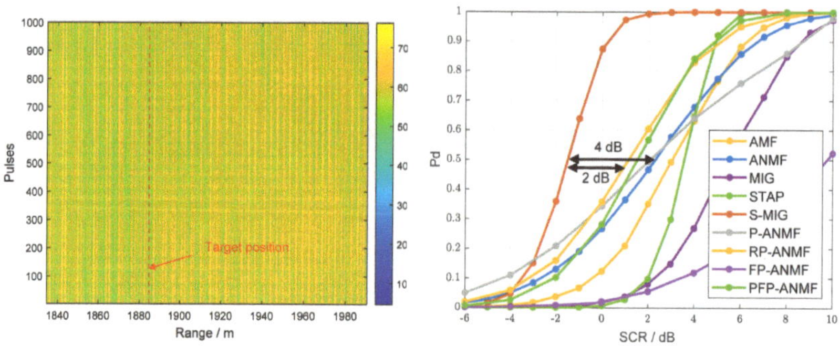

(**a**) Raw data (**b**) Probability of detection

Figure 20. Ground clutter and Pd curves.

5. Conclusions

This paper focuses on the weak target detection problem in heterogeneous clutter for airborne radars. Based on the GRLT and information geometry, a novel detector that makes use of prior information is proposed to extend the MIG detector to airborne multi-channel radar. According to the theoretical deduction result, the decision boundary on the manifold for the GRLT ought to be a hyperbola rather than a circle when prior knowledge is known. Then, it is revealed that the proposed S-MIG detector has better directional sensitivity than traditional methods and better robustness that the MIG detector; furthermore, its computational efficiency maintains the same level as the comparison methods. The simulated results validate that the performance of the proposed method is better than others.

Author Contributions: Writing—original draft preparation, R.Z.; writing—review and editing, R.Z., Y.C., H.W. (Hao Wu), Z.Y., and X.H.; supervision, Y.C.; conceptualization and methodology, R.Z.; investigation, Y.C.; resources and software, H.W. (Hanjie Wu). All authors have read and agreed to the published version of the manuscript.

Funding: This work was supported by the National Key Research and Development Program of China under grant 2022YFB3902400, the Distinguished Youth Science Foundation of Hunan Province under grant 2022JJ10063, and the National Natural Science Foundation of China under grants 61921001 and 62371458.

Data Availability Statement: Data is contained within the article.

Conflicts of Interest: The authors declare no conflicts of interest.

Appendix A. Proof of Theorem

Assume that there are K reference cells valuable for detectors; among them, m reference cells are homogeneous clutter and the other $n = K - m$ are heterogeneous.

The influence factors of the MIG detector and S-MIG detector, respectively, are

$$IF_{\mathrm{MIG}} = E[\frac{|\mathcal{D}_{\mathrm{KL}}(\mathbf{R}_{\mathrm{CUT}}, \hat{\mathbf{R}}) - \mathcal{D}_{\mathrm{KL}}(\mathbf{R}_{\mathrm{CUT}}, \mathbf{R}_{\mathrm{c}})|}{mean(\mathcal{D}_{\mathrm{KL}}(\mathbf{R}_{\mathrm{CUT}}, \mathbf{R}_{\mathrm{c}}), \mathcal{D}_{\mathrm{KL}}(\mathbf{R}_{\mathrm{CUT}}, \mathbf{R}_{\mathrm{J}}))}] \tag{A1}$$

$$IF_{\mathrm{S-MIG}} = E[\frac{|\mathcal{D}_{\mathrm{KL}}(\mathbf{R}_{\mathrm{CUT}}, \hat{\mathbf{R}}) - \mathcal{D}_{\mathrm{KL}}(\mathbf{R}_{\mathrm{CUT}}, \mathbf{R}_{\mathrm{c}}) - \mathcal{D}_{\mathrm{KL}}(\mathbf{R}_{\mathrm{CUT}}, \mathbf{R}_{\mathrm{t1}} + \hat{\mathbf{R}}) + \mathcal{D}_{\mathrm{KL}}(\mathbf{R}_{\mathrm{CUT}}, \mathbf{R}_{\mathrm{t0}} + \mathbf{R}_{\mathrm{c}})|}{mean(\mathcal{D}_{\mathrm{KL}}(\mathbf{R}_{\mathrm{CUT}}, \mathbf{R}_{\mathrm{c}}), \mathcal{D}_{\mathrm{KL}}(\mathbf{R}_{\mathrm{CUT}}, \mathbf{R}_{\mathrm{J}}))}] \tag{A2}$$

where $\mathbf{R}_{\mathrm{t1}} = \frac{<\mathbf{R}_{\mathrm{CUT}} - \hat{\mathbf{R}}, \mathbf{s}_{\mathrm{t}}\mathbf{s}_{\mathrm{t}}^{\mathrm{H}}>}{<\mathbf{s}_{\mathrm{t}}\mathbf{s}_{\mathrm{t}}^{\mathrm{H}}, \mathbf{s}_{\mathrm{t}}\mathbf{s}_{\mathrm{t}}^{\mathrm{H}}>}\mathbf{s}_{\mathrm{t}}\mathbf{s}_{\mathrm{t}}^{\mathrm{H}}$ and $\mathbf{R}_{\mathrm{t0}} = \frac{<\mathbf{R}_{\mathrm{CUT}} - \mathbf{R}_{\mathrm{c}}, \mathbf{s}_{\mathrm{t}}\mathbf{s}_{\mathrm{t}}^{\mathrm{H}}>}{<\mathbf{s}_{\mathrm{t}}\mathbf{s}_{\mathrm{t}}^{\mathrm{H}}, \mathbf{s}_{\mathrm{t}}\mathbf{s}_{\mathrm{t}}^{\mathrm{H}}>}\mathbf{s}_{\mathrm{t}}\mathbf{s}_{\mathrm{t}}^{\mathrm{H}}$.

Their difference is

$$\begin{aligned} \Delta(IF_{\mathrm{S-MIG}}, IF_{\mathrm{MIG}}) &= IF_{\mathrm{S-MIG}} - IF_{\mathrm{MIG}} \\ &= \frac{-\mathcal{D}_{\mathrm{KL}}(\mathbf{R}_{\mathrm{CUT}}, \mathbf{R}_{\mathrm{t1}} + \hat{\mathbf{R}}) + \mathcal{D}_{\mathrm{KL}}(\mathbf{R}_{\mathrm{CUT}}, \mathbf{R}_{\mathrm{t0}} + \mathbf{R}_{\mathrm{c}})}{mean(\mathcal{D}_{\mathrm{KL}}(\mathbf{R}_{\mathrm{CUT}}, \mathbf{R}_{\mathrm{c}}), \mathcal{D}_{\mathrm{KL}}(\mathbf{R}_{\mathrm{CUT}}, \mathbf{R}_{\mathrm{J}}))} \end{aligned} \tag{A3}$$

In the following proof, we focus on the numerator of (A3) foremost and deduce in two cases that the clutter is localized and distributed.

- In case I, according to Formula (30), we have

$$\begin{aligned} E[\hat{\mathbf{R}}] &= \frac{1}{K}\sum_{i=1}^{K}\mathbf{R}_i \\ &= \frac{m}{m+n}\mathbf{R}_{\mathrm{c}} + \frac{n}{m+n}\mathbf{R}_{\mathrm{J}} \\ &= \mathbf{R}_{\mathrm{c}} + \frac{n}{m+n}\sum_{i=1}^{n_{\mathrm{J}}}\alpha_{\mathrm{J}i}\alpha_{\mathrm{J}i}^{\mathrm{H}}\mathbf{s}_{\mathrm{J}i}\mathbf{s}_{\mathrm{J}i}^{\mathrm{H}} \end{aligned} \tag{A4}$$

where n_{J} is the number of interference clutter patches. Therefore, we have

$$\mathbf{R}_{\mathrm{t1}} = \frac{<\mathbf{R}_{\mathrm{CUT}} - \mathbf{R}_{\mathrm{c}} - \frac{n}{m+n}\sum_{i=1}^{n_{\mathrm{J}}}\alpha_{\mathrm{J}i}\alpha_{\mathrm{J}i}^{\mathrm{H}}\mathbf{s}_{\mathrm{J}i}\mathbf{s}_{\mathrm{J}i}^{\mathrm{H}}, \mathbf{s}_{\mathrm{t}}\mathbf{s}_{\mathrm{t}}^{\mathrm{H}}>}{<\mathbf{s}_{\mathrm{t}}\mathbf{s}_{\mathrm{t}}^{\mathrm{H}}, \mathbf{s}_{\mathrm{t}}\mathbf{s}_{\mathrm{t}}^{\mathrm{H}}>}\mathbf{s}_{\mathrm{t}}\mathbf{s}_{\mathrm{t}}^{\mathrm{H}} \overset{\mathrm{a}}{=} \mathbf{R}_{\mathrm{t0}} \tag{A5}$$

where Equation (a) is established on both the property of the matrix inner product that $<\mathbf{X}+\mathbf{Y}, \mathbf{Z}> = <\mathbf{X}, \mathbf{Z}> + <\mathbf{Y}, \mathbf{Z}>$ and the equation that $<\mathbf{s}_{\mathrm{J}i}\mathbf{s}_{\mathrm{J}i}^{\mathrm{H}}, \mathbf{s}_{\mathrm{t}}\mathbf{s}_{\mathrm{t}}^{\mathrm{H}}> = 0$ because $\mathbf{s}_{\mathrm{J}i}$ is orthogonal to $\mathbf{s}_{\mathrm{t}}$.

Based on Formula (A5), the numerator of (A3) is

$$\begin{aligned} \Delta_{\mathrm{num}}(IF_{\mathrm{S-MIG}}, IF_{\mathrm{MIG}}) &= -\mathcal{D}_{\mathrm{KL}}(\mathbf{R}_{\mathrm{CUT}}, \mathbf{R}_{\mathrm{t1}} + \hat{\mathbf{R}}) + \mathcal{D}_{\mathrm{KL}}(\mathbf{R}_{\mathrm{CUT}}, \mathbf{R}_{\mathrm{t0}} + \mathbf{R}_{\mathrm{c}}) \\ &= -\mathcal{D}_{\mathrm{KL}}(\mathbf{R}_{\mathrm{CUT}}, \mathbf{R}_{\mathrm{t1}} + \mathbf{R}_{\mathrm{c}} + \frac{n}{m+n}\sum_{i=1}^{n_{\mathrm{J}}}\alpha_{\mathrm{J}i}\alpha_{\mathrm{J}i}^{\mathrm{H}}\mathbf{s}_{\mathrm{J}i}\mathbf{s}_{\mathrm{J}i}^{\mathrm{H}}) \\ &\quad + \mathcal{D}_{\mathrm{KL}}(\mathbf{R}_{\mathrm{CUT}}, \mathbf{R}_{\mathrm{t1}} + \mathbf{R}_{\mathrm{c}}) \end{aligned} \tag{A6}$$

- In case II, as analyzed in Section 3.3.2, the hetero-clutter shows entirely different characteristics than the homo-clutter. We denote that

$$\mathbf{R}_{\mathrm{J}} = \mathbf{R}_{\mathrm{D}} \neq \mathbf{R}_{\mathrm{c}}; \tag{A7}$$

then

$$\begin{aligned} E[\hat{\mathbf{R}}] &= \frac{1}{K}\sum_{i=1}^{K}\mathbf{R}_i = \frac{m}{m+n}\mathbf{R}_{\mathrm{c}} + \frac{n}{m+n}\mathbf{R}_{\mathrm{D}} \\ &= \mathbf{R}_{\mathrm{c}} + \frac{n}{m+n}(\mathbf{R}_{\mathrm{D}} - \mathbf{R}_{\mathrm{c}}). \end{aligned} \tag{A8}$$

Similar to (A5), in case II, we have

$$\begin{aligned} \mathbf{R}_{\mathrm{t1}} &= \frac{< \mathbf{R}_{\mathrm{CUT}} - \hat{\mathbf{R}}, \mathbf{s}_{\mathrm{t}}\mathbf{s}_{\mathrm{t}}^{\mathrm{H}} >}{< \mathbf{s}_{\mathrm{t}}\mathbf{s}_{\mathrm{t}}^{\mathrm{H}}, \mathbf{s}_{\mathrm{t}}\mathbf{s}_{\mathrm{t}}^{\mathrm{H}} >}\mathbf{s}_{\mathrm{t}}\mathbf{s}_{\mathrm{t}}^{\mathrm{H}} \\ &= \frac{< \mathbf{R}_{\mathrm{CUT}} - \mathbf{R}_{\mathrm{c}} - \frac{n}{m+n}(\mathbf{R}_{\mathrm{D}} - \mathbf{R}_{\mathrm{c}}), \mathbf{s}_{\mathrm{t}}\mathbf{s}_{\mathrm{t}}^{\mathrm{H}} >}{< \mathbf{s}_{\mathrm{t}}\mathbf{s}_{\mathrm{t}}^{\mathrm{H}}, \mathbf{s}_{\mathrm{t}}\mathbf{s}_{\mathrm{t}}^{\mathrm{H}} >}\mathbf{s}_{\mathrm{t}}\mathbf{s}_{\mathrm{t}}^{\mathrm{H}} \quad = \mathbf{R}_{\mathrm{t0}} \end{aligned} \tag{A9}$$

Furthermore, applying (A8) and (A9) to (A3),

$$\begin{aligned} \Delta_{\mathrm{num}}(IF_{\mathrm{S-MIG}}, IF_{\mathrm{MIG}}) &= -\mathcal{D}_{\mathrm{KL}}(\mathbf{R}_{\mathrm{CUT}}, \mathbf{R}_{\mathrm{t1}} + \hat{\mathbf{R}}) + \mathcal{D}_{\mathrm{KL}}(\mathbf{R}_{\mathrm{CUT}}, \mathbf{R}_{\mathrm{t1}} + \mathbf{R}_{\mathrm{c}}) \\ &= -\mathcal{D}_{\mathrm{KL}}(\mathbf{R}_{\mathrm{CUT}}, \mathbf{R}_{\mathrm{t1}} + \mathbf{R}_{\mathrm{c}} + \frac{n}{m+n}(\mathbf{R}_{\mathrm{D}} - \mathbf{R}_{\mathrm{c}}))) \\ &+ \mathcal{D}_{\mathrm{KL}}(\mathbf{R}_{\mathrm{CUT}}, \mathbf{R}_{\mathrm{t1}} + \mathbf{R}_{\mathrm{c}}) \end{aligned} \tag{A10}$$

It can be seen that both Formulas (A6) and (A10) obey the following mathematical structure:

$$\Delta_{\mathrm{num}}(IF_{\mathrm{S-MIG}}, IF_{\mathrm{MIG}}) = -\mathcal{D}_{\mathrm{KL}}(\mathbf{A}, \mathbf{B} + \mathbf{C}) + \mathcal{D}_{\mathrm{KL}}(\mathbf{A}, \mathbf{B}) \tag{A11}$$

where $\mathbf{A} = \mathbf{R}_{\mathrm{CUT}}$, and $\mathbf{B} = \mathbf{R}_{\mathrm{t1}} + \mathbf{R}_{\mathrm{c}}$. In case I, $\mathbf{C} = \frac{n}{m+n}\sum_{i=1}^{n_{\mathrm{J}}} \alpha_{\mathrm{J}i}\alpha_{\mathrm{J}i}^{\mathrm{H}}\mathbf{s}_{\mathrm{J}i}\mathbf{s}_{\mathrm{J}i}^{\mathrm{H}}$. In case II, $\mathbf{C} = \frac{n}{m+n}(\mathbf{R}_{\mathrm{D}} - \mathbf{R}_{\mathrm{c}})$.

According to (10), formula (A11) can be further written as

$$\begin{aligned} \Delta_{\mathrm{num}}(IF_{\mathrm{S-MIG}}, IF_{\mathrm{MIG}}) &= -\mathrm{tr}(\mathbf{A}^{-1}(\mathbf{B} + \mathbf{C})) + \log(|\mathbf{A}^{-1}(\mathbf{B} + \mathbf{C})|) \\ &+ \mathrm{tr}(\mathbf{A}^{-1}\mathbf{B}) - \log(|\mathbf{A}^{-1}\mathbf{B}|) \\ &\overset{\mathrm{b}}{=} \log(\frac{|\mathbf{A}^{-1}(\mathbf{B} + \mathbf{C})|}{|\mathbf{A}^{-1}\mathbf{B}|}) - \mathrm{tr}(\mathbf{A}^{-1}\mathbf{C}) \end{aligned} \tag{A12}$$

where (b) relays on the property of matrix trace $\mathrm{tr}(\mathbf{M} + \mathbf{N}) = \mathrm{tr}(\mathbf{M}) + \mathrm{tr}(\mathbf{N})$.

Noted that in the binary hypothesis problem (6), $E[\mathbf{R}_{\mathrm{CUT}}] = \mathbf{R}_{\mathrm{c}} + \mathbf{R}_{\mathrm{t1}}$, $\mathbf{A} = \mathbf{B}$ holds for both hypotheses (it should be emphasized that the noise of an adaptive detector is usually contained in the clutter). Therefore,

$$\begin{aligned} \Delta_{\mathrm{num}}(IF_{\mathrm{S-MIG}}, IF_{\mathrm{MIG}}) &= \log(|\mathbf{I} + \mathbf{A}^{-1}\mathbf{C}|) - \mathrm{tr}(\mathbf{A}^{-1}\mathbf{C}) \\ &\overset{\mathrm{c}}{=} \log(\prod_{i=1}^{M}\lambda_i(\mathbf{I} + \mathbf{A}^{-1}\mathbf{C})) - \sum_{i=1}^{M}\lambda_i(\mathbf{A}^{-1}\mathbf{C}) \\ &= \sum_{i=1}^{M}\log(\lambda_i(\mathbf{I} + \mathbf{A}^{-1}\mathbf{C})) - \sum_{i=1}^{M}\lambda_i(\mathbf{A}^{-1}\mathbf{C}) \\ &\overset{\mathrm{d}}{=} \sum_{i=1}^{M}(\log(1 + \lambda_i(\mathbf{A}^{-1}\mathbf{C})) - \lambda_i(\mathbf{A}^{-1}\mathbf{C})) \end{aligned} \tag{A13}$$

where Equation (c) relies on the properties of the matrix determinant and matrix trace. Equation (d) is established on the formula $\lambda_i(\mathbf{I} + \mathbf{M}) = 1 + \lambda_i(\mathbf{M})$.

Because $\log(1 + x) \leq x$, we have $\Delta(IF_{\mathrm{S-MIG}}, IF_{\mathrm{MIG}}) \leq 0$: namely, $IF_{\mathrm{S-MIG}} \leq IF_{\mathrm{MIG}}$.

References

1. Guerci, J.R. *Space-Time Adaptive Processing for Radar*, 2nd ed.; Artech House: Fitchburg, MA, USA, 2015; pp. 1–9.
2. Reed, I.S.; Mallett, J.D.; Brennan, L.E. Rapid Convergence Rate in Adaptive Arrays. *IEEE Trans. Aerosp. Electron. Syst.* **1974**, *10*, 853–863. [CrossRef]
3. Melvin, W.L. A STAP Overview. *IEEE Aerosp. Electron. Syst. Mag.* **2004**, *19*, 19–35. [CrossRef]
4. Liu, W.; Xie, W.; Wang, Y. Diagonally Loaded Space-time Adaptive Detection. In Proceedings of the IEEE International Conference on Radar, Chengdu, China, 24–27 October 2011.
5. Liu, W.; Xie, W.; Wang Y. AMF and ACE Detectors Based on Diagonal Loading. *Syst. Eng. Electron.* **2013**, *35*, 463–468.
6. Liu, W.; Liu, J.; Cao, C. Multichannel Adaptive Signal Detection: Basic Theory and Literature Review. *Sci. China Inf. Sci.* **2022**, *65*, 121301. [CrossRef]
7. Wang, Y.; Liu, W.; Xie, W. Research Progress of Space-Time Adaptive Detection for Airborne Radar. *J. Radars* **2014**, *3*, 201–207. [CrossRef]
8. Kelly, E.J. An Adaptive Detection Algorithm. *IEEE Trans. Aerosp. Electron. Syst.* **1986**, *22*, 112–127. [CrossRef]
9. Chen, W.; Reed, I.S. A New CFAR Detection Test for Radar. *Digit. Signal Process.* **1991**, *1*, 198–214. [CrossRef]
10. Robey, F.C.; Fuhrmann, D.R.; Kelly, E.J. A CFAR Adaptive Matched Filter Detector. *IEEE Trans. Aerosp. Electron. Syst.* **1992**, *28*, 208–216. [CrossRef]
11. De Maio, A. Rao Test for Adaptive Detection in Gaussian Interference with Unknown Covariance Matrix. *IEEE Trans. Signal Process.* **2007**, *55*, 3577–3584. [CrossRef]
12. De Maio, A. A New Derivation of The Adaptive Matched Filter. *IEEE Signal Process. Lett.* **2004**, *11*, 792–793. [CrossRef]
13. Liu, W.; Xie, W.; Liu, J. Adaptive double subspace signal detection in Gaussian background—part. I: Homogeneous environments. *IEEE Trans. Signal Process.* **2014**, *62*, 2345–2357. [CrossRef]
14. Liu, W.; Xie, W.; Gao, Y. Multichannel signal detection in interference and noise when signal mismatch happens. *Signal Process.* **2020**, *166*, 107268. [CrossRef]
15. Liu, W.; Xie, W.; Liu, J. Adaptive double subspace signal detection in Gaussian background—part. II: Partially homogeneous environments. *IEEE Trans. Signal Process.* **2014**, *62*, 2358–2369. [CrossRef]
16. Liu, W.; Gao, F.; Luo, Y. GLRT-based generalized direction detector in partially homogeneous environment. *Sci. China Inf. Sci.* **2019**, *62*, 209303. [CrossRef]
17. Liu, W.; Wang, Y.; Liu, J. Adaptive detection without training data in colocated MIMO radar. *IEEE Trans. Aerosp. Electron. Syst.* **2015**, *51*, 2469–2479. [CrossRef]
18. Liu, W.; Liu J.; Huang L. Rao tests for distributed target detection in interference and noise. *Signal Process.* **2015**, *117*, 333–342. [CrossRef]
19. *Introduction to Airborne Radar[M]. The Institution of Engineering and Technology*; Geroge, W.S., Ed.; SciTech Publishing: Raleigh, NC, USA, 2010.
20. He, C.; Zhang, R.; Huang, B. Moving-Target Detection for FDA-MIMO Radar in Partially Homogeneous Environments. *Electronics* **2024**, *13*, 851. [CrossRef]
21. Chen, X.; Cheng, Y.; Wu, H. Heterogeneous Clutter Suppression via Affine Transformation on Riemannian Manifold of HPD Matrices. *IEEE Trans. Geosci. Remote Sens.* **2022**, *60*, 5109816. [CrossRef]
22. Yang, Y.; Yang, B. Overview of Radar Detection Methods for Low Altitude Targets in Marine Environments. *J. Syst. Eng. Electron.* **2024**, *35*, 1–13. [CrossRef]
23. Kononov, A.A.; Ka, M.-H. Rapid Adaptive Matched Filter for Detecting Radar Targets with Unknown Velocity. *IEEE Access* **2024**, *12*, 25411-25428. [CrossRef]
24. Conte, E.; De Maio, A. Mitigation Techniques for Non-Gaussian Sea Clutter. *IEEE J. Ocean. Eng.* **2003**, *29*, 284–302. [CrossRef]
25. Gini, F.; Greco, M. Covariance matrix estimation for CFAR detection in correlated heavy tailed clutter. *Signal Process.* **2002**, *82*, 1847–1859. [CrossRef]
26. Conte, E.; De Maio, A.; Ricci, G. Recursive estimation of the covariance matrix of a compound-Gaussian process and its application to adaptive CFAR detection. *IEEE Trans. Signal Process.* **2002**, *50*, 1908–1915. [CrossRef]
27. Pailloux, G.; Forster, P.; Ovarlez, J.P. On Persymmetric Covariance Matrices in Adaptive Detection. In Proceedings of the IEEE International Conference on Acoustics, Speech and Signal Processing, Las Vegas, NV, USA, 31 March–4 April 2008.
28. Pascal, F.; Chitour, Y.; Ovarlez, J.P. Covariance structure maximum likelihood estimates in compound-Gaussian noise: Existence and algorithm analysis. *IEEE Trans. Signal Process.* **2008**, *56*, 34–48. [CrossRef]
29. Barbaresco, F. Innovative tools for radar signal processing based on Cartan's geometry of SPD matrices & information geometry. In Proceedings of the IEEE Radar Conference, Rome, Italy, 26–30 May 2008.
30. Arnaudon, M.; Yang, L.; Barbaresco, F. Stochastic algorithms for computing p-means of probability measures, geometry of radar Toeplitz covariance matrices and applications to HR Doppler processing. In Proceedings of the International Radar Symposium, Leipzig, Germany, 6–9 September 2011.
31. Balaji, B.; Barbaresco, B. Application of Riemannian mean of covariance matrices to space-time adaptive processing. In Proceedings of the 9th European Radar Conference, Amsterdam, The Netherlands, 31 October–2 November 2012.
32. Chen, X.; Cheng, Y.; Wang, H. Geometric Estimation on Manifolds Based Training Data Selection. In Proceedings of the 4th IEEE International Conference on Electronic Information and Communication Technology, Xi'an, China, 18–20 August 2021.

33. Chen, X.; Cheng, Y.; Wu, H. Heterogeneous Clutter Suppression for Airborne Radar STAP Based on Matrix Manifolds. *Remote Sens.* **2021**, *13*, 3195. [CrossRef]
34. Wu, H.; Cheng, Y.; Chen, X. Power Spectrum Information Geometry-Based Radar Target Detection in Heterogeneous Clutter. *IEEE Trans. Geosci. Remote Sens.* **2024**, *62*, 1–16. [CrossRef]
35. Xie, Z.; Cheng, Y.; Wu, H. Ship Detection in SAR Images via Kullback-Leibler Divergence-Based Matrix Information Geometry Detector. *J. Phys. Conf. Ser.* **2023**, *69*, 4326. [CrossRef]
36. Wu, H.; Cheng, Y.; Chen, X. Geodesic Normal Coordinate-Based Manifold Filtering for Target Detection. *IEEE Trans. Geosci. Remote Sens.* **2022**, *60*, 1–15. [CrossRef]
37. Yang, Z.; Cheng, Y.; Wu, H. Enhanced Matrix CFAR Detection with Dimensionality Reduction of Riemannian Manifold. *IEEE Signal Process. Lett.* **2020**, *27*, 2084–2088. [CrossRef]
38. Ye, L.; Yang, Q.; Chen, Q. Multidimensional Joint Domain Localized Matrix Constant False Alarm Rate Detector Based on Information Geometry Method with Applications to High Frequency Surface Wave Radar. *IEEE Access* **2019**, *7*, 28080–28088. [CrossRef]
39. Ye, L.; Yang, Q.; Deng, W. Matrix Constant False Alarm Rate(MCFAR) Detector Based on Information Geometry. In Proceedings of the 2016 IEEE Information Technology, Networking, Electronic and Automation Control Conference, Chongqing, China, 20–22 May 2016.
40. Yang, J.; Chen, Y.; Dirk, T.M. Signal Detection for Ultra-Massive MIMO: An Information Geometry Approach. *IEEE Trans. Signal Process.* **2024**, *72*, 824–838. [CrossRef]
41. Wu, H.; Cheng, Y.; Chen, X. Adaptive Matrix Information Geometry Detector with Local Metric Tensor. *IEEE Trans. Signal Process.* **2020**, *70*, 3758–3773. [CrossRef]
42. Yang, Z.; Cheng, Y.; Wu, H. Manifold Projection-Based Subband Matrix Information Geometry Detection for Radar Targets in Sea Clutter. *IEEE Trans. Geosci. Remote Sens.* **2023**, *61*, 1–15. [CrossRef]
43. Wu, H.; Cheng, Y.; Chen, X. Information Geometry-Based Track-Before-Detect Algorithm for Slow-Moving Fluctuating Target Detection. *IEEE Trans. Signal Process.* **2023**, *71*, 1833–1848. [CrossRef]
44. Chen, X.; Wu, H.; Cheng, Y. Joint Design of Transmit Sequence and Receive Filter Based on Riemannian Manifold of Gaussian Mixture Distribution for MIMO Radar. *IEEE Trans. Geosci. Remote Sens.* **2023**, *61*, 1–13. [CrossRef]
45. Lapuyade-Lahorgue, J.; Barbaresco, F. Radar Detection Using Siegel Distance between Autoregressive Processes, Application to HF and X-band Radar. In Proceedings of the 2008 IEEE Radar Conference, Rome, Italy, 26–30 May 2008; pp. 1–6.
46. Cheng, Y.; Hua, X.; Wang, H. The Geometry of Signal Detection with Applications to Radar Signal Processing. *Entropy* **2016**, *18*, 381. [CrossRef]
47. Hua, X.; Peng, L.; Liu, W. LDA-MIG Detectors for Maritime Targets in Nonhomogeneous Sea Clutter. *IEEE Trans. Geosci. Remote Sens.* **2023**, *61*, 5101815. [CrossRef]
48. Arnaudon, M.; Barbaresco, F.; Yang, L. Riemannian Medians and Means with Applications to Radar Signal Processing. *IEEE J. Sel. Top. Signal Process.* **2013**, *7*, 595–604. [CrossRef]
49. Hua, X.; Yusuke, O.; Peng, L. Target Detection within Nonhomogeneous Clutter Via Total Bregman Divergence-Based Matrix Information Geometry Detectors. *IEEE Trans. Signal Process.* **2021**, *69*, 4326. [CrossRef]
50. Hua, X.; Cheng, Y.; Wang, H. Matrix CFAR detectors based on symmetrized Kullback-Leibler and total Kullback-Leibler divergences. *Digit. Signal Process.* **2017**, *69*, 106–116. [CrossRef]
51. Zhou, Z.; Hua, X.; Guan, C. Target Detection Based on Matrix Information Geometry for Continuous Active Sonar. *J. Shaanxi Norm. Univ.* **2019**, *47*, 23–29.
52. Shi, Y.; Li, J. Target Detecting Algorithm Based on Subband Matrix for Slow Target in Sea Clutter. *J. Electron. Inf. Technol.* **2021**, *43*, 2703–2710.
53. Zhao, W.; Jin, M.; Liu, W. Matrix CFAR Detector Based on Matrix Spectral Norm. *Acta Electron. Sin.* **2019**, *47*, 1951–1956.
54. Ward, J. *Space-Time Adaptive Processing for Airborne Radar*; Lincoln Lab: Lexington, MA, USA, 1994; pp. 20–51.
55. Conte, E.; Lops, M.; Ricci, G. Asymptotically Optimum Radar Detection in Compound-Gaussian Clutter. *IEEE Trans. Aerosp. Electron. Syst.* **1995**, *31*, 617–625. [CrossRef]
56. Adve, R.; Hale, T.; Wicks, M. Knowledge Based Adaptive Processing for Ground Moving Target Indication. *Digit. Signal Process.* **2007**, *17*, 495–541. [CrossRef]
57. Guan, J.; Liu, N.; Wang, G. Sea-detecting radar experiment and target feature data acquisition for dual polarization multistate scattering dataset of marine targets. *J. Radars* **2023**, *12*, 456–469.
58. Liu, N.; Dong, Y.; Wang, G. Sea-detecting X-band radar and data acquisition program. *J. Radars* **2019**, *8*, 656–667.
59. Hu, C.; Yan, Y.J.; Wang, R. High-resolution, Multi-frequency and Full-polarization Radar Database of Small and Group Targets in Clutter Environment. *Sci. China Inf. Sci.* **2023**, *66*, 227301. Available online: https://yjyan99.github.io/ (accessed on 1 May 2024). [CrossRef]
60. Titi, G.W.; Marshall, D.F. The ARPA/NAVY Mountaintop Program: Adaptive Signal Processing for Airborne Early Warning Radar. In Proceedings of the 1996 IEEE International Conference on Acoustics, Speech and Signal Processing, Atlanta, GA, USA, 9 May 1996.
61. Titi, G.W. An Overview of the ARPA/NAVY Mountaintop Program. In Proceedings of the 1994 IEEE Long Island Section Adaptive Antenna Systems Symposium, Melville, NY, USA, 19 November 1992.

62. Carlson, B.D.; Goodman, L.M.; Asutin, J. An Ultralow-Sidelobe Adaptive Array Antenna. *Linc. Lab. J.* **1990**, *3*, 291–310.
63. The Development of E-2D Radar from the Radar Monitoring Technology Experimental Radar RSTER to the E-2C RMP Project. Available online: http://www.360doc.com/content/18/0126/21/13664199_725368484.shtml (accessed on 1 May 2024).

remote sensing

MDPI

Article

Efficient and Robust Adaptive Beamforming Based on Coprime Array Interpolation

Siming Chen, Xiaochuan Wu, Shujie Li, Weibo Deng and Xin Zhang *

The School of Electronics and Information Engineering, Harbin Institute of Technology, Harbin 150001, China; chsmhit@stu.hit.edu.cn (S.C.); wxc@hit.edu.cn (X.W.); 23b905057@stu.hit.edu.cn (S.L.); dengweibo@hit.edu.cn (W.D.)

* Correspondence: zhangxinhit@hit.edu.cn

Abstract: Unlike uniform linear arrays (ULAs), coprime arrays require fewer physical sensors yet provide higher degrees of freedom (DOF) and larger array apertures. However, due to the existence of "holes" in the differential co-array, the target detection performance deteriorates, especially in adaptive beamforming. To address these challenges, this paper proposes an efficient and robust adaptive beamforming algorithm leveraging coprime array interpolation. The algorithm eliminates unwanted signals and uses the Gauss–Legendre quadrature method to reconstruct an Interference-plus-Noise Covariance Matrix (INCM), thereby obtaining the beamforming coefficients. Unlike previous techniques, we utilize a virtual interpolated ULA to expand the aperture, enabling the acquisition of a high-dimensional covariance matrix. Additionally, a projection matrix is constructed to eliminate unwanted signals from the received data, greatly enhancing the accuracy of INCM reconstruction. To address the high computational complexity of integral operations used in most INCM reconstruction algorithms, we propose an approximation based on the Gauss–Legendre quadrature, which reduces the computational load while maintaining accuracy. This algorithm avoids the array aperture loss caused by using only the ULA segment in the difference co-array and improves the accuracy of INCM reconstruction. Simulation and experimental results show that the performance of the proposed algorithm is superior to the compared beamformers and is closer to the optimal beamformer in various scenarios.

Keywords: coprime array; robust adaptive beamforming; covariance matrix reconstruction; virtual uniform linear array; signal interpolation; degrees of freedom; Gauss–Legendre quadrature

1. Introduction

Adaptive beamforming is a widely used technology in array signal processing, with applications in radar, sonar, wireless communications, and medical fields [1–7]. It dynamically adjusts the weights using the received data to isolate the signal of interest (SOI) and minimize interference and noise. The Minimum Variance Distortionless Response (MVDR) is a powerful adaptive beamforming technique that relies on precise knowledge of the desired signal and array configuration, providing high resolution and anti-interference performance [8,9]. However, its performance degrades significantly when there are mismatches in the steering vector of the signal of interest due to non-ideal factors such as look direction errors and wavefront distortions. Furthermore, the performance drops drastically if the received data contain information about the SOI, which is unavoidable in practical scenarios. Enhancing the robustness of adaptive beamformers has been the focus of extensive research over the past few decades. Typically, robust adaptive beamforming techniques focus on several categories of methods: diagonal loading methods [10–16], eigenstructure-based methods [17–23], uncertainty set-based methods [24–34], and covariance matrix reconstruction-based methods [35–45].

The performance of adaptive beamforming is heavily influenced by the degrees of freedom (DOF) and array aperture, which determine its spatial resolution and ability to counteract interference. Due to Nyquist sampling constraints, most commonly used beamformers rely on traditional uniform linear arrays (ULAs), with element spacing set to within half a wavelength to prevent aliasing. In a ULA, as the number of array elements increases, the array aperture grows proportionally, thereby increasing the number of distinguishable DOA in space. However, the increase in the number of physical sensors leads to higher hardware costs and greater computational complexity [46–48]. Fortunately, non-uniform linear arrays (NLAs) provide a viable solution to these challenges [49]. Utilizing mathematical techniques to derive difference arrays, NLAs can attain significantly higher DOF with fewer physical sensors [50,51].

In recent years, the coprime array, a notable type of NLA, has attracted considerable attention. Compared to traditional ULAs with the same number of sensors, the coprime array configuration achieves a larger array aperture, offers significant advantages in terms of DOF, and substantially reduces the coupling effect between sensors. Research on coprime arrays has predominantly concentrated on direction of arrival (DOA) estimation techniques [52–58], while studies on adaptive beamforming methods utilizing coprime arrays are comparatively scarce. NLA improves the robustness of beamforming by adjusting the spacing between array elements, making it better suited to complex environments. The traditional MVDR adaptive beamforming algorithm for coprime arrays first derives the differential array of the coprime array through mathematical methods and then uses the MVDR algorithm to obtain the beamformer. In [47], an adaptive beamforming algorithm based on virtual array spatial spectrum estimation was proposed. In this method, the signal received by the array is transformed into the virtual domain through mathematical techniques. Spectrum estimation is performed in the virtual domain, resulting in robust beamformers obtained through both integration and summation methods. Consequently, this approach yields two robust beamformers. Nevertheless, errors in the virtual domain Capon spectrum lead to some performance degradation relative to the optimal signal-to-interference-plus-noise ratio (SINR). In [59], an effective adaptive beamforming algorithm for coprime arrays was introduced. This algorithm divides the coprime array into subarrays for DOA estimation and then reconstructs the INCM using these DOA estimates and the power of the input signals. However, these methods only leverage the continuous ULA segment in the difference co-array, which causes a loss of certain degrees of freedom and aperture, resulting in reduced performance of the adaptive beamformer. The aforementioned method does not address the issue of differential array aperture loss. In recent research [60], a robust adaptive beamforming method using coprime co-array interpolation was introduced. The proposed method interpolates the missing points of the physical sensor to preserve the coaxial array aperture. The weight coefficients for beamforming are then obtained by reconstructing the INCM. However, due to its dependence on DOA estimation, errors in the sample covariance matrix (SCM) at low snapshot counts can result in some SOI information remaining in the INCM, leading to a decline in beamforming performance.

It is noteworthy that there is a gap in the precise reconstruction of the INCM matrix in the field of coprime arrays. Research in this area is highly beneficial for the performance of adaptive beamformers. In this paper, to further improve the performance of adaptive beamformers, reduce computational complexity, and reconstruct a more accurate INCM, we propose a robust adaptive beamforming method based on coprime array interpolation that enables unwanted signal removal and employs the Gauss–Legendre quadrature (URG) for covariance matrix reconstruction. In most existing algorithms, the INCM is reconstructed using the received signals, which results in the INCM containing some SOI information [35,60–63]. Since beamformers are highly sensitive to SOI information, their anti-interference performance decreases. Therefore, after reconstructing the covariance matrix of the coprime array using the interpolated virtual ULA, it eliminates the unwanted signals by constructing a specific projection matrix. The resulting pseudo-covariance matrix has less desired signal content compared to the original data, thereby enhancing the accuracy of INCM reconstruction. To mitigate the high computational cost of traditional integral calculation methods, this paper employs the Gauss–Legendre quadrature (GLQ) approximation method to reduce computational complexity and streamline the integration process. The algorithm expands the effective aperture of the coprime array, reduces computational complexity, and maintains high algebraic accuracy, enabling more precise INCM reconstruction. Both simulation and experimental results confirm the algorithm's effectiveness. In summary, the primary contributions of this paper are:

1. We expand the coprime array's effective aperture through coaxial array interpolation, constructing and interpolating it to achieve a higher-dimensional covariance matrix;
2. We construct a projection matrix using the eigen-subspace method to eliminate unwanted signal information from the received signals of the array, resulting in a matrix with minimal SOI information.
3. We propose an efficient and robust adaptive beamforming algorithm for coprime arrays, which employs GLQ approximation for the integral computation of the power spectrum. This approach achieves an accurate INCM and significantly reduces computational complexity;

The remainder of this paper is organized as follows: Section 2 encompasses the signal model for coprime arrays and adaptive beamforming, discusses an interpolation technique for coprime arrays, and describes the new INCM reconstruction algorithm. Additionally, it evaluates the computational complexity of the proposed method. Section 3 illustrates the beamformer's performance in various scenarios through simulation and experimental results. Section 4 discusses the performance of the algorithm and its potential drawbacks. Finally, Section 5 summarizes the findings of the paper.

The notations and their meanings mentioned in the text are listed in Table 1.

Table 1. Notations and their meanings.

Notation	Meaning
$\mathbf{I}$	Identity matrix
$\mathbf{0}$	Zero matrix
$*$	Complex conjugate
T	Transpose
H	Conjugate transpose
$\mathbb{E}\{\cdot\}$	Statistical expectation
$\|\|\cdot\|\|_F$	Frobenius norm
$\|\|\cdot\|\|_*$	Nuclear norm
$\text{rank}(\cdot)$	Rank of a matrix
$\text{vec}(\cdot)$	Vectorization operator
$\text{diag}(\mathbf{a})$	Diagonal matrix formed from the elements of the vector $\mathbf{a}$
$\mathbf{A} \succ 0$	$\mathbf{A}$ is positive definite
$[\mathbf{x}]_i$	ith component of the vector $\mathbf{x}$
$\langle \mathbf{x} \rangle_n$	Value of $\mathbf{x}$ at index n
$\langle \mathbf{R} \rangle_{a,b}$	$\mathbb{E}[\langle \mathbf{x} \rangle_a \langle \mathbf{x} \rangle_b^*]$
$\|\mathcal{A}\|$	Cardinality of the set $\mathcal{A}$
$\otimes$	Kronecker product

2. Signal Model and Proposed Algorithm

2.1. Signal Model for Adaptive Beamforming

The coprime array configuration integrates two sparse ULAs as depicted in Figure 1a. The sensor elements are strategically located at $\{0, Md, 2Md, \ldots, (N-1)Md\}$ and $\{0, Nd, 2Nd, \ldots, (M-1)Nd\}$ within the respective ULAs. Specifically, the first ULA comprises N elements with an inter-element spacing of Md, while the second ULA consists of M elements with an inter-element spacing of Nd, where M and N are coprime integers. The coprime nature of M and N ensures that the two ULAs have a single overlap point at the origin. For generality, we assume $M < N$. The inter-element spacing d is set to half the carrier wavelength, i.e., $d = \lambda/2$, where λ denotes the carrier wavelength. Consequently, the coprime array is composed of $M + N - 1$ physical sensors, with the sensor positions given by $p_i d$, where $i = 1, 2, \ldots, |\mathbb{S}|$, forming a discrete integer set:

$$P_{\mathbb{S}} = \{Mn | n \in \mathbb{N}, n \leq N - 1\} \bigcup \{Nm | m \in \mathbb{N}, m \leq M - 1\} \quad (1)$$

which represents a non-uniform linear sparse array with $|\mathbb{S}| = M + N - 1$ sensors, where $|\cdot|$ denotes the cardinality of the set. Thus, the effective aperture of the array is determined by $\max((N-1)Md, (M-1)Nd)$.

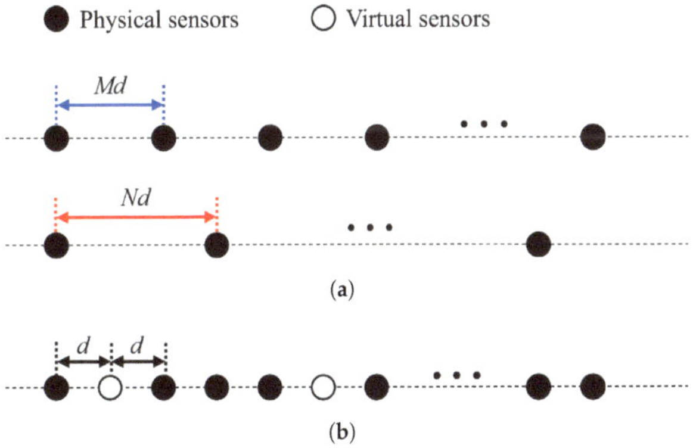

Figure 1. Illustration of array configurations. (**a**) Non-uniform coprime array configurations; (**b**) The interpolated array.

Considering a narrowband SOI and $\mathcal{J}$ narrowband interference signals. The DOA for the signal and interference are $\theta_0, \theta_1, \ldots, \theta_{\mathcal{J}}$, and the observation vector of the array can be mathematically expressed as follows:

$$\mathbf{y}(k) = \mathbf{d}(k) + \mathbf{j}(k) + \mathbf{v}(k) \quad (2)$$

the observation vector of the array, which at time k can be decomposed into components corresponding to the SOI, interference, and noise. Specifically, $\mathbf{d}(k) = \mathbf{a}_0 x_0(k)$ represents the SOI component, $\mathbf{j}(k) = \sum_{l=1}^{\mathcal{J}} \mathbf{a}_l x_l(k)$ represents the interference component, where $\mathbf{a}_l = \mathbf{a}(\theta_l) \in \mathbb{C}^{|\mathbb{S}| \times 1}$ is the steering vector associated with the l-th source, and $\mathbf{v}(k) \in \mathbb{C}^{|\mathbb{S}| \times 1}$ represents the noise component. Here, $x_l(k)$ denotes the waveform of the l-th source signal, and $\mathbf{a}_l = \mathbf{a}(\theta_l) \in \mathbb{C}^{|\mathbb{S}| \times 1}$ is the steering vector associated with the l-th source. The steering vector of a coprime array with respect to different directions is defined as follows,

$$\mathbf{a}(\theta_l) = \left[1, e^{jp_2\beta_l}, \ldots, e^{jp_{|\mathbb{S}|}\beta_l}\right]^T,$$

where $\beta_l = \frac{2\pi}{\lambda} d \sin\theta_l$. The noise components $\mathbf{v}(k)$ are modeled as independent and identically distributed (i.i.d.) additive white Gaussian noise, expressed as $\mathbf{v}(k) \sim \mathcal{CN}(0, \sigma_n^2 \mathbf{I})$, where σ_n^2 denotes the noise power and $\mathbf{I}$ is the identity matrix of the appropriate dimensions.

Utilizing adaptive beamforming techniques in coprime arrays effectively removes interference. The resulting output of the adaptive beamformer can be represented as

$$\mathbf{z}(k) = \boldsymbol{w}^H \mathbf{y}(k), \quad (3)$$

where the beamformer weight vector $w = [w_1, w_2, \ldots, w_{|\mathbb{S}|}] \in \mathbb{C}^{|\mathbb{S}| \times 1}$ is utilized to define the output SINR for the coprime array. The SINR for this configuration is mathematically expressed as follows:

$$\begin{aligned} \text{SINR} &= \frac{\mathcal{E}\{w^{\text{H}}\mathbf{d}(k)\mathbf{d}(k)^{\text{H}}(k)w\}}{\mathcal{E}\{w^{\text{H}}(\mathbf{j}(k)+\mathbf{v}(k))(\mathbf{j}(k)+\mathbf{v}(k))^{\text{H}}w\}} \\ &= \frac{\sigma_0^2|w^{\text{H}}\mathbf{a}_0|^2}{w^{\text{H}}\mathbf{R}_{\text{INF}}w} \end{aligned} \tag{4}$$

where the parameter $\sigma_0^2 = \mathcal{E}|s_0(k)|^2$ denotes the power of the SOI, while $\mathbf{R}_{\text{INF}} \in \mathbb{C}^{|\mathbb{S}| \times |\mathbb{S}|}$ represents the INCM, defined as

$$\begin{aligned} \mathbf{R}_{\text{INF}} &= E\Big\{(\mathbf{j}(k)+\mathbf{v}(k))(\mathbf{j}(k)+\mathbf{v}(k))^H\Big\} \\ &= \sum_{l=1}^{\mathcal{J}} \sigma_l^2 \mathbf{a}_l \mathbf{a}_l^H + \sigma_n^2 \mathbf{I}, \end{aligned} \tag{5}$$

where σ_l^2 denotes the power of the l-th interference.

To maximize the SINR, it is essential to ensure a distortion-free response of the SOI while minimizing the output power of interference and noise. Consequently, the SINR maximization optimization problem in (4) can be transformed into an interference and noise minimization problem, expressed as follows:

$$\begin{aligned} \min_{w} \quad & w^H \mathbf{R}_{\text{INF}} w \\ \text{s.t.} \quad & w^H \mathbf{a}_0 = 1. \end{aligned} \tag{6}$$

By solving this optimization problem using the Lagrange multiplier method, the optimal weight vector is determined as follows:

$$w_{\text{opt}} = \frac{\mathbf{R}_{\text{INF}}^{-1}\mathbf{a}_0}{\mathbf{a}_0^H \mathbf{R}_{\text{INF}}^{-1}\mathbf{a}_0}. \tag{7}$$

In practical scenarios, obtaining the theoretical $\mathbf{R}_{\text{INF}}$ is challenging. Therefore, it is typically estimated using the sample covariance matrix (SCM), which is calculated directly from the received signal. The SCM for $\mathcal{K}$ snapshots is obtained by the following equation:

$$\hat{\mathbf{R}} = \frac{1}{\mathcal{K}} \sum_{k=1}^{\mathcal{K}} \mathbf{y}(k)\mathbf{y}^H(k), \tag{8}$$

The beamformer (7) can be reformulated as the SMI beamformer:

$$w_{\text{SMI}} = \frac{\hat{\mathbf{R}}^{-1}\mathbf{a}_0}{\mathbf{a}_0^H \hat{\mathbf{R}}^{-1}\mathbf{a}_0}, \tag{9}$$

which is the SCM beamformer, and is closely related to the steering vector of the SOI. Therefore, any inaccuracies in the steering vector obtained from the actual coprime array can affect the performance of the beamformer. However, there is a notable discrepancy between the SCM and the INCM. Therefore, this paper focuses on accurately estimating the INCM $\mathbf{R}_{\text{INF}}$ and the ideal signal steering vector $\mathbf{a}_0$ from the received signals $\mathbf{y}(k)$ in the coprime array.

2.2. Coprime Array Interpolation Algorithm

Coprime array interpolation entails placing extra sensors into the 'holes' at positions that are integer multiples of half-wavelengths, where no physical sensors are present in the coprime array $\mathbb{S}$.

Vectorizing $\hat{\mathbf{R}}$ yields a new vector

$$\nu = \text{vec}(\hat{\mathbf{R}}) = \tilde{\mathbf{A}}\mathbf{p} + \sigma_n^2 \mathbf{1}_n \tag{10}$$

where $\tilde{\mathbf{A}} = [\tilde{\mathbf{a}}(\theta_0), \tilde{\mathbf{a}}(\theta_1), \ldots, \tilde{\mathbf{a}}(\theta_{\mathcal{J}})]$, with $\tilde{\mathbf{a}}(\theta_j) = \mathbf{a}^*(\theta_j) \otimes \mathbf{a}(\theta_j)$ for $j = 0, 1, \ldots, \mathcal{J}$, $\mathbf{p} = [\sigma_0^2, \sigma_1^2, \ldots, \sigma_{\mathcal{J}}^2]^T$, and

$$\mathbf{1}_n = \text{vec}(\mathbf{I}) = \begin{bmatrix} \mathbf{e}_1 \\ \mathbf{e}_2 \\ \vdots \\ \mathbf{e}_{M+N-1}, \end{bmatrix}$$

where $\mathbf{e}_i$ is a column vector with all elements zero except for a one at the i-th position. The vector ν represents the received data from the virtual array extending the resonant aperture, and its corresponding steering vector is denoted as $\tilde{\mathbf{A}}$. The sensor positions of this virtual array can all be described by integers in the set $\mathbb{D} = \{a - b | a, b \in \mathbb{S}\}$. Nevertheless, the virtual source signal reduces to a single snapshot of $\mathbf{p}$, and $\sigma_n^2 \mathbf{1}_n$ constitutes a constant vector.

Vector ν has some duplicate items, and by eliminating them and reordering them we obtain a new vector ν_1,

$$\langle \nu_1 \rangle_n = \frac{1}{|\Gamma(m)|} \sum_{(a,b) \in \Gamma(m)} \langle \hat{\mathbf{R}} \rangle_{a,b}, \quad n \in \mathbb{D} \tag{11}$$

where $\Gamma(m)$ represents the co-array matrix of (a, b). Specifically, $\Gamma(m) = \{(a, b) \in \mathbb{S}^2 | a - b = m\}$.

According to the properties of coprime numbers, the resulting difference co-array $\mathbb{D}$ is not an ideal ULA, meaning the integer indices are not a continuous sequence of integers and there are 'holes'. Therefore, algorithms that derive the covariance matrix of the virtual array, such as spatial smoothing, cannot directly use the array output ν_1. To address this issue, an interpolated vector $\bar{\nu}_1$ is constructed as

$$\langle \bar{\nu}_1 \rangle_n = \begin{cases} \langle \nu_1 \rangle_n, & n \in \mathbb{D}, \\ 0, & n \in \mathbb{U} \backslash \mathbb{D} \end{cases} \tag{12}$$

where $\mathbb{U} = \{n | \min(\mathbb{D}) \leq n \leq \max(\mathbb{D})\}$, $\bar{\nu}_1 \in \mathbb{C}^{2MN-2M+1}$.

It is important to note that the interpolated sensors $\mathbb{U} \backslash \mathbb{D}$ exist mathematically rather than physically. Consequently, the received signals associated with these interpolated sensors should be regarded as unknown.

Given that ν_1 is Hermitian symmetric, it follows that $\bar{\nu}_1$ is also Hermitian symmetric [55], i.e.,

$$\begin{aligned} \langle \bar{\nu}_1 \rangle_n &= \langle \bar{\nu}_1 \rangle_{-n}^* \\ [\bar{\nu}_1]_i &= [\bar{\nu}_1]_{2MN-2M+2-i}^* \end{aligned} \tag{13}$$

where $n \in [0, MN - M]$ and $i \in [1, 2MN - 2M + 1]$.

Notably, the theoretical covariance matrix for the uncorrelated signals received by the ULA exhibits a Hermitian Toeplitz structure. This characteristic can be leveraged as prior information for structured matrix recovery. Utilizing the properties of the vector $\bar{\nu}1$, the resulting Hermitian Toeplitz matrix $\mathbf{R}_{\mathbb{U}} \in \mathbb{C}^{|\mathbb{U}^+| \times |\mathbb{U}^+|}$ is given by

$$\mathbf{R}_{\mathbb{U}} = \begin{bmatrix} [\bar{\nu}_1]_{|\mathbb{U}^+|} & [\bar{\nu}_1]_{|\mathbb{U}^+|-1} & \cdots & [\bar{\nu}_1]_1 \\ [\bar{\nu}_1]_{|\mathbb{U}^+|+1} & [\bar{\nu}_1]_{|\mathbb{U}^+|} & \cdots & [\bar{\nu}_1]_2 \\ \vdots & \vdots & \ddots & \vdots \\ [\bar{\nu}_1]_{2|\mathbb{U}^+|-1} & [\bar{\nu}_1]_{2|\mathbb{U}^+|-2} & \cdots & [\bar{\nu}_1]_{|\mathbb{U}^+|} \end{bmatrix} \tag{14}$$

where $\mathbb{U}^{+} = n|n \in \mathbb{U}, n \geq 0$ represents the subset of $\mathbb{U}$ containing non-negative integers.

The presence of zero elements in $\bar{v}_1$ indicates corresponding zero elements in $\mathbf{R}_{\mathbb{U}}$, necessitating the recovery of this missing information. The covariance matrix features a Hermitian Toeplitz structure. Furthermore, because the number of incident sources is relatively small compared to the dimension of the covariance matrix, the noise-free covariance matrix exhibits low-rank characteristics. Given these conditions, the interpolated ULA covariance matrix can be recovered by solving the following optimization problem:

$$\begin{aligned} \min_{\bar{\mathbf{R}}_{\mathbb{U}} \in \mathbb{C}^{|\mathbb{U}^{+}| \times |\mathbb{U}^{+}|}} \quad & \operatorname{rank}(\bar{\mathbf{R}}_{\mathbb{U}}) \\ \text{s.t.} \quad & \bar{\mathbf{R}}_{\mathbb{U}} = \bar{\mathbf{R}}_{\mathbb{U}}^{H} \\ & \langle \bar{\mathbf{R}}_{\mathbb{U}} \rangle_{a,b} = \langle \mathbf{R}_{\mathbb{U}} \rangle_{a,b} \\ & \bar{\mathbf{R}}_{\mathbb{U}} \succeq 0, \end{aligned} \tag{15}$$

where $a, b \in \mathbb{U}^{+}, a - b \in \mathbb{D}$. The goal of the aforementioned optimization problem is to retrieve the low-rank Hermitian PSD Toeplitz covariance matrix of the interpolated ULA. This is done by ensuring that the corresponding elements of the sample covariance matrix $\mathbf{R}_{\mathbb{U}}$ match those of the recovered covariance matrix $\bar{\mathbf{R}}_{\mathbb{U}}$. However, (15) represents a non-convex NP-hard problem. By applying a nuclear norm convex relaxation, the optimization function can be reformulated as

$$\begin{aligned} \min_{\bar{\mathbf{R}}_{\mathbb{U}} \in \mathbb{C}^{|\mathbb{U}^{+}| \times |\mathbb{U}^{+}|}} \quad & ||\bar{\mathbf{R}}_{\mathbb{U}}||_{*} \\ \text{s.t.} \quad & \bar{\mathbf{R}}_{\mathbb{U}} = \bar{\mathbf{R}}_{\mathbb{U}}^{H} \\ & \langle \bar{\mathbf{R}}_{\mathbb{U}} \rangle_{a,b} = \langle \mathbf{R}_{\mathbb{U}} \rangle_{a,b} \\ & \bar{\mathbf{R}}_{\mathbb{U}} \succeq 0, \end{aligned} \tag{16}$$

The optimization problem (16) is convex and can be effectively addressed using interior point methods. For clarity, the covariance matrix reconstruction algorithm based on coprime array interpolation is outlined in Algorithm 1 below:

Algorithm 1 Covariance matrix reconstruction based on coprime array interpolation

1: **Input:** Array received data $\{x(k)\}_{k=1}^{\mathcal{K}}$
2: Calculate the SCM $\hat{\mathbf{R}} = \frac{1}{\mathcal{K}} \sum_{k=1}^{\mathcal{K}} x(k)x(k)^{\mathrm{H}}$
3: Transform the SCM $\hat{\mathbf{R}}$ into a vector to acquire the difference co-array output $\boldsymbol{v}$. Using (11), $\boldsymbol{v}_1$ is obtained.
4: Interpolate $\boldsymbol{v}_1$ using (12) to $\bar{v}_1$.
5: Reconstruct the Hermitian Toeplitz matrix $\mathbf{R}_{\mathbb{U}}$ using $\bar{v}_1$.
6: Through (16), the nuclear norm minimization results in $\bar{\mathbf{R}}_{\mathbb{U}}$.
7: **Output:** Proposed interpolated covariance matrix $\bar{\mathbf{R}}_{\mathbb{U}}$ for the coprime array.

2.3. *Pre-Estimated INCM*

For accurate INCM estimation, this paper employs an efficient reconstruction method grounded in the Capon spatial spectrum. The INCM is determined by integrating across an angular range that excludes the desired signal's direction. By reintroducing the MVDR beamformer (9) into the objective function (6), the Capon spatial spectrum is derived as

$$\mathbf{P}(\theta) = \frac{1}{\mathbf{a}^{\mathrm{H}}(\theta)\hat{\mathbf{R}}^{-1}\mathbf{a}(\theta)}. \tag{17}$$

To enhance the precision of $\bar{\mathbf{R}}_{\mathbb{U}}$, we first reconstruct the coprime array covariance matrix using Equation (16). This reconstructed matrix is subsequently substituted into Equation (17) to refine the results. Additionally, the steering vector originally associated with $\bar{\mathbf{R}}_{\mathbb{U}}$ should be replaced with the steering vector corresponding to the interpolated ULA. Consequently, the Capon spatial spectrum can be reformulated as follows:

$$\mathbf{P}_{\mathbb{U}}(\theta) = \frac{1}{\mathbf{a}_{\mathbb{U}}^{\mathrm{H}}(\theta)\bar{\mathbf{R}}_{\mathbb{U}}^{-1}\mathbf{a}_{\mathbb{U}}(\theta)}, \tag{18}$$

where $\mathbf{a}_{\mathbb{U}}(\theta) = \left[1, e^{j\frac{2\pi}{\lambda}d\sin\theta}, e^{j\frac{2\pi}{\lambda}2d\sin\theta}, \ldots, e^{j\frac{2\pi}{\lambda}|\mathbb{U}^{+}|d\sin\theta_l}\right]^T$. At this stage, $P_{\mathbb{U}}(\theta)$ serves as an approximation for the Capon spatial spectrum of the ULA. Following this, the INCM is reconstructed by integrating the Capon spatial spectrum.

$$\begin{aligned}\hat{\mathbf{R}}_{\mathrm{INF}} &= \oint_{\Theta_{\mathrm{i}}} \mathbf{P}_{\mathbb{U}}(\theta)\mathbf{a}_{\mathbb{U}}(\theta)\mathbf{a}_{\mathbb{U}}^{\mathrm{H}}(\theta)d\theta \\ &= \oint_{\Theta_{\mathrm{i}}} \frac{\mathbf{a}_{\mathbb{U}}(\theta)\mathbf{a}_{\mathbb{U}}^{\mathrm{H}}(\theta)}{\mathbf{a}_{\mathbb{U}}^{\mathrm{H}}(\theta)\bar{\mathbf{R}}_{\mathbb{U}}^{-1}\mathbf{a}_{\mathbb{U}}(\theta)}d\theta,\end{aligned} \tag{19}$$

where the integral range of the angle, Θ_{i}, includes only the angles of the interference signals, excluding the SOI signal. $\hat{\mathbf{R}}_{\mathrm{INF}}$ includes all the information about interference and noise and is not affected by the SOI. However, conventional integration methods lead to a sharp increase in computational complexity, necessitating the adoption of an efficient solution method. Therefore, the integral operation can be approximately simplified into a discrete form [36]. In other words, the integration can be approximated by summation as follows:

$$\hat{\mathbf{R}}_{\mathrm{INF}} \approx \sum_{l=1}^{L} \frac{\mathbf{a}_{\mathbb{U}}(\theta)\mathbf{a}_{\mathbb{U}}^{\mathrm{H}}(\theta)}{\mathbf{a}_{\mathbb{U}}^{\mathrm{H}}(\theta)\bar{\mathbf{R}}_{\mathbb{U}}^{-1}\mathbf{a}_{\mathbb{U}}(\theta)}d\theta \tag{20}$$

where the original integral range is discretized into L angles within Θ_{i}. To simplify the notation, θ_l is written as θ. The precision of the approximate integral expression increases with the number of discretized angles L. Additionally, the computational complexity of this approach is $\mathcal{O}((MN - M + 1)^2 L)$.

Nonetheless, the estimated $\hat{\mathbf{R}}_{\mathrm{INF}}$ is obtained from $\bar{\mathbf{R}}_{\mathbb{U}}$, which includes data related to the desired signal. Additionally, $\hat{\mathbf{R}}_{\mathrm{INF}}$ through summation poses problems that need urgent resolution. To accurately approximate the integral operation, a large number of discretized angles L are required, resulting in significant computational complexity.

2.4. INCM Reconstruction

In this section, an adaptive beamforming algorithm for coprime arrays is introduced, leveraging INCM reconstruction. Initially, the method employs a projection matrix to remove the SOI from the received signals. The projection matrix-processed approximate covariance matrix will then replace $\bar{\mathbf{R}}_{\mathbb{U}}$ in Equation (20). Then, to reduce computational complexity, a GLQ algorithm is introduced to propose a new matrix reconstruction method as an alternative to the integral operation. Finally, the steering vector is updated by maximizing the beamformer's output power, addressing the model mismatch issue.

The goal is to achieve $\mathbf{By}(k) \to \mathbf{j}(k) + \mathbf{v}(k)$. Furthermore, the function of the projection matrix $\mathbf{B}$ is to extract the signal that only contains $\mathbf{j}(k) + \mathbf{v}(k)$ from the received signal $\mathbf{y}(k)$. This reduces the impact of the desired signal on the INCM, resulting in a more accurate INCM. First, a pseudo-covariance matrix is constructed as

$$C = \alpha\mathbf{a}_{\mathbb{U}}(\theta_0)\mathbf{a}_{\mathbb{U}}^{\mathrm{H}}(\theta_0) + \mathbf{I}_{MN-M+1} \tag{21}$$

where $\alpha \gg 1$ is a construction parameter. The value of α affects the orthogonality between the constructed matrix and the desired signal and should be chosen to be relatively large. In the simulations of this paper, α is set to $101\mathrm{Tr}\{\bar{\mathbf{R}}_{\mathbb{U}}\}$. In practical applications, to reduce computational load, α should be set to 10^4. C can be expressed in terms of its eigenvalues and eigenvectors, with the eigenvalues μ_i arranged in descending order, as follows:

$$C = \sum_{i=1}^{MN-M+1} \mu_i \xi_i \xi_i^{\mathrm{H}} \tag{22}$$

where $\mu_1 = (MN - M + 1)\alpha + 1$ and $\mu_i = 1$ for $i \neq 1, \mu_i$. The eigenvectors ξ_i correspond to the eigenvalues μ_i. Consequently, the inverse of the constructed pseudo-covariance matrix C is given by

$$C^{-1} = \sum_{i=1}^{MN-M+1} \frac{\xi_i \xi_i^{\mathrm{H}}}{\mu_i} = \frac{p_1 p_1^{\mathrm{H}}}{M\alpha + 1} + \sum_{i=2}^{MN-M+1} \xi_i \xi_i^{\mathrm{H}}. \tag{23}$$

Based on the condition $\sum_{i=1}^{MN-M+1} p_i p_i^{\mathrm{H}} = \mathbf{I}_{MN-M+1}$ being satisfied and considering $\alpha \gg 1$, C^{-1} can be approximated as

$$C^{-1} \approx \sum_{i=2}^{MN-M+1} \xi_i \xi_i^{\mathrm{H}} = \mathbf{I}_{MN-M+1} - \xi_i \xi_i^{\mathrm{H}}. \tag{24}$$

According to the mathematical derivation, C^{-1} can serve as the projection matrix $\mathbf{B}$, which is given by

$$\mathbf{B} = \mathbf{I}_{MN-M+1} - \xi_i \xi_i^{\mathrm{H}}. \tag{25}$$

Figure 2 illustrates the function $\| B^{\mathrm{H}} a(\theta) \|_2^2$ as it varies with θ. In practical applications, a 12 dB threshold is commonly used for signal detection. When the signal is below 12 dB, it is evident that the projection matrix $\mathbf{B}$ can effectively remove the target signal. This is true even in the presence of certain angular errors, where the matrix still maintains its signal elimination capability. In this scenario, a coprime array with $M = 3$ and $N = 5$ is examined. The desired signal comes from $\theta_0 = 0°$, while three interferers originate from $\theta_1 = -30°$, $\theta_2 = 30°$, and $\theta_3 = 45°$. Each interferer has a signal-to-noise ratio (SNR) and interference-to-noise ratio (INR) of 20 dB. As illustrated in Figure 2, when θ aligns with the desired signal direction θ_0, $\| \mathbf{B}^{\mathrm{H}} a(\theta) \|_2^2$ is significantly reduced. Hence, the SOI is effectively suppressed.

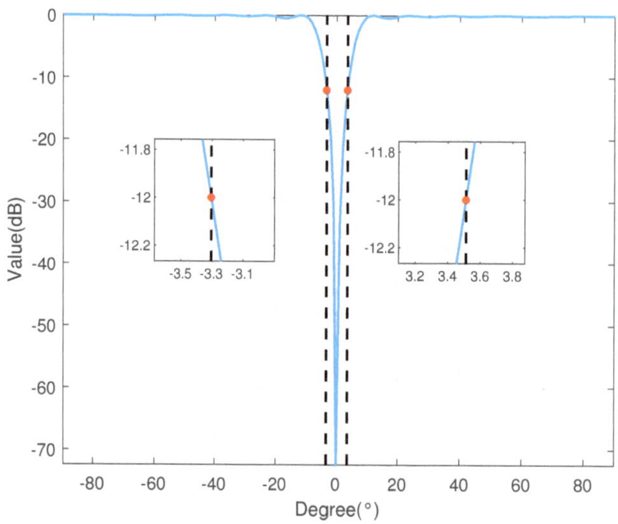

Figure 2. The value of $\| B^{\mathrm{H}} a(\theta) \|_2^2$ as a function of θ.

Next, a covariance matrix $\tilde{\mathbf{R}}_{\mathbb{U}}$ processed by the projection matrix $\mathbf{B}$ can be obtained. In this covariance matrix, the SOI information is eliminated, retaining only the interference and noise information, as follows:

$$\tilde{\mathbf{R}}_{\mathbb{U}} = \mathbf{B}^{\mathrm{H}}\bar{\mathbf{R}}_{\mathbb{U}}\mathbf{B} + \hat{\sigma}_{\mathrm{n}}^2\mathbf{I}_{MN-M+1} \tag{26}$$

Here, $\hat{\sigma}_{\mathrm{n}}^2$ denotes the estimated noise power, which is commonly derived from the smallest eigenvalue of $\bar{\mathbf{R}}_{\mathbb{U}}$. By substituting $\bar{\mathbf{R}}_{\mathbb{U}}$ in Equation (20) with $\tilde{\mathbf{R}}_{\mathbb{U}}$, we obtain

$$\hat{\mathbf{R}}_{\mathrm{INF}} \approx \sum_{l=1}^{L} \frac{\mathbf{a}_{\mathbb{U}}(\theta)\mathbf{a}_{\mathbb{U}}^{\mathrm{H}}(\theta)}{\mathbf{a}_{\mathbb{U}}^{\mathrm{H}}(\theta)\tilde{\mathbf{R}}_{\mathbb{U}}^{-1}\mathbf{a}_{\mathbb{U}}(\theta)} d\theta \tag{27}$$

The quadrature formula with L discrete points, also known as the mechanical quadrature formula, is used for numerical integration. When the discrete points for integration are equally spaced, using L nodes can approximate the integration accuracy up to the $(2L-1)$-th order. Nodes that achieve this level of accuracy are referred to as Gaussian points, and the quadrature formula that attains this accuracy is known as the Gaussian quadrature formula. Currently, the Gaussian quadrature formula is the most accurate quadrature formula available.

The necessary and sufficient condition for nodes to be Gaussian points is that the polynomial having these nodes as its zeros must be orthogonal to any polynomial of degree not exceeding $L-1$ over the interval $[\tau_0, \tau_1]$. That is,

$$\int_{\tau_0}^{\tau_1} P(x)Q(x)\,dx = 0$$

for all polynomials $Q(x)$ of degree not exceeding $L-1$, where $P(x)$ is the polynomial having the Gaussian points as its zeros.

In practical applications of numerical integration, computational complexity is a critical evaluation criterion. The computational complexity of the trapezoidal rule is $O(n)$, requiring $n+1$ evaluations of the function $f(q)$ to complete the integration. Similarly, Simpson's rule has a computational complexity of $O(n)$, also requiring $n+1$ function evaluations. However, despite having the same computational complexity, both methods require a large number of nodes to achieve high accuracy, significantly increasing the computational load. In contrast, the Gauss–Legendre quadrature (GLQ) method also has a computational complexity of $O(n)$, but due to its high polynomial accuracy, it can achieve the same precision with fewer nodes. For example, to achieve an accuracy of 10^{-6}, the trapezoidal rule may need thousands of nodes, Simpson's rule may require hundreds of nodes, whereas GLQ may only need around 10 nodes or fewer. Therefore, to achieve an optimal balance between computational complexity and accuracy, the GLQ method is the most efficient choice.

The N-th standard Legendre polynomial is expressed as

$$F_N(q) = \frac{1}{2^N N!}\frac{\partial^N (q^2-1)^N}{\partial q^N}. \tag{28}$$

where it is orthogonal to any polynomial of degree not exceeding $N-1$ over the interval $[-1,1]$, and all its distinct roots lie within the interval $[-1,1]$. Therefore, the N zeros of the Legendre polynomial can be used as the Gaussian points for the Gaussian quadrature formula.

To mitigate the problem of increasing computational complexity in solving (27) as the number of discretizations L grows, this paper introduces an effective matrix reconstruction technique utilizing the GLQ. GLQ, which is the Gaussian quadrature formula using the Legendre polynomials, is a highly accurate method for numerical integration, known for its algebraic precision. When the integrand is polynomial, GLQ provides a precise approximation. The generalized interpolatory Gaussian quadrature is formulated as

$$\int_{\tau_0}^{\tau_1} f(q)dq = \sum_{n=0}^{N-1} A_n f(q_n) + \epsilon \tag{29}$$

where $q_n \in [\tau_0, \tau_1]$ are the Gaussian quadrature nodes, and τ_0 and τ_1 represent the lower and upper limits of the integration, respectively. ϵ represents the error between the actual value of the integral and its approximation.

A_n are the weights, calculated as follows:

$$A_n = \int_{\tau_0}^{\tau_1} \frac{h(q)}{(q-q_n)\frac{\partial h(q)}{\partial q}} dq \tag{30}$$

Here, $h(q) = (q-q_0)(q-q_1)\cdots(q-q_n)$ is a polynomial in q. According to Gauss's theorem, $h(q)$ must be orthogonal to any polynomial of lower degree than N.

For ease of computation, the nodes of the Gaussian quadrature in (29) are chosen to be the roots of $F_N(q) = 0$, which are used to construct the GLQ. To balance computational complexity and algebraic accuracy, this paper employs third-order GLQ to calculate the integral [45]. The Gaussian integration nodes required for third-order GLQ are determined by $F_3(q) = \frac{1}{2}(5q^3 - 3q) = 0$, i.e., $q_0 = -\rho$, $q_1 = 0$, and $q_2 = \rho, \rho = \sqrt{15}/5$.

The integration interval is conveniently chosen as $[-1, 1]$. Therefore, (29) is rewritten as

$$\int_{-1}^{1} f(q)dq = A_0 f(q_0) + A_1 f(q_1) + A_2 f(q_2) + E \tag{31}$$

Let $h(q) = F_N(q)$ in (30), and the resulting weights A_n are

$$A_n = \int_{-1}^{1} \frac{F_N(q)}{(q-q_n)\frac{\partial F_N(q_n)}{\partial q}} dz = \frac{2}{(1-q_n^2)\left[\frac{\partial F_N(q_n)}{\partial q}\right]^2}. \tag{32}$$

Using the third-order Legendre polynomial, the weights are obtained as $A_0 = 5/9$, $A_1 = 8/9$, and $A_2 = 5/9$. Combining the roots q_n and weights A_n of the third-order Legendre polynomial and substituting them into (31), we obtain

$$\int_{-1}^{1} f(q)dq \approx \frac{5}{9} f(\rho) + \frac{8}{9} f(0) + \frac{5}{9} f(\rho). \tag{33}$$

And by using a linear transformation of the integral, the third-order GLQ can be generalized to a general integral as

$$\int_{\tau_0}^{\tau_1} f(q)dz \approx \frac{\tau_1 - \tau_0}{2}\left[\frac{5}{9} f(l_0) + \frac{8}{9} f(l_1) + \frac{5}{9} f(l_2)\right] \tag{34}$$

where $l_n = (\tau_0 + \tau_1)/2 + q_n(\tau_1 - \tau_0)/2$. Therefore, the third-order GLQ can also be used to compute the INCM (27) as

$$\begin{aligned} \hat{\mathbf{R}}_{\mathrm{INF}} &= \oint_{\Theta_{\mathrm{i}}} \frac{\mathbf{a}_{\mathbb{U}}(\theta)\mathbf{a}_{\mathbb{U}}^{\mathrm{H}}(\theta)}{\mathbf{a}_{\mathbb{U}}^{\mathrm{H}}(\theta)\tilde{\boldsymbol{R}}_{\mathbb{U}}^{-1}\mathbf{a}_{\mathbb{U}}(\theta)} d\theta \\ &= \frac{\theta_b - \theta_a}{2}\left[\frac{5}{9} f(l_0) + \frac{8}{9} f(l_1) + \frac{5}{9} f(l_2)\right] \end{aligned} \tag{35}$$

where $f(\theta) = (\mathbf{a}_{\mathbb{U}}(\theta)\mathbf{a}_{\mathbb{U}}^{\mathrm{H}}(\theta))/(\mathbf{a}_{\mathbb{U}}^{\mathrm{H}}(\theta)\tilde{\boldsymbol{R}}_{\mathbb{U}}^{-1}\mathbf{a}_{\mathbb{U}}(\theta))$, and the integral range is given by $\Theta_{\mathrm{i}} = [\theta_a, \theta_b]$. It should be noted that the integration range Θ_{i} represents the interference range.

By obtaining $\hat{R}_{\text{INF}}$ from (35), which excludes the desired signal information, and substituting it into (20) in place of $\bar{\mathbf{R}}_{\mathbb{U}}$, the output power of the beamformer can be expressed as

$$\mathbf{P}_{\mathbb{U}}(\theta) = \frac{1}{\mathbf{a}_{\mathbb{U}}^{\text{H}}(\theta)\hat{\mathbf{R}}_{\text{INF}}^{-1}\mathbf{a}_{\mathbb{U}}(\theta)} \tag{36}$$

In the Capon spatial spectrum, the peak position corresponds to the DOA estimate $\hat{\theta_0}$ of the desired signal. Furthermore, the steering vector of the SOI can be estimated as $\mathbf{a}_{\mathbb{U}}(\hat{\theta_0})$.

Substituting the reconstructed INCM $\hat{\mathbf{R}}_{\text{INF}}$ and $\mathbf{a}_{\mathbb{U}}(\hat{\theta_0})$ to (9), we obtain

$$w_{\text{prop}} = \frac{\hat{\mathbf{R}}_{\text{INF}}^{-1}\mathbf{a}_{\mathbb{U}}(\hat{\theta_0})}{\mathbf{a}_{\mathbb{U}}^{\text{H}}(\hat{\theta_0})\hat{\mathbf{R}}_{\text{INF}}^{-1}\mathbf{a}_{\mathbb{U}}(\hat{\theta_0})}. \tag{37}$$

The obtained beamformer, combined with the corresponding channel information, can achieve good anti-interference performance. Based on the previous discussion, Algorithm 2 outlines the steps for implementing the proposed robust adaptive beamforming algorithm for coprime arrays.

Algorithm 2 The implementation process of the proposed robust adaptive beamforming

1: **Input:** The interpolated covariance matrix $\bar{\mathbf{R}}_{\mathbb{U}}$
2: Construct the pseudo-covariance matrix. $C = \alpha\mathbf{a}_{\mathbb{U}}(\theta_0)\mathbf{a}_{\mathbb{U}}^{\text{H}}(\theta_0) + \mathbf{I}_{MN-M+1}$, $\alpha = 101\text{Tr}\{\bar{\mathbf{R}}_{\mathbb{U}}\}$
3: Use (24) to obtain the approximate matrix C^{-1}, and construct the projection matrix $\mathbf{B} = C^{-1}$.
4: Use $\mathbf{B}$ to obtain $\tilde{\mathbf{R}}_{\mathbb{U}}$ through (26).
5: Reconstruction the INCM $\hat{\mathbf{R}}_{\text{INF}}$ based on GLQ, $\hat{\mathbf{R}}_{\text{INF}} = \frac{\theta_b-\theta_a}{2}[A_0 f(l_0) + A_1 f(l_1) + A_2 f(l_2)]$, $f(\theta) = \frac{\mathbf{a}_{\mathbb{U}}(\theta)\mathbf{a}_{\mathbb{U}}^{\text{H}}(\theta)}{\mathbf{a}_{\mathbb{U}}^{\text{H}}(\theta)\tilde{R}_{\mathbb{U}}^{-1}\mathbf{a}_{\mathbb{U}}(\theta)}, \theta \in \Theta_i$.
6: Obtain $\hat{\theta_0}$ from the peak values in the spectrum of (36).
7: Design the adaptive beamformer $w_{\text{prop}} = \frac{\hat{\mathbf{R}}_{\text{INF}}^{-1}\mathbf{a}_{\mathbb{U}}(\hat{\theta_0})}{\mathbf{a}_{\mathbb{U}}^{\text{H}}(\hat{\theta_0})\hat{\mathbf{R}}_{\text{INF}}^{-1}\mathbf{a}_{\mathbb{U}}(\hat{\theta_0})}$.
8: **Output:** Proposed robust adaptive beamforming weight vector w_{prop} for the coprime array.

2.5. Algorithm Performance Analysis

The computational complexity of the proposed robust beamforming algorithm is mainly determined by the construction of the projection matrix, INCM reconstruction, and spectral peak search. However, the INCM reconstruction based on GLQ requires only three additions, resulting in very low computational complexity. Therefore, the primary factors affecting the computational complexity are the construction of the projection matrix and the spectral peak search. The construction of the projection matrix $\mathbf{B}$ requires obtaining the eigenvalues and eigenvectors of C as described in (23). The eigen decomposition of C is the primary factor affecting the computational complexity, with a complexity of $\mathcal{O}((MN - M + 1)^3)$. The computational burden for the spectral peak search is $\mathcal{O}((MN - M + 1)^2 P)$, where P represents the number of grid points in the angular domain. Therefore, the overall complexity of the algorithm is approximately $\mathcal{O}(max((MN - M + 1)^3, (MN - M + 1)^2 P))$.

3. Simulation and Experimental Results

The effectiveness of the proposed algorithm is verified through simulation and experimental results. In the example of this section, we assume a coprime array with coprime integers 3 and 5, i.e., $(M, N) = (3, 5)$. This means that the sensor array consists of $M + N - 1 = 7$ physical sensors. Assume the SOI impinges on the array from $\theta_0 = 0°$, and three interferers come from $\theta_1 = -30°$, $\theta_2 = 30°$, and $\theta_3 = 45°$, respectively. In all simulation experiments, the INR of each interferer is set to 30 dB. When the experiment studies the impact of input SNR variation on SINR, the number of snapshots is fixed at $\mathcal{K} = 50$. When the experiment studies the impact of snapshot number variation on SINR, the input SNR is fixed at 20 dB. Each figure represents the results of 500 Monte Carlo experiments. The performance of the proposed beamformer is evaluated in comparison to the following beamformers:

(i) integration beamformer [47];
(ii) summation beamformer [47];
(iii) coprime co-array interpolation beamformer [60];
(iv) reconstruction-based beamformer [35];
(v) worst-case SINR maximization beamformer [34].

We assume the angular sector of the SOI is $\Theta = [-3°, 3°]$, then $\bar{\Theta} = [-90°, -3°) \cup (3°, 90°]$.

3.1. Example 1: Known Signal Steering Vector

In the first example, we examine the scenario where the steering vector of the input signal is known. It is worth noting that, due to the presence of SOI information in the snapshots, the adaptive beamformer is still affected, leading to a performance degradation.

The relationship between the output SINR of the beamformer and the input SNR is shown in Figure 3. From this figure, we can intuitively see that the proposed beamformer achieves nearly optimal SINR within the range of -30 dB to 30 dB when the signal steering vector is known. Furthermore, it outperforms all other beamformers compared within the defined SNR range.

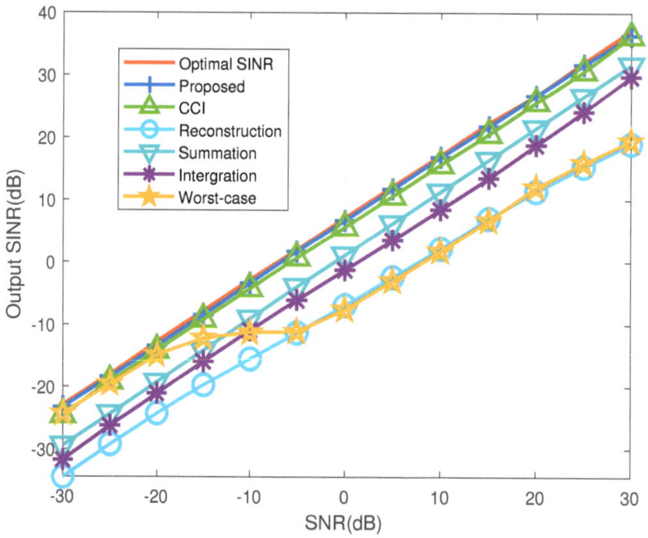

Figure 3. Output SINR versus input SNR in example 1 (Snapshots: 50, INR: 30 dB).

The beamformer based on coprime co-array interpolation also performs well within the defined SNR range. However, because the SOI is not completely removed within the snapshots, its performance is somewhat lacking compared to the proposed beamformer. Based on the worst-case beamformer, the performance is close to that of the optimal beamformer under low SNR conditions. However, its performance degrades at high SNR because the covariance matrix used does not effectively remove the SOI information. The proposed beamformer surpasses existing beamformers by fundamentally removing the SOI information through the application of a projection matrix. This approach enables the construction of a more precise INCM. The relationship between the output SINR of the beamformer and the number of snapshots is shown in Figure 4. It is evident that as the number of snapshots varies from 10 to 100, the proposed beamformer consistently outperforms the other beamformers, demonstrating particularly superior performance at lower snapshot numbers.

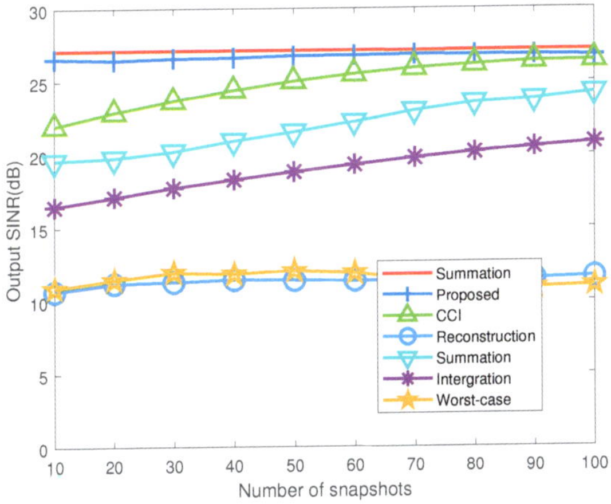

Figure 4. Output SINR versus the number of snapshots in example 1 (SNR: 20 dB, INR: 30 dB).

3.2. Example 2: Mismatch Due to Signal Look Direction Error

In the second example, the effect of signal and interference DOA mismatch on the beamformer's performance is examined. Assume the random DOA errors of the input signals are uniformly distributed within $[-3^{\circ}, 3^{\circ}]$. Furthermore, the actual DOA of the SOI is distributed within $[-3^{\circ}, 3^{\circ}]$, and the DOAs of the interferers are distributed within $[-33^{\circ}, -27^{\circ}]$, $[27^{\circ}, 33^{\circ}]$, and $[42^{\circ}, 48^{\circ}]$, respectively. In each Monte Carlo simulation, the DOA of the input signals changes but stays constant for the same set of snapshots during an experiment.

Figure 5 presents the output SINR of the evaluated beamformers as a function of the SNR. It is clear that both the proposed beamformer and the coprime co-array interpolation beamformer operate effectively within the specified SNR range. Additionally, the proposed beamformer surpasses the coprime co-array interpolation beamformer due to its superior INCM reconstruction accuracy. The worst-case beamformer performs well at low SNR but its performance degrades at high SNR. The correlation between the beamformer's output SINR and the number of snapshots is illustrated in Figure 6. The proposed beamformer demonstrates better performance compared to other algorithms, particularly in the range of 10 to 100 snapshots, with a more notable advantage at lower snapshot counts.

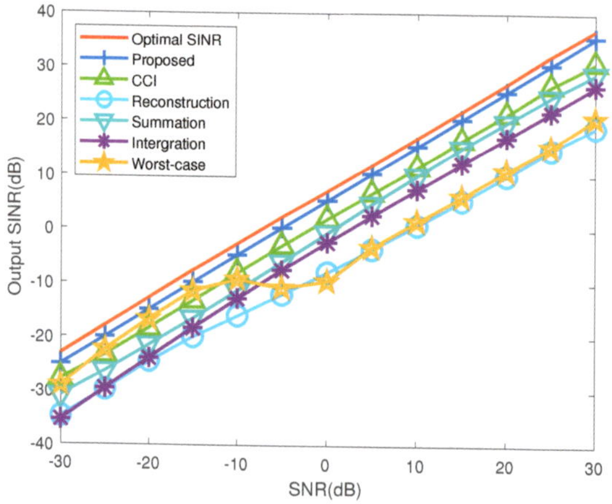

Figure 5. Output SINR versus input SNR in example 2 (Snapshots: 50, INR: 30 dB).

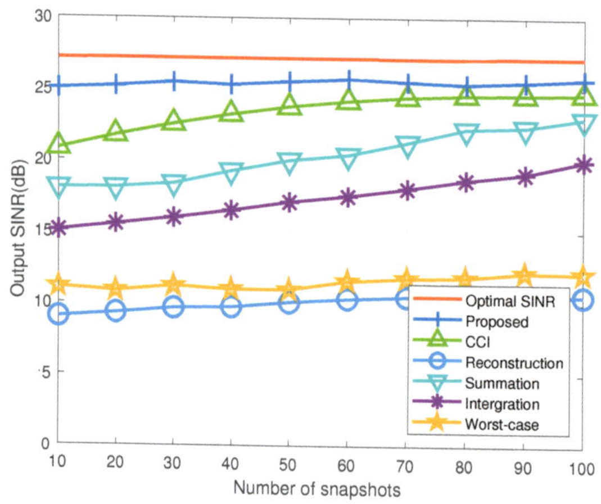

Figure 6. Output SINR versus the number of snapshots in example 2 (SNR: 20 dB, INR: 30 dB).

3.3. *Example 3: Gain and Phase Mismatch*

The third example examines how gain and phase mismatches affect beamformer performance. The amplitude and phase error of the i-th sensor is

$$a_i(\theta) = (1 + \gamma_i)e^{j(\pi \sin\theta(i-1)+\delta_i)} \tag{38}$$

where $\gamma_i \in \mathcal{N}(0, 0.05^2)$ denotes the gain variation at the i-th sensor, characterized by a zero mean and randomness, and $\delta_i \in \mathcal{N}(0, (0.02\pi)^2)$ denotes the phase variation at the same sensor, also random with a zero mean.

Figure 7 shows the relationship between the input SNR and the SINR of the beamformer, and Figure 8 shows the relationship between the number of snapshots used to generate the SCM from the received signals and SINR of the resulting beamformer. The worst-case beamformer exhibits superior performance compared to the other beamformers under low SNR conditions. However, its performance deteriorates at high SNR. In Figure 7, it can be seen that the proposed beamformer performs well even in the presence of gain and phase perturbation mismatches, and it is the closest to the optimal performance. Figure 8 shows that the proposed algorithm nearly achieves optimal performance with fewer snapshots and surpasses other existing methods.

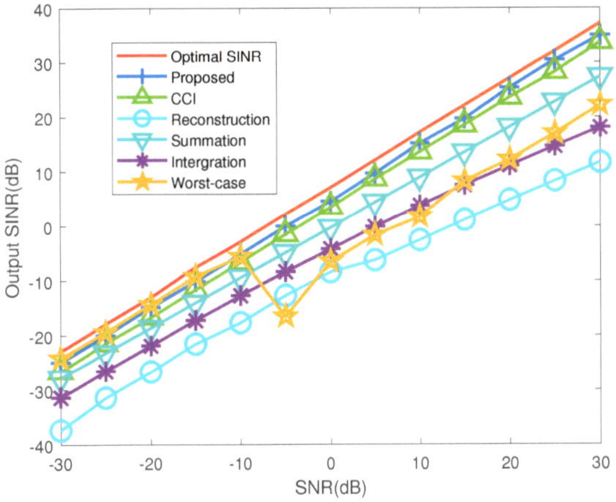

Figure 7. Output SINR versus input SNR in example 3 (Snapshots: 50, INR: 30 dB).

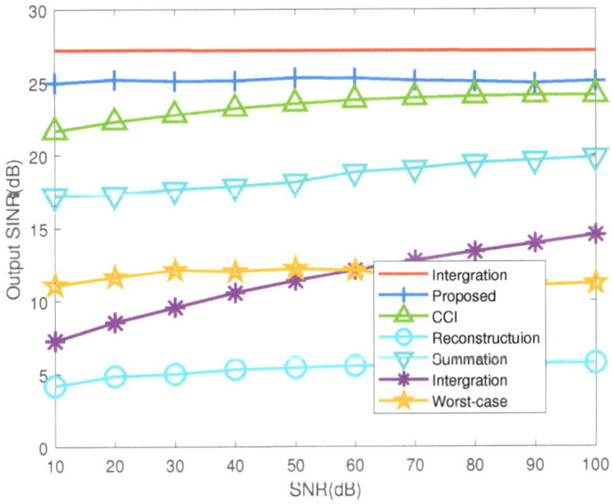

Figure 8. Output SINR versus the number of snapshots in example 3 (SNR: 20 dB, INR: 30 dB).

4. Discussion

The simulation results presented in Section 3 demonstrate the robustness and effectiveness of the proposed beamformer under various conditions, including known signal steering vectors, DOA mismatches, and gain and phase mismatches. These scenarios are critical in practical applications where perfect knowledge of the environment is often unattainable.

When the steering vector of the SOI is known, the proposed beamformer achieves nearly optimal performance across a wide range of SNR values, consistently outperforming other methods. This highlights the algorithm's ability to accurately reconstruct the INCM by effectively eliminating the SOI component through projection matrices.

In the presence of DOA mismatches, the proposed beamformer still maintains superior performance compared to other beamformers. The improved INCM reconstruction accuracy accounts for this robustness, making the algorithm suitable for environments where signal and interference directions are not perfectly known.

In applying the GLQ method for power spectrum integration to obtain the covariance matrix, gain and phase mismatches significantly affect the numerical integration accuracy, leading to performance degradation. Gain mismatch implies that the actual values of the integrand function deviate from the true values due to measurement inaccuracies, data noise, or model imperfections. As GLQ nodes and weights are designed for ideally smooth functions, these errors accumulate directly at each node, causing the overall integration result to deviate from the true value. Additionally, for power spectra with significant variations, gain mismatch amplifies because GLQ nodes might fall within regions of large errors, further exacerbating the integration inaccuracy.

Phase mismatch affects the phase component of the power spectrum, which is critical in calculating the covariance matrix, such as in the term $\cos(2\pi f(t_i - t_j))$. Phase mismatch causes deviations in these phase relationships, leading to systematic biases in the integration results and introducing periodic disturbances. These disturbances result in additional oscillatory terms in the integration result, further reducing accuracy. Therefore, in the presence of gain and phase mismatches, the performance of the GLQ method significantly degrades, highlighting the need for careful consideration of these error sources in practical applications.

5. Conclusions

This paper proposes a novel robust adaptive beamforming algorithm for coprime arrays based on URG. In contrast to existing methods, this approach expands the array's effective aperture via coprime array interpolation and efficiently removes SOI information using a projection matrix, thus allowing for a more precise INCM reconstruction. To reduce the computational complexity of the algorithm, GLQ is introduced. Compared with existing algorithms, this method exhibits greater robustness in scenarios with known steering vectors, DOA mismatches between signals and interferers, and gain and phase perturbation mismatches. Simulation results show that the proposed beamformer achieves superior performance with less computational effort compared to other beamformers and consistently remains close to optimal. Additionally, the algorithm outperforms existing methods, particularly with fewer snapshots. Future work will investigate novel INCM reconstruction techniques aimed at increasing the beamformer's robustness and decreasing computational complexity.

Author Contributions: Conceptualization, S.C. and X.Z.; methodology, S.C.; software, S.C.; validation, S.C., X.Z. and X.W.; formal analysis, S.C.; investigation, S.C.; resources, S.C.; data curation, S.C.; writing—original draft preparation, S.C.; writing—review and editing, S.C.; visualization, S.C.; supervision, X.Z.; project administration, X.Z.; funding acquisition, X.Z., S.L. and W.D. contributed to the validation and review processes. All authors have read and agreed to the published version of the manuscript.

Funding: This work was supported in part by the National Natural Science Foundation of China under Grant 61701140, and research project fund of Songjiang Laboratory (No. SL20230204).

Data Availability Statement: Data are contained within the article.

Conflicts of Interest: The authors declare no conflicts of interest.

Abbreviations

The following abbreviations are used in this manuscript:

ULAs	Uniform Linear Arrays
DOF	Degrees of Freedom
DOA	Direction-of-arrival
INCM	Interference-plus-Noise Covariance Matrix
MVDR	Minimum Variance Distortionless Response
SOI	Signal of Interest
NLAs	Non-Uniform Linear Arrays
SCM	Sample Covariance Matrix
GLQ	Gauss–Legendre Quadrature
URG	Unwanted Signal Removal and Employs Gauss–Legendre Quadrature
SMI	Sample Matrix Inversion
CCI	Coprime Coarray Interpolation
SINR	Signal-to-Interference-plus-Noise Ratio
SNR	Signal-to-Noise Ratio
INR	Interference-to-Noise Ratio

References

1. Fabrizio, G.A.; Gershman, A.B.; Turley, M.D. Robust adaptive beamforming for HF surface wave over-the-horizon radar. *IEEE Trans. Aerosp. Electron. Syst.* **2004**, *40*, 510–525. [CrossRef]
2. Li, J.; Stoica, P. *Robust Adaptive Beamforming*; John Wiley & Sons: Hoboken, NJ, USA, 2005.
3. Xiang, C.; Feng, D.Z.; Lv, H.; He, J.; Cao, Y. Robust adaptive beamforming for MIMO radar. *Signal Process.* **2010**, *90*, 3185–3196. [CrossRef]
4. Xu, J.; Liao, G.; Huang, L.; So, H.C. Robust adaptive beamforming for fast-moving target detection with FDA-STAP radar. *IEEE Trans. Signal Process.* **2016**, *65*, 973–984. [CrossRef]
5. Lin, Z.; Chen, Y.; Liu, X.; Jiang, R.; Shen, B. Adaptive beamforming design of planar arrays based on Bayesian compressive sensing. *IEEE Sens. J.* **2020**, *21*, 5185–5194. [CrossRef]
6. Liu, C.; Li, M.; Zhao, L.; Whiting, P.; Hanly, S.V.; Collings, I.B.; Zhao, M. Robust adaptive beam tracking for mobile millimetre wave communications. *IEEE Trans. Wirel. Commun.* **2020**, *20*, 1918–1934. [CrossRef]
7. Chang, S.; Deng, Y.; Zhang, Y.; Zhao, Q.; Wang, R.; Zhang, K. An advanced scheme for range ambiguity suppression of spaceborne SAR based on blind source separation. *IEEE Trans. Geosci. Remote Sens.* **2022**, *60*, 5230112. [CrossRef]
8. Huang, Y.; Zhou, M.; Vorobyov, S.A. New designs on MVDR robust adaptive beamforming based on optimal steering vector estimation. *IEEE Trans. Signal Process.* **2019**, *67*, 3624–3638. [CrossRef]
9. Yazdi, N.; Todros, K. Measure-transformed MVDR beamforming. *IEEE Signal Process. Lett.* **2020**, *27*, 1959–1963. [CrossRef]
10. Shahbazpanahi, S.; Gershman, A.B.; Luo, Z.Q.; Wong, K.M. Robust adaptive beamforming using worst-case SINR optimization: A new diagonal loading-type solution for general-rank signal models. In Proceedings of the 2003 IEEE International Conference on Acoustics, Speech, and Signal Processing, 2003 (ICASSP'03), Hong Kong, China, 6–10 April 2003; Volume 5, p. V-333. [CrossRef]
11. Elnashar, A.; Elnoubi, S.M.; El-Mikati, H.A. Further study on robust adaptive beamforming with optimum diagonal loading. *IEEE Trans. Antennas Propag.* **2006**, *54*, 3647–3658. [CrossRef]
12. Lin, J.R.; Peng, Q.C.; Shao, H.Z. On Diagonal Loading for Robust Adaptive Beamforming Based on Worst-Case Performance Optimization. *ETRI J.* **2007**, *29*, 50–58. [CrossRef]
13. Du, L.; Li, J.; Stoica, P. Fully automatic computation of diagonal loading levels for robust adaptive beamforming. *IEEE Trans. Aerosp. Electron. Syst.* **2010**, *46*, 449–458. [CrossRef]
14. Zhang, M.; Chen, X.; Zhang, A. A simple tridiagonal loading method for robust adaptive beamforming. *Signal Process.* **2019**, *157*, 103–107. [CrossRef]
15. Gao, J.; Zhen, J.; Lv, Y.; Guo, B. Beamforming technique based on adaptive diagonal loading in wireless access networks. *Ad Hoc Netw.* **2020**, *107*, 102249. [CrossRef]
16. Yang, Y.; Xu, X.; Yang, H.; Li, W. Robust adaptive beamforming via covariance matrix reconstruction with diagonal loading on interference sources covariance matrix. *Digit. Signal Process.* **2023**, *136*, 103977. [CrossRef]
17. Youn, W.S.; Un, C.K. Robust adaptive beamforming based on the eigenstructure method. *IEEE Trans. Signal Process.* **1994**, *42*, 1543–1547. [CrossRef]
18. Choi, Y.H. Eigenstructure-based adaptive beamforming for coherent and incoherent interference cancellation. *IEICE Trans. Commun.* **2002**, *85*, 633–640.
19. Huang, F.; Sheng, W.; Ma, X. Modified projection approach for robust adaptive array beamforming. *Signal Process.* **2012**, *92*, 1758–1763. [CrossRef]
20. Huang, L.; Zhang, B.; Ye, Z. Robust adaptive beamforming using a new projection approach. In Proceedings of the 2015 IEEE International Conference on Digital Signal Processing (DSP), Singapore, 21–24 July 2015; pp. 1181–1185.
21. Yi, S.; Wu, Y.; Wang, Y. Projection-based robust adaptive beamforming with quadratic constraint. *Signal Process.* **2016**, *122*, 65–74. [CrossRef]

22. Yang, X.; Li, Y.; Liu, F.; Lan, T.; Long, T.; Sarkar, T.K. Robust adaptive beamforming based on covariance matrix reconstruction with annular uncertainty set and vector space projection. *IEEE Antennas Wirel. Propag. Lett.* **2020**, *20*, 130–134. [CrossRef]
23. Guo, J.; Yang, H.; Ye, Z. A novel robust adaptive beamforming algorithm based on subspace orthogonality and projection. *IEEE Sens. J.* **2023**, *23*, 12076–12083. [CrossRef]
24. Vorobyov, S.A.; Gershman, A.B.; Luo, Z.Q.; Ma, N. Adaptive beamforming with joint robustness against mismatched signal steering vector and interference nonstationarity. *IEEE Signal Process. Lett.* **2004**, *11*, 108–111. [CrossRef]
25. Hassanien, A.; Vorobyov, S.A.; Wong, K.M. Robust adaptive beamforming using sequential quadratic programming: An iterative solution to the mismatch problem. *IEEE Signal Process. Lett.* **2008**, *15*, 733–736. [CrossRef]
26. Vorobyov, S.A. Principles of minimum variance robust adaptive beamforming design. *Signal Process.* **2013**, *93*, 3264–3277. [CrossRef]
27. Somasundaram, S.D.; Butt, N.R.; Jakobsson, A.; Hart, L. Low-complexity uncertainty-set-based robust adaptive beamforming for passive sonar. *IEEE J. Ocean. Eng.* **2015**, *Early Access*. [CrossRef]
28. Huang, L.; Zhang, J.; Xu, X.; Ye, Z. Robust adaptive beamforming with a novel interference-plus-noise covariance matrix reconstruction method. *IEEE Trans. Signal Process.* **2015**, *63*, 1643–1650. [CrossRef]
29. Liao, B.; Guo, C.; Huang, L.; Li, Q.; Liao, G.; So, H.C. Robust adaptive beamforming with random steering vector mismatch. *Signal Process.* **2016**, *129*, 190–194. [CrossRef]
30. Zhuang, J.; Shi, B.; Zuo, X.; Ali, A.H. Robust adaptive beamforming with minimum sensitivity to correlated random errors. *Signal Process.* **2017**, *131*, 92–98. [CrossRef]
31. Liao, B.; Guo, C.; Huang, L.; Li, Q.; So, H.C. Robust adaptive beamforming with precise main beam control. *IEEE Trans. Aerosp. Electron. Syst.* **2017**, *53*, 345–356. [CrossRef]
32. Feng, Y.; Liao, G.; Xu, J.; Zhu, S.; Zeng, C. Robust adaptive beamforming against large steering vector mismatch using multiple uncertainty sets. *Signal Process.* **2018**, *152*, 320–330. [CrossRef]
33. Chen, P.; Yang, Y. An uncertainty-set-shrinkage-based covariance matrix reconstruction algorithm for robust adaptive beamforming. *Multidimens. Syst. Signal Process.* **2021**, *32*, 263–279. [CrossRef]
34. Huang, Y.; Fu, H.; Vorobyov, S.A.; Luo, Z.Q. Robust adaptive beamforming via worst-case SINR maximization with nonconvex uncertainty sets. *IEEE Trans. Signal Process.* **2023**, *71*, 218–232. [CrossRef]
35. Gu, Y.; Leshem, A. Robust adaptive beamforming based on interference covariance matrix reconstruction and steering vector estimation. *IEEE Trans. Signal Process.* **2012**, *60*, 3881–3885.
36. Gu, Y.; Goodman, N.A.; Hong, S.; Li, Y. Robust adaptive beamforming based on interference covariance matrix sparse reconstruction. *Signal Process.* **2014**, *96*, 375–381. [CrossRef]
37. Shen, F.; Chen, F.; Song, J. Robust adaptive beamforming based on steering vector estimation and covariance matrix reconstruction. *IEEE Commun. Lett.* **2015**, *19*, 1636–1639. [CrossRef]
38. Yuan, X.; Gan, L. Robust adaptive beamforming via a novel subspace method for interference covariance matrix reconstruction. *Signal Process.* **2017**, *130*, 233–242. [CrossRef]
39. Zheng, Z.; Zheng, Y.; Wang, W.Q.; Zhang, H. Covariance matrix reconstruction with interference steering vector and power estimation for robust adaptive beamforming. *IEEE Trans. Veh. Technol.* **2018**, *67*, 8495–8503. [CrossRef]
40. Zhu, X.; Xu, X.; Ye, Z. Robust adaptive beamforming via subspace for interference covariance matrix reconstruction. *Signal Process.* **2020**, *167*, 107289. [CrossRef]
41. Yang, H.; Wang, P.; Ye, Z. Robust adaptive beamforming via covariance matrix reconstruction under colored noise. *IEEE Signal Process. Lett.* **2021**, *28*, 1759–1763. [CrossRef]
42. Li, H.; Geng, J.; Xie, J. Robust adaptive beamforming based on covariance matrix reconstruction with RCB principle. *Digit. Signal Process.* **2022**, *127*, 103565. [CrossRef]
43. Yang, H.; Ye, Z. Robust adaptive beamforming based on covariance matrix reconstruction via steering vector estimation. *IEEE Sens. J.* **2022**, *23*, 2932–2939. [CrossRef]
44. Ge, S.; Fan, C.; Wang, J.; Huang, X. Robust adaptive beamforming based on sparse Bayesian learning and covariance matrix reconstruction. *IEEE Commun. Lett.* **2022**, *26*, 1893–1897. [CrossRef]
45. Luo, T.; Chen, P.; Cao, Z.; Zheng, L.; Wang, Z. URGLQ: An efficient covariance matrix reconstruction method for robust adaptive beamforming. *IEEE Trans. Aerosp. Electron. Syst.* **2023**, *59*, 5634–5645. [CrossRef]
46. Liu, C.L.; Vaidyanathan, P. Super nested arrays: Linear sparse arrays with reduced mutual coupling—Part I: Fundamentals. *IEEE Trans. Signal Process.* **2016**, *64*, 3997–4012. [CrossRef]
47. Zhou, C.; Shi, Z.; Gu, Y. Coprime array adaptive beamforming with enhanced degrees-of-freedom capability. In Proceedings of the 2017 IEEE Radar Conference (RadarConf), Seattle, WA, USA, 8–12 May 2017; pp. 1357–1361.
48. Zheng, Z.; Wang, W.Q.; Kong, Y.; Zhang, Y.D. MISC array: A new sparse array design achieving increased degrees of freedom and reduced mutual coupling effect. *IEEE Trans. Signal Process.* **2019**, *67*, 1728–1741. [CrossRef]
49. Wagner, M.; Park, Y.; Gerstoft, P. Gridless DOA estimation and root-MUSIC for non-uniform linear arrays. *IEEE Trans. Signal Process.* **2021**, *69*, 2144–2157. [CrossRef]
50. Pal, P.; Vaidyanathan, P.P. Nested arrays: A novel approach to array processing with enhanced degrees of freedom. *IEEE Trans. Signal Process.* **2010**, *58*, 4167–4181. [CrossRef]

51. Li, S.; Zhang, X.P. A new approach to construct virtual array with increased degrees of freedom for moving sparse arrays. *IEEE Signal Process. Lett.* **2020**, *27*, 805–809. [CrossRef]
52. Qin, S.; Zhang, Y.D.; Amin, M.G. Generalized coprime array configurations for direction-of-arrival estimation. *IEEE Trans. Signal Process.* **2015**, *63*, 1377–1390. [CrossRef]
53. Zhou, C.; Gu, Y.; Fan, X.; Shi, Z.; Mao, G.; Zhang, Y.D. Direction-of-arrival estimation for coprime array via virtual array interpolation. *IEEE Trans. Signal Process.* **2018**, *66*, 5956–5971. [CrossRef]
54. Zhou, C.; Gu, Y.; Shi, Z.; Zhang, Y.D. Off-grid direction-of-arrival estimation using coprime array interpolation. *IEEE Signal Process. Lett.* **2018**, *25*, 1710–1714. [CrossRef]
55. Zheng, Z.; Huang, Y.; Wang, W.Q.; So, H.C. Direction-of-arrival estimation of coherent signals via coprime array interpolation. *IEEE Signal Process. Lett.* **2020**, *27*, 585–589. [CrossRef]
56. Zhou, C.; Gu, Y.; Shi, Z.; Haardt, M. Direction-of-arrival estimation for coprime arrays via coarray correlation reconstruction: A one-bit perspective. In Proceedings of the 2020 IEEE 11th Sensor Array and Multichannel Signal Processing Workshop (SAM), Hangzhou, China, 8–11 June 2020; pp. 1–4.
57. Shi, J.; Wen, F.; Liu, Y.; Liu, Z.; Hu, P. Enhanced and generalized coprime array for direction of arrival estimation. *IEEE Trans. Aerosp. Electron. Syst.* **2022**, *59*, 1327–1339. [CrossRef]
58. Li, W.; Xu, X.; Huang, X.; Yang, Y. Direction-of-arrival estimation for coherent signals exploiting moving coprime array. *IEEE Signal Process. Lett.* **2023**, *30*, 304–308. [CrossRef]
59. Zhou, C.; Gu, Y.; He, S.; Shi, Z. A robust and efficient algorithm for coprime array adaptive beamforming. *IEEE Trans. Veh. Technol.* **2017**, *67*, 1099–1112. [CrossRef]
60. Zheng, Z.; Yang, T.; Wang, W.Q.; Zhang, S. Robust adaptive beamforming via coprime coarray interpolation. *Signal Process.* **2020**, *169*, 107382. [CrossRef]
61. Zhang, Z.; Liu, W.; Leng, W.; Wang, A.; Shi, H. Interference-plus-noise covariance matrix reconstruction via spatial power spectrum sampling for robust adaptive beamforming. *IEEE Signal Process. Lett.* **2015**, *23*, 121–125. [CrossRef]
62. Ruan, H.; de Lamare, R.C. Robust adaptive beamforming based on low-rank and cross-correlation techniques. *IEEE Trans. Signal Process.* **2016**, *64*, 3919–3932. [CrossRef]
63. Duan, Y.; Yu, X.; Mei, L.; Cao, W. Low-complexity robust adaptive beamforming based on INCM reconstruction via subspace projection. *Sensors* **2021**, *21*, 7783. [CrossRef]

remote sensing

Communication

Delay-Doppler Map Shaping through Oversampled Complementary Sets for High-Speed Target Detection

Jiahua Zhu [1,2], Zhuang Xie [3,*], Nan Jiang [4], Yongping Song [3], Sudan Han [5], Weijian Liu [6] and Xiaotao Huang [3]

1 College of Meteorology and Oceanography, National University of Defense Technology, Changsha 410073, China; zhujiahua1019@nudt.edu.cn
2 Hunan Key Laboratory for Marine Detection Technology, Changsha 410073, China
3 College of Electronic Science and Technology, National University of Defense Technology, Changsha 410073, China
4 School of Automation, Central South University, Changsha 410083, China
5 Defense Innovation Institute, Beijing 100071, China
6 Wuhan Electronic Information Institute, Wuhan 430010, China
* Correspondence: xiezhuang18@nudt.edu.cn

Abstract: Advanced waveform design schemes have been widely employed for radar and sonar remote sensing analysis such as target detection and separation, where significant range sidelobe is a main factor that limits the improvement of analysis performance. As an extensional type of Golay complementary waveforms, complementary sets are a waveform design scenario of concern that shows more diversity in the design of transmission order, and results in a different distribution of range sidelobes. This work proposes an oversampled generalized Prouhet–Thue–Morse (OGPTM) method for the transmitted signal design of complementary sets, with comprehensive analysis to the influence on the sidelobe distribution. Based on this idea and our previous work, we further put forward a pointwise multiplication processor (PMuP) to integrate two delay-Doppler maps of oversampled complementary sets, which achieve much better sidelobe suppression performance on high-speed target detection with range migration.

Keywords: delay-Doppler map; oversampled complementary sets; pointwise multiplication processor; high-speed target detection; sidelobe suppression

1. Introduction

A delay-Doppler map is a kind of remote sensing image that indicates the range and velocity information of targets obtained by radar, sonar, and/or other sensors, which are two highly important parameters during the whole remote sensing process, while range sidelobes always exist to influence the correct extraction of targets. To address this drawback, various advanced waveform schemes are proposed, in which "complementary sets" is a particular waveform pattern that is free of sidelobe in theory. As a generalized type of Golay complementary waveforms (Golay pair) [1] proposed by Tseng and Liu [2], complementary sets has been widely researched in terms of target detection and separation [3] based on the ambiguity function (or delay-Doppler map) [4], as well as the application in different fields, during the past decades. For instance, the meteorological and marine remote sensing of complementary sets have been investigated in [5,6], respectively. In [7], complementary sets are used to identify the acoustic signals of two autonomous underwater vehicles (AUVs). In recent years, complementary sets support the transmission of synthetic aperture radar (SAR) for better autocorrelation and peak sidelobe ratio (PSLR) [8,9]. Additionally, ambiguity function shaping and optimization approaches of complementary sets have also been proposed [10–13].

Notice that the above studies all rely on the complementarity of this waveform scheme to achieve a theoretical impulse along the zero-Doppler shift, but significant range sidelobes are still induced at other Doppler values in the delay-Doppler map. On the other hand, when the targets are moving fast enough, the range migration effect also turns out to be obvious and brings non-ignorable Doppler sidelobes. Both kinds of sidelobes may submerge the weak targets of interest and lead to a drop in detection performance.

The range sidelobes were first introduced by Calderbank et al. [14,15], who developed a transmission order schedule of a Golay pair named Prouhet–Thue–Morse (PTM) design to dramatically decrease the range sidelobes near the zero-Doppler axis. In a parallel direction, Dang et al. [16,17] produced a visually larger range sidelobe blanking area than the PTM through a Binomial Design (BD) on the weights of received returns, at the cost of extremely worsened Doppler resolution. Later, from 2017 to 2023, we separately researched three different pointwise processors [18–21] to combine the advantages of the two ideas for lower range sidelobes without sacrificing Doppler resolution. Unfortunately, both of these works are unable to be directly employed on complementary sets since they have higher design degrees than the Golay sequences studied before [22], and the only contributions to the range sidelobe suppression of complementary sets come from Tang et al. and Nguyen et al. [23,24], who pushed ahead the work of Calderbank et al. through a Generalized PTM (GPTM) design just with close-range sidelobe suppression effect compared to that of the PTM. Additionally, all the above studies cannot reduce the Doppler sidelobes generated by range migration, and the detection of weak targets with high speed is still difficult.

In order to further enhance the elimination ability of range sidelobes and solve the suppression of Doppler sidelobes, we employ an oversampled complementary sets on the scheme of transmission order, named oversampled GPTM (OGPTM) in this paper, and perform a sufficient investigation on the redistribution pattern of sidelobes and the blanking windows, which gives better illumination of the targets and sidelobes in the delay-Doppler map. Next, to address the Doppler sidelobe rise due to range migration, a pointwise multiplication processor (PMuP) is applied to combine the advantages of two delay-Doppler maps of oversampled complementary sets, with an almost sidelobe-free output result for high-speed targets.

The remainder of this paper is organized as follows. Section 2 first gives a signal model of complementary sets under range migration. The technical approaches of OGPTM and PMuP are proposed and analyzed in Section 3. Typical simulation results compared to traditional methods for complementary sets are presented in Section 4, followed by the conclusion in Section 5.

2. Signal Model

The complementarity of complementary sets consist of D unimodular binary sequences $[a_0(l), a_1(l), \ldots, a_d(l), \ldots, a_D(l)]$ $(l = 0, 1, \ldots, L-1)$, usually expressed as

$$\sum_{d=0}^{D-1} C_{a_d}(k) = DL\delta(k) \tag{1}$$

where $\delta(k)$ stands for a Kronecker impulse function, $C_{a_d}(k)$ is the k-th value of the autocorrelation of sequence $a_d(l)$, L means the number of sequence elements (or chips), and each element occupies a pulse width of T_c. Due to this complementarity, the complementary sets can potentially be employed in research fields such as multi-input–multi-output (MIMO) radar and sonar, imaging, communication, and target detection and recognition [25–30], etc.

Next, a baseband modulation is operated to obtain the following time domain waveforms

$$\begin{cases} a_d(t) = \sum_{l=0}^{L-1} a_d(l)\Omega(t - lT_c) \\ \int_{-T_c/2}^{T_c/2} |\Omega(t)|^2 \mathrm{d}t = 1 \end{cases} \tag{2}$$

the shape of $\Omega(t)$ can be selected as rectangle in theory or raised cosine, but it is Gaussian in practice. In order to transmit the complementary sets, two pulse trains $P = \{p(n)\}_{n=0}^{N-1}$ and $Q = \{q(n)\}_{n=0}^{N-1}$ are employed to control the transmission order and received weights, respectively, which could achieve the ideal range sidelobe level on the delay-Doppler map after careful design. The transmitted and received signal are then written as follows:

$$\begin{cases} z_P(t) = \sum_{n=0}^{N-1} a_{p(n)}(t - nT) \\ z_Q(t) = \sum_{n=0}^{N-1} q(n) a_{p(n)}(t - nT) \end{cases} \tag{3}$$

in which N equal to a power of 2 is the pulse number under a periodic interval T. The D-ary sequence P with a cyclic form $P_{\text{std}} = \{0, 1, \ldots, D-1, 0, 1, \ldots, D-1, \ldots\}$ shows the standard order of transmission, and the positive sequence Q represents the classical matched filtering when the elements are all 1 (standard weighting). The element values of (P, Q) are variable for other existing designs, like the aforementioned PTM, GPTM, and BD; readers may refer to [14,16,23] for detailed schemes. Therefore, the delay-Doppler map of complementary sets is given by the following formula:

$$\chi(t, F_D) = \int_{-\infty}^{+\infty} z_P(s) \exp(j2\pi F_D s) z_Q^*(t - s) ds \tag{4}$$

where "$*$" represents the complex conjugation. Without loss of generality, the delay-Doppler map of 2-ary complementary sets (i.e., Golay pair) is deduced as follows:

$$\begin{aligned} \chi(t, F_D) = & \frac{1}{2} \sum_{k=-L+1}^{L-1} [C_{a_0}(k) + C_{a_1}(k)] \sum_{n=0}^{N-1} q(n) C_\Omega(t - kT_c - nT) \\ & - \frac{1}{2} \sum_{k=-L+1}^{L-1} [C_{a_0}(k) - C_{a_1}(k)] \sum_{n=0}^{N-1} (-1)^{p(n)} q(n) C_\Omega(t - kT_c - nT) \end{aligned} \tag{5}$$

It is easy to find that the first term is an impulse function, which makes the waveform free of range sidelobe when $F_D = 0$, and the second term indicates the energy of range sidelobes that is only influenced by the transmission order during the standard weighting.

Note that in our previous work [18,20,21], the targets are assumed to be stabled during the whole signal illumination. However, for a high-speed target whose velocity is considered to be non-negligible to the transmission speed of a signal (e.g., electromagnetic wave or sound), this assumption is invalid due to the range migration. Under this situation, Equation (3) is modified to be

$$\begin{cases} z_P(t) = \sum_{n=0}^{N-1} a_{p(n)}(t - nT) \\ z_Q(t) = \sum_{n=0}^{N-1} q(n) a_{p(n)} \left[t \left(1 + \frac{F_D \lambda}{c} \right) - nT \right] \end{cases} \tag{6}$$

where λ is the wave length of complementary sets, and F_D denotes the Doppler shift of target. This modification will induce a remainder term $O(C_\Omega(t - kT_c - nT))$ due to the mismatch of autocorrelation after the first term of Equation (4)

$$\begin{aligned} \chi(t, F_D) = & \frac{1}{2} \sum_{k=-L+1}^{L-1} [C_{a_0}(k) + C_{a_1}(k)] \sum_{n=0}^{N-1} q(n) C_\Omega(t - kT_c - nT) + O(C_\Omega(t - kT_c - nT)) \\ & - \frac{1}{2} \sum_{k=-L+1}^{L-1} [C_{a_0}(k) - C_{a_1}(k)] \sum_{n=0}^{N-1} (-1)^{p(n)} q(n) C_\Omega(t - kT_c - nT) \end{aligned} \tag{7}$$

which results in significant Doppler sidelobes, as shown in the right column of Figure 1b.

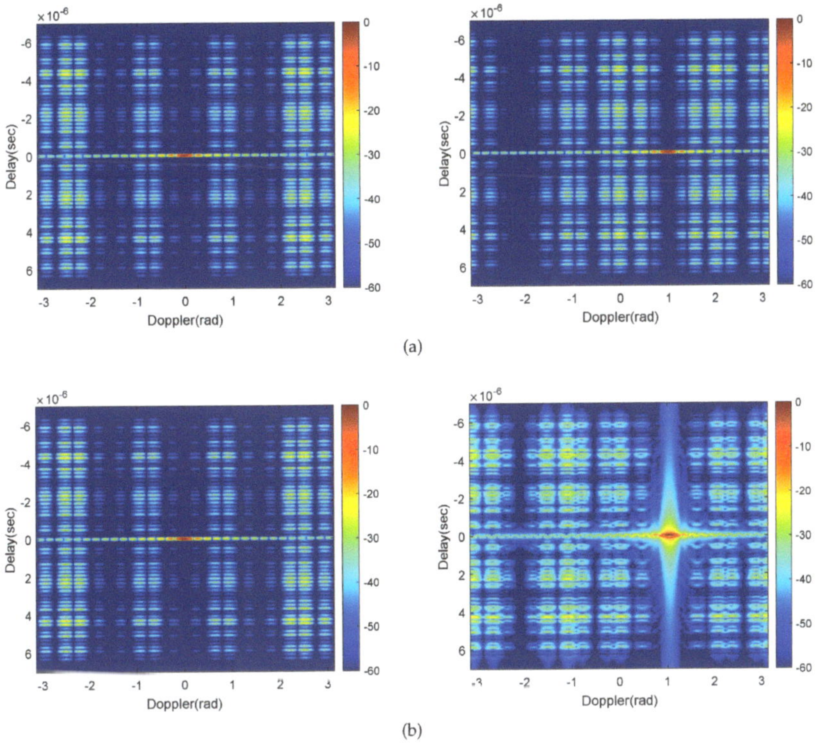

Figure 1. The delay-Doppler map of (**a**) theoretical GPTM algorithm without range migration and (**b**) GPTM algorithm with range migration. (the Doppler of the target equals 0 and 1 rad, respectively, and the unit of the color bar is dB).

3. Technical Approach

Inspired by the idea of our recent publication [21], the process flow of the approach in this paper is demonstrated as Figure 2.

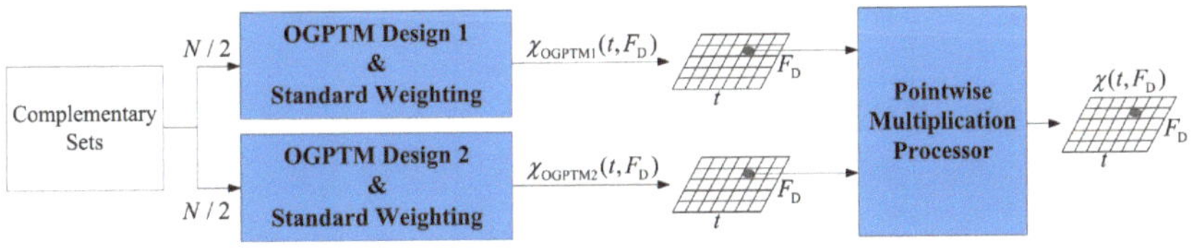

Figure 2. Process flow of OGPTM scheme for complementary sets.

In this proposed procedure, the transmission order of complementary sets is designed by two oversampled GPTM (OGPTM) schemes on the first $N/2$ pulses and the last $N/2$ pulses, respectively, and followed by a pointwise multiplication processor (PMuP) that has not been investigated in our early works to integrate the results of two oversampled sequences as usual. The final output delay-Doppler map would show a satisfied sidelobe suppression performance, particularly the elimination of migration (Doppler) sidelobes generated by the high-speed target. Details of the approaches in the procedure are given in the following section.

3.1. OGPTM

Early in 2007, Pezeshki et al., as aforementioned, put forward the well-known PTM design method, whose transmission order is illustrated as follows [14]:

$$\begin{cases} P(2k) = P(k) \\ P(2k+1) = 1 - P(k) \quad , \quad k \in N \\ P(0) = 0 \end{cases} \tag{8}$$

while Q holds the standard weights.

Seven years later, Tang et al. [23] published a paper and considered extending the PTM on complementary sets for the first time, creating the so-called GPTM algorithm. The detailed process of GPTM contains the following three steps:

1. Assume a sequence $S = [0, 1, \ldots n, \ldots, N-1]$;
2. Change the elements in S to base D, denoted as S_D;
3. The GPTM sequence $P_{\text{GPTM}} = \text{mod}[c_d(S_D(n)), D]$, where $c_d(\cdot)$ represents a function that adds each digit of $S_D(n)$, e.g., $c_d(128) = 1 + 2 + 8 = 11$.

OGPTM, as illustrated by its name, is a generalized version of GPTM with several rounds of oversampling, in order to achieve the redistribution of sidelobes. Without loss of generality, the times of oversampling should be no more than the pulse number, and their values are both limited to a power of 2. An example is now introduced for a clear expression of OGPTM.

Example:

Set a GPTM sequence for a 4-ary complementary sets with $N = 16$; then, the 1-time oversampling, also the conventional GPTM, is given as follows:

$$\begin{aligned} P_{\text{OGPTM1}} &= [0,1,2,3,1,2,3,0,2,3,0,1,3,0,1,2] \\ &= P_{\text{GPTM}} \end{aligned}$$

The twice oversampling GPTM is

$$P_{\text{OGPTM2}} = [0,0,1,1,2,2,3,3,1,1,2,2,3,3,0,0]$$

The 4-times oversampling GPTM is

$$P_{\text{OGPTM4}} = [0,0,0,0,1,1,1,1,2,2,2,2,3,3,3,3]$$

The 8-times oversampling GPTM is

$$P_{\text{OGPTM8}} = [0,0,0,0,0,0,0,0,1,1,1,1,1,1,1,1]$$

The 16-times oversampling GPTM is

$$P_{\text{OGPTM16}} = [0,0,0,0,0,0,0,0,0,0,0,0,0,0,0,0]$$

■

In the following section, we will perform an analysis of the influence of oversampling times in terms of the distribution of sidelobes and reveal the detailed scheme employed in Figure 2.

3.2. PMuP

The pointwise multiplication processor (PMuP) is an alternative method to the previous proposed pointwise minimization processor (PMP) [18] and pointwise thresholding processor (PTP) [21]. This processor is induced as follows:

$$\chi(t, F_D) = \chi_{OGPTM1}(t, F_D) \times \chi_{OGPTM2}(t, F_D) \tag{9}$$

Since the magnitude and location of targets will theoretically stay the same in two maps, but the counterparts of the sidelobe are much different, this provides a chance for PMuP to make more intuitive enhancement to the targets than the sidelobe. Actually, the output of the targets after PMuP is squared, while it only exerts an linear increase to the larger sidelobe of the two, which means as long as the magnitude of target is not far less than the larger value of sidelobe, it would be exposed on the sidelobe background. This discussion will be further evaluated by the later simulations.

4. Simulation Discussion and Results

Before simulation and discussion, the global parameters during the analysis are first listed in Table 1.

Table 1. The global parameters of the simulation.

Variable Name	Parameter	Value (Unit)
Sequences of complementary sets	D	4
Chip number	L	64
Chip length	T_c	0.1 µs
Carrier frequency	f_c	1 GHz
Bandwidth	B	50 MHz
Delay sampling rate	f_{ts}	$2B$
Doppler sampling rate	f_{ds}	0.01 rad
Pulse number	N	32
Pulse repetition interval (PRI)	T	50 µs

Complex Gaussian zero-mean white noise is simulated, satisfying $E \sim \mathcal{CN}(0,1)$ and $SNR = 10$ dB. The Swerling II model is applied to the generation of targets with 10% fluctuation in the radar cross-section (RCS).

4.1. Evaluation of OGPTM

First, a visual comparison of the standard design of Golay and complementary sets, PTM and GPTM, is given in Figure 3, in which $N = 32$ and $D = 4$. The results show that both standard and PTM series designs obtain equal Doppler resolution of the target and similar sidelobe suppression performance, particularly around the zero-Doppler.

The delay-Doppler map of GPTM with range migration is correspondingly plotted in Figure 1b, in which significant spread sidelobes (migration tails) along the delay/range domain are observed caused by the range migration, which unfortunately just fill in the sidelobe blank and neutralize the suppression effect brought by the GPTM design. On the other hand, the shape of target is worsened, becoming irregular. Note that the migration sidelobes only exist on moving targets (since the left columns of Figure 1a,b are identical), especially on fast-moving targets.

Next, the sidelobe distributions of OGPTM with different oversampling rates are exhibited in Figure 4.

For a better illustration, a zero-Doppler target is plotted in this simulation. As can be observed from these figures, with more rounds of oversampling, the GPTM gradually gathers the energy of range sidelobes together, and finally to the Doppler of target. Specifically, when the oversampling rate is N_{over}, the Doppler values of all the range sidelobes decrease to their $1/N_{over}$ compared to that of the traditional GPTM. The OGPTM also pulls the range sidelobes that should be located at $\pm[N_{over} - 1, N_{over}]\pi$rad into the interval $[-\pi, \pi]$rad.

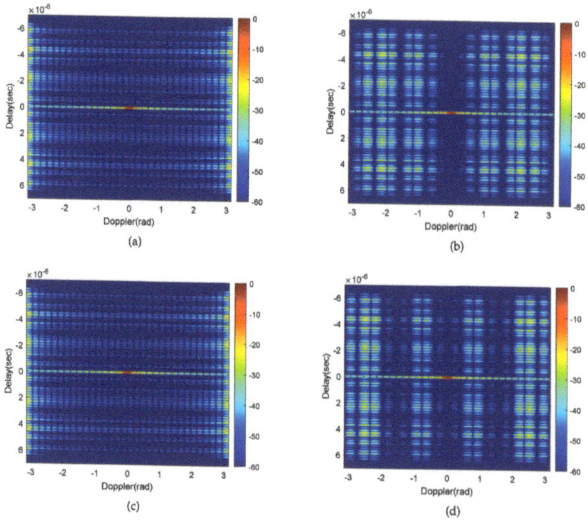

Figure 3. The comparison results of (**a**) standard Golay pair; (**b**) PTM design; (**c**) standard complementary sets; and (**d**) GPTM algorithm (the unit of the color bar is dB).

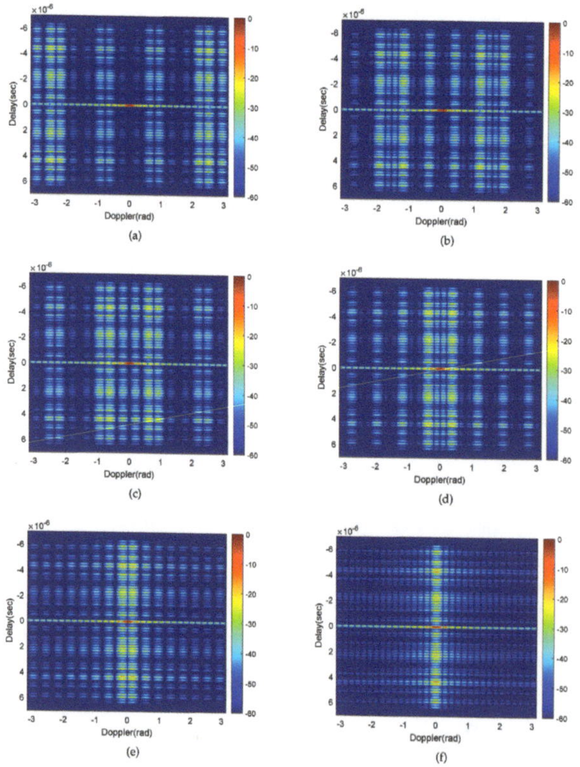

Figure 4. The delay-Doppler maps of OGPTM with different oversampling rates: (**a**) 1 time; (**b**) 2 times; (**c**) 4 times; (**d**) 8 times; (**e**) 16 times; and (**f**) 32 times ($N = 32$, the unit of the color bar is dB).

For comparison, we double the pulse number to 64 and obtain the following range sidelobe distributions in Figure 5. Increasing the pulse number does not change the distribution of OGPTM, but improves the Doppler resolution of the target $1/N_{\text{times}}$, where N_{times} is the multiple of pulse number enlargement (say, N_{times} is 2 when the pulse number increases from 32 to 64), and vice versa.

Let $\Theta = \sum_{n=0}^{N-1}(-1)^{p(n)} q_{(}n) C_{\Omega}(t - kT_{\text{c}} - nT)$ in Equation (5); it equals to 0 when $F_{\text{D}} = \frac{1}{NT}$ or $\theta = 2\pi F_{\text{D}} T = \frac{2\pi}{N}$, which means that the range sidelobes are cut into grids in the delay-Doppler map, and the resolution of each range sidelobe grid is $2T_{\text{c}}\mu\text{s} \times \frac{4\pi}{N}\text{rad}$ ($N_{\text{over}} < N$) and $2T_{\text{c}}\mu\text{s} \times \frac{2\pi}{N}\text{rad}$ ($N_{\text{over}} = N$), but the target always holds the resolution $2T_{\text{c}}\mu\text{s} \times \frac{4\pi}{N}\text{rad}$, and it is surrounded by a sinc-shape range sidelobe group along the target delay axis.

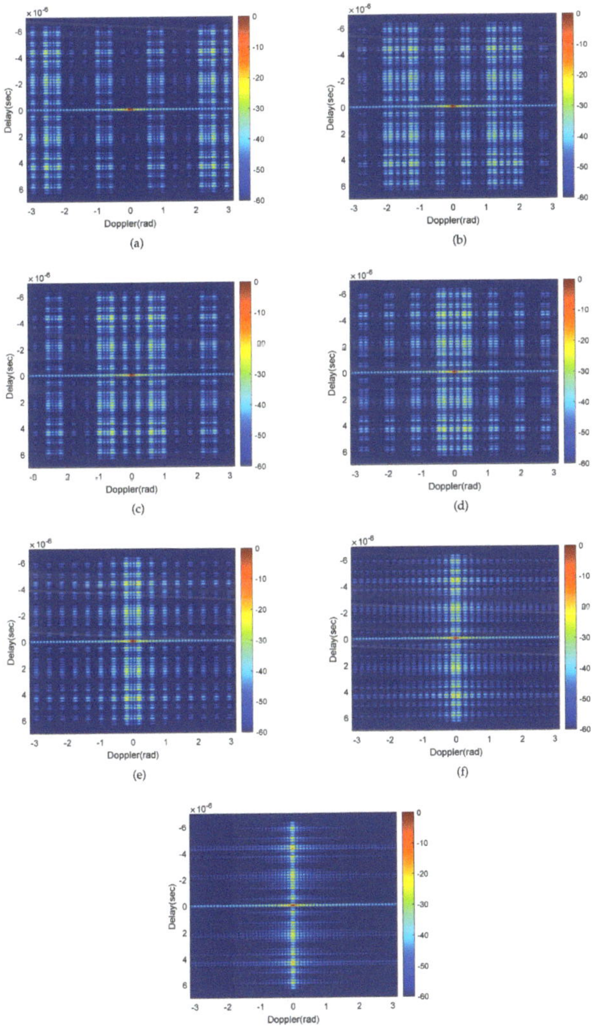

Figure 5. The delay-Doppler maps of OGPTM with different oversampling rates: (**a**) 1 time; (**b**) 2 times; (**c**) 4 times; (**d**) 8 times; (**e**) 16 times; (**f**) 32 times; and (**g**) 64 times ($N = 64$, the unit of the color bar is dB).

Based on these figures, we further investigate the range sidelobe blanking windows (≤ -60 dB) of OGPTM. Using Figure 4 as an example, the correspondence of the oversampling rate and the blanking intervals is

- 1 time: $\pm\left[0, \frac{2\pi}{16}\right] \cup \left[\frac{6\pi}{16}, \frac{10\pi}{16}\right]$rad;
- 2 times: $\pm\left[0, \frac{1\pi}{16}\right] \cup \left[\frac{3\pi}{16}, \frac{5\pi}{16}\right] \cup \left[\frac{11\pi}{16}, \frac{16\pi}{16}\right]$rad;
- 4 times: $\pm\left[\frac{6\pi}{16}, \frac{10\pi}{16}\right] \cup \left[\frac{14\pi}{16}, \frac{16\pi}{16}\right]$rad;
- 8 times: $\pm\left[\frac{3\pi}{16}, \frac{5\pi}{16}\right] \cup \left[\frac{7\pi}{16}, \frac{9\pi}{16}\right] \cup \left[\frac{11\pi}{16}, \frac{13\pi}{16}\right]$rad.

Since the Doppler resolution of the range sidelobe grid is $2\pi/N$rad, the blanking window narrower than the grid is not calculated. This statistical result shows that when N_{over} reaches half of the pulse number, all the blanking window widths would be smaller than the grid resolution, so it makes no sense to calculate the intervals. However, the windows of 16-times oversampling can be observed from the figure of $N = 64$, i.e.,

- 16 times:
$\pm\left[\frac{3\pi}{32}, \frac{5\pi}{32}\right] \cup \left[\frac{7\pi}{32}, \frac{9\pi}{32}\right] \cup \left[\frac{11\pi}{32}, \frac{13\pi}{32}\right] \cup \left[\frac{15\pi}{32}, \frac{17\pi}{32}\right] \cup \left[\frac{19\pi}{32}, \frac{21\pi}{32}\right] \cup \left[\frac{23\pi}{32}, \frac{25\pi}{32}\right] \cup \left[\frac{27\pi}{32}, \frac{29\pi}{32}\right]$rad.

Therefore, we summarize the sidelobe blanking bars of $N/4$ oversampling when $N > 64$ as follows:

$$\pm\left[\frac{3\pi}{N/2}, \frac{5\pi}{N/2}\right] \cup \left[\frac{7\pi}{N/2}, \frac{9\pi}{N/2}\right] \cup \left[\frac{11\pi}{N/2}, \frac{13\pi}{N/2}\right] \cup \cdots \cup \left[\frac{(N/2-5)\pi}{N/2}, \frac{(N/2-3)\pi}{N/2}\right] \text{rad}.$$

With the increase in pulse number and oversampling rate, each window holds more precise Doppler indication to the target, but it also become increasingly sensitive to the Doppler fluctuation, and there is a much longer accumulation time. Therefore, it is not acceptable to blindly increase the oversampling rate of GPTM.

It is worth pointing that the blanking area of the sidelobe using 1-time and 2-times oversampled complementary sets shows almost perfect complementarity in $[-\pi, \pi]$rad, which means if a target is submerged in one of these delay-Doppler maps, it would emerge in the other. This gives us an opportunity to combine the two figures through the following PMuP and further suppress the unwanted range and migration sidelobes.

4.2. Performance of PMuP

As discussed before, the PMuP has the potential to realize improved sidelobe suppression, especially migration-induced sidelobe. Since this issue has also been studied in our previous studies with other pointwise processors—pointwise minimization processor (PMP) [18], pointwise addition processor (PAP) [20] and pointwise thresholding processor (PTP; the threshold was selected to be 2 dB for simulation) [21]—we then make a comparison to these processors, listed in Figure 6, for a better illustration of the performance of PMuP.

In these subfigures, one strong target (0 dB) and one weak target (-20 dB) are presented, whose locations are [1 µs,1 rad] and [3.2 µs, -1.4 rad], respectively. The weak target is covered by the GPTM-generated sidelobe of the strong target (see Figure 6a), but it can be observed in Figure 6b after twice oversampling. Additionally, two tails exists on the targets due to range migration. The results show that PMuP almost eliminates all the sidelobes (including the range sidelobe and migration tails) in the map and two targets are clearly visible. On the other hand, PTP exerts the second best effect on the sidelobe suppression followed by PMP and PAP, while they all preserve unsatisfied tails that are still significant to influence the weak target detection. More seriously, the overall sidelobe of PAP remains comparable to the weak targets, which definitely increases the false alarm, and the PTP even blanked the weak target and led to misdetection. Additionally, it is worth mentioning that the computational complexity of these pointwise processors is comparable due to different point-by-point logical operations, which are also the main difference between these processors, and do not cause a significant increase in processing burden in the approach.

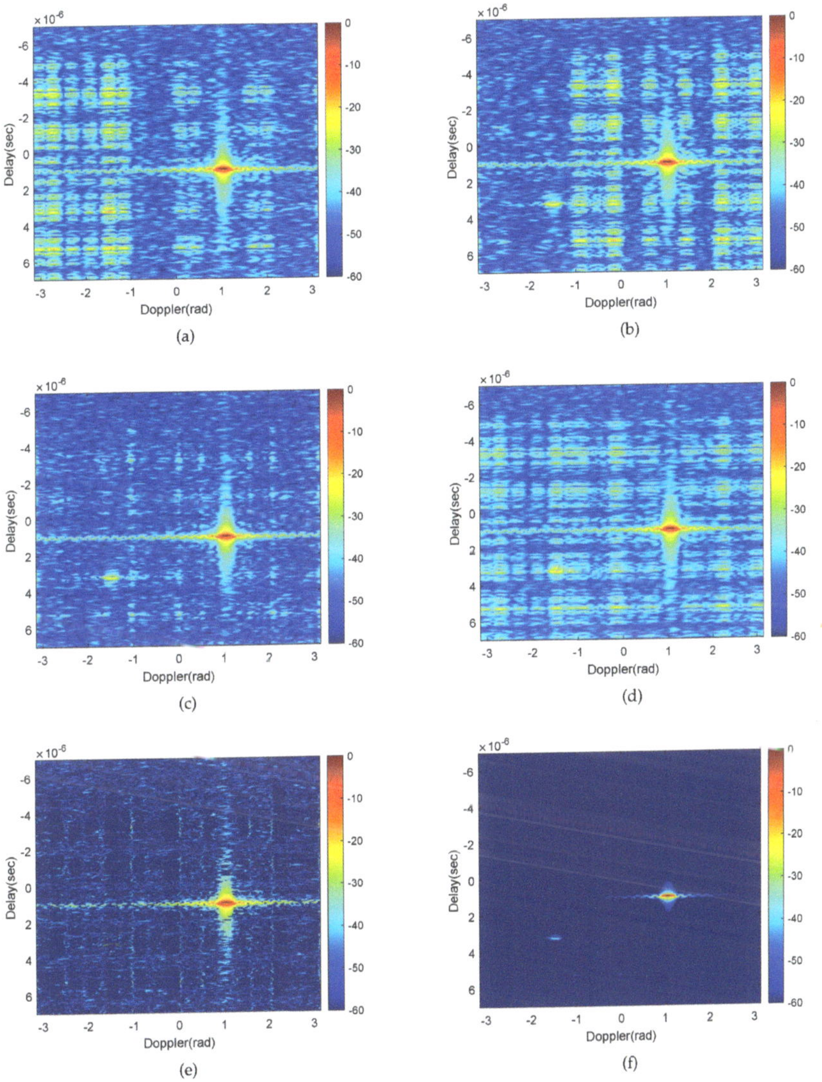

Figure 6. The delay-Doppler maps of (**a**) $\chi_{\mathrm{OGPTM1}}(t, F_{\mathrm{D}})$ (1 time OGPTM, i.e., GPTM); (**b**) $\chi_{\mathrm{OGPTM2}}(t, F_{\mathrm{D}})$ (2 times OGPTM); (**c**) PMP; (**d**) PAP; (**e**) PTP ($thr = 2$dB); and (**f**) PMuP ($N = 32$, the unit of the color bar is dB).

5. Conclusions

This work proposes an oversampled transmission order called OGPTM design for complementary sets, and a PMuP is put forward to exploit the advantages of this scheme for high-speed target detection. Comprehensive theoretical analysis and simulations on the influence of the delay-Doppler map in terms of different sidelobe distributions are given, and a comparison of sidelobe suppression on PMuP and the past-researched PMP, PAP, and PTP shows that this current processor has much better performance on the elimination of overall sidelobe and the migration tails than the others. Practical experiments for the validation of our methods will be considered in future work.

Author Contributions: Conceptualization and supervision, X.H. and W.L.; methodology, J.Z. and Z.X.; software validation, J.Z., N.J. and Y.S.; writing—original draft preparation, J.Z.; writing—review and editing, N.J., S.H. and Z.X.; funding acquisition, J.Z. and S.H. All authors have read and agreed to the published version of the manuscript.

Funding: This work was partly supported by the National Natural Science Foundation of China under grant numbers 62101573 and 62201607.

Data Availability Statement: The data presented in this study are available on request from the first author. The data are not publicly available due to privacy.

Acknowledgments: The authors would like to thank the editors and reviewers for their valuable suggestions to the improvement of the manuscript.

Conflicts of Interest: The authors declare no conflicts of interest.

References

1. Golay, M. Complementary series. *IRE Trans. Inform. Theory* **1961**, *7*, 82–87. [CrossRef]
2. Tseng, C.C.; Liu, C. Complementary sets of sequences. *IEEE Trans. Inform. Theory* **1972**, *18*, 644–652. [CrossRef]
3. Aparicio, J.; Shimura, T. Asynchronous detection and identification of multiple users by multi-carrier modulated complementary set of sequences. *IEEE Access* **2018**, *6*, 22054–22069. [CrossRef]
4. Farnane, K.; Minaoui, K. Analysis of the radar ambiguity function for a new Golay sets shaping. In Proceedings of the 2020 10th International Symposium on Signal, Image, Video and Communications (ISIVC), Tokyo, Japan, 7–9 April 2021; pp. 1–4.
5. Urkowitz, H.; Bucci, N.J. Complementary sequences with Doppler tolerance for radar sidelobe suppression in meteorological radar. In Proceedings of 1992 IEEE International Geoscience and Remote Sensing Symposium (IGARSS), Houston, TX, USA, 26–29 May 1992; pp. 1737–1738.
6. Koshevyy, V.; Popova, V. Complementary coded waveforms sets in marine radar application. In Proceedings of the 2017 IEEE First Ukraine Conference on Electrical and Computer Engineering (UKRCON), Kyiv, Ukraine, 29 May–2 June 2017; pp. 215–220.
7. Aparicio, J.; Shimura, T. Detection of acoustic signals based on OFDM modulated complementary set of sequences. In Proceedings of the OCEANS 2017, Aberdeen, UK, 19–22 June 2017; pp. 1–7.
8. Zhang, Q.; Zhang, Y.; Qi, X.; Li, H. A novel phase coding design for MIMO SAR based on Golay complementary sequences. In Proceedings of the 2022 IEEE International Geoscience and Remote Sensing Symposium (IGARSS), Kuala Lumpur, Malaysia, 17–22 July 2022; pp. 563–566.
9. Zhu, Y.; Chen, J.; Zhang, H.; Wang, P.B.; Yang, W. A novel SAR scheme using CC-S based phase coding waveform for ultra-low range PSLR performance. In Proceedings of the 2013 IEEE International Geoscience and Remote Sensing Symposium (IGARSS), Melbourne, Australia, 21–26 July 2013; pp. 2039–2042.
10. Wang, F.; Feng, S.; Yin, J.; Pang, C.; Li, Y.; Wang, X. Unimodular sequence and receiving filter design for local ambiguity function shaping. *IEEE Trans. Geosci. Remote Sens.* **2022**, *60*, 5113012. [CrossRef]
11. Xie, Z.; Xu, Z.; Fan, C.; Han, S.; Huang, X. Robust radar waveform optimization under target interpulse fluctuation and practical constraints via sequential Lagrange dual approximation. *IEEE Trans. Aerosp. Electron. Sys.* **2023**, *59*, 9711–9721. [CrossRef]
12. Xie, Z.; Wu, L.; Zhu, J.; Lops, M.; Huang, X.; Shankar, B. RIS-Aided radar for target detection clutter region analysis and joint active-passive design. *IEEE Trans. Signal Process.* **2024**, *72*, 1706–1723. [CrossRef]
13. Wang, F.; Pang, C.; Zhou, J.; Li, Y.; Wang, X. Design of complete complementary sequences for ambiguity functions optimization with a PAR constraint. *IEEE Geosci. Remote Sens. Lett.* **2022**, *19*, 3505705. [CrossRef]
14. Pezeshki, A.; Calderbank, A.R.; Moran, W.; Howard, S.D. Doppler resilient Golay complementary waveforms. *IEEE Trans. Inform. Theory* **2008**, *54*, 4254–4266. [CrossRef]
15. Calderbank, R.; Howard, S.D.; Moran, B. Waveform diversity in radar signal processing. *IEEE Signal Process. Mag.* **2009**, *26*, 32–41. [CrossRef]
16. Dang, W.; Pezeshki, A.; Howard, S.; Moran, W.; Calderbank, R. Coordinating complementary waveforms for sidelobe suppression. In Proceedings of the 45th Asilomar Conference Signals, Systems and Computers, Pacific Grove, CA, USA, 6–9 November 2011; pp. 2096–2100.
17. Dang, W. Signal Design for Active Sensing. Ph.D. Thesis, Colorado State University, Fort Collins, CO, USA, 2014.
18. Zhu, J.; Wang, X.; Huang, X.; Suvorova, S.; Moran, B. Range sidelobe suppression for using Golay complementary waveforms in multiple moving target detection. *Signal Process.* **2017**, *141*, 28–31. [CrossRef]
19. Zhu, J.; Wang, X.; Huang, X.; Suvorova, S.; Moran, B. Detection of moving targets in sea clutter using complementary waveforms. *Signal Process.* **2018**, *146*, 15–21. [CrossRef]
20. Zhu, J.; Chu, N.; Song, Y.; Yi, S.; Wang, X.; Huang, X.; Moran, B. Alternative signal processing of complementary waveform returns for range sidelobe suppression. *Signal Process.* **2019**, *159*, 187–192. [CrossRef]

21. Zhu, J.; Song, Y.; Jiang, N.; Xie, Z.; Fan, C.; Huang, X. Enhanced Doppler resolution and sidelobe suppression performance for Golay complementary waveforms. *Remote Sens.* **2023**, *15*, 2452. [CrossRef]
22. Zhu, J.; Fan, C.; Song, Y.; Huang, X.; Zhang, B.; Ma, Y. Coordination of complementary sets for low Doppler-induced Sidelobes. *Remote Sens.* **2022**, *14*, 1549. [CrossRef]
23. Tang, J.; Zhang, N.; Ma, Z.; Tang, B. Construction of Doppler resilient complete complementary code in MIMO radar. *IEEE Trans. Signal Process.* **2014**, *62*, 4704–4712. [CrossRef]
24. Nguyen, H.D.; Coxson, G.E. Doppler tolerance, complementary code sets, and generalised Thue-Morse sequences. *IET Radar Sonar Navi.* **2016**, *10*, 1603–1610. [CrossRef]
25. Zhang, R.; Cheng, L.; Wang, S.; Lou, Y.; Gao, Y.; Wu, W.; Ng, D.W.K. Integrated sensing and communication with massive MIMO: A unified tensor approach for channel and target parameter estimation. *IEEE Trans. Wireless Commun.* **2024**. [CrossRef]
26. Zhang, X.; Yang, P.; Wang, Y.; Shen, W.; Yang, J.; Ye, K.; Zhou, M.; Sun, H. LBF-based CS algorithm for multireceiver SAS. *IEEE Geosci. Remote Sens. Lett.* **2024**, *21*, 1502505. [CrossRef]
27. Zhang, X.; Yang, P.; Wang, Y.; Shen, W.; Yang, J.; Wang, J.; Ye, K.; Zhou, M.; Sun, H. A novel multireceiver SAS RD processor. *IEEE Trans. Geosci. Remote Sens.* **2024**, *62*, 4203611. [CrossRef]
28. Zhang, X. An efficient method for the simulation of multireceiver SAS raw signal. *Mutimedia Tools Appl.* **2023**, *83*, 37351–37368. [CrossRef]
29. Zhang, X.; Yang, P.; Sun, H. Frequency-domain multireceiver synthetic aperture sonar imagery with Chebyshev polynomials. *Electron. Lett.* **2022**, *58*, 995–998. [CrossRef]
30. Zhang, X.; Yang, P.; Feng, X.; Sun, H. Efficient imaging method for multireceiver SAS. *IET Radar Sonar Navi.* **2022**, *16*, 1470–1483. [CrossRef]

remote sensing

MDPI

Article

Infrared Camera Array System and Self-Calibration Method for Enhanced Dim Target Perception

Yaning Zhang, Tianhao Wu, Jungang Yang * and Wei An

College of Electronic Science and Technology, National University of Defense Technology, Changsha 410073, China; zhangyaning@nudt.edu.cn (Y.Z.); wutianhao16@nudt.edu.cn (T.W.); anwei@nudt.edu.cn (W.A.)
* Correspondence: yangjungang@nudt.edu.cn

Abstract: Camera arrays can enhance the signal-to-noise ratio (SNR) between dim targets and backgrounds through multi-view synthesis. This is crucial for the detection of dim targets. To this end, we design and develop an infrared camera array system with a large baseline. The multi-view synthesis of camera arrays relies heavily on the calibration accuracy of relative poses in the sub-cameras. However, the sub-cameras within a camera array lack strict geometric constraints. Therefore, most current calibration methods still consider the camera array as multiple pinhole cameras for calibration. Moreover, when detecting distant targets, the camera array usually needs to adjust the focal length to maintain a larger depth of field (DoF), so that the distant targets are located on the camera's focal plane. This means that the calibration scene should be selected within this DoF range to obtain clear images. Nevertheless, the small parallax between the distant sub-aperture views limits the calibration. To address these issues, we propose a calibration model for camera arrays in distant scenes. In this model, we first extend the parallax by employing dual-array frames (i.e., recording a scene at two spatial locations). Secondly, we investigate the linear constraints between the dual-array frames, to maintain the minimum degrees of freedom of the model. We develop a real-world light field dataset called NUDT-Dual-Array using an infrared camera array to evaluate our method. Experimental results on our self-developed datasets demonstrate the effectiveness of our method. Using the calibrated model, we improve the SNR of distant dim targets, which ultimately enhances the detection and perception of dim targets.

Keywords: camera arrays; self-calibration; light field; structure from motion

1. Introduction

Camera arrays capture both the intensity and direction of light rays [1] and enable the synthesis of multiple perspectives to enhance the signal-to-noise ratio (SNR) between spatial dim targets and backgrounds [2], thereby enhancing the detection capability for dim targets [3]. This method addresses the limitations of traditional techniques in adapting to changes in illumination and complex backgrounds [4], while also circumventing the data requirements and computational resource dependencies associated with deep-learning-based approaches [5,6].

The commercialization of light field (LF) imaging devices has accelerated the development of LF technology. Common LF imaging devices can be categorized into three distinct types based on their structural design: scanning LF cameras [7–9], microlens-array-based LF cameras (also known as plenoptic cameras) [10–12], and camera arrays [13–15]. Scanning LF cameras require stability of the LF throughout the capture process, which limits their applicability to dynamic scenes [8]. Plenoptic cameras have a limited overall resolution, and therefore must trade-off between spatial and angular resolution of captured LF images [16]. Existing camera arrays are typically bulky and expensive [17], and most still operate in the visible spectral range.

To address the structural limitations of conventional LF imaging devices for tasks such as dim target detection, this paper introduces a novel infrared camera array system (as shown in Figure 1). The infrared camera array we designed offers excellent mobility, highly synchronized image acquisition accuracy, impressive baseline depth resolution, and the ability to detect dim targets in the long-wave infrared spectrum (LWIR). The camera array fulfills the requirement to capture moving dim targets under a wide range of conditions, which eliminates the limitations of traditional LF imaging equipment.

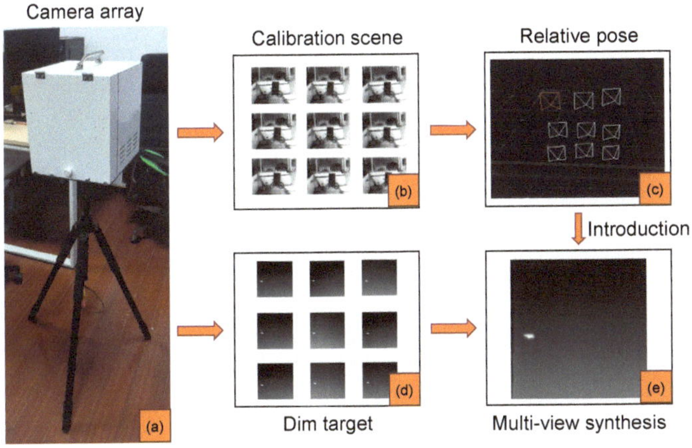

Figure 1. Infrared camera array self-calibration and dim target perception framework. (**a**) Designed camera array. (**b**) Captured calibration scene. (**c**) Calibrated relative pose. (**d**) Obtained dim target (**e**) Multi-view synthesis result.

The calibration of camera arrays is critical for LF imaging and spatial dim target perception. Extensive research has been conducted across various domains, including 3D target localization [18], full optical camera calibration [19], spatial position and pose measurement [20], target position measurement [21], and robot visual measurement [22]. The calibration of camera arrays involves considerations of pose [23,24] and disparity [25,26] for each camera viewpoint. Numerous calibration methods have been developed to meet the diverse requirements of different applications. While current methods, such as metric calibration [27] and structure-from-motion (SfM)-based method [28], have made considerable progress, they still treat the camera array as a combination of pinhole cameras, and the geometrical constraints that exist between sub-cameras have not been investigated, which leads to unsatisfactory calibration accuracy. In addition, the infrared camera arrays we designed faced unique challenges. Since the camera array is used to detect dim targets at long distances, its sub-cameras capture clear images at infinity by adjusting their focal lengths. Therefore, during the calibration process, the calibration scene should also be chosen far enough away so that the camera array can capture a calibrated scene with sharp texture details. However, choosing a calibration scene that is too far away can lead to parallax deficiency problems, so the rays reconstructed from a single array frame are very dense and easily corrupted by noise.

To address this problem, we propose a comprehensive calibration model. The model enhances the discrepancy by capturing the scene twice at different spatial locations, thus extending the spatial resolution of the LF. In addition, the model derives a set of linear constraints regarding the ray space correspondence between dual-array frames. This maintains the minimum degrees of freedom of the model by adding orderly sub-aperture images (SAIs) and effectively reduces the cumulative reprojection error due to the addition of LF images.

In this paper, our goal is to provide a high-precision self-calibration method for infrared camera arrays to achieve dim target perception, as shown in Figure 1. The main contributions of this paper can be briefly described as follows:

(1) We design an innovative infrared camera array system to provide a novel solution for open world dim target detection.
(2) We propose a comprehensive calibration model for the infrared camera array. This model establishes linear constraints on the ray space correspondence within the camera array at two different spatial positions.
(3) We develop a real-world LF dataset for comprehensive performance evaluation. In addition, guided by the obtained camera poses, we achieve dim target SNR enhancement based on multi-view synthesis.

The remaining sections of the paper are structured as follows. Section 2 provides an overview of previous related work. In Section 3, we introduce our own designed and developed infrared camera array, which demonstrates its advantages in visual perception and target detection. Section 4 proposes a self-calibration method for the infrared camera array. Section 5 presents experimental results on real-world data, which demonstrate the accuracy and robustness of the proposed self-calibration method compared to state-of-the-art methods. Section 6 discusses the application prospects and existing problems of the designed infrared camera array and its self-calibration model in practical tasks, which provides direction for our next research efforts. Finally, Section 7 concludes the paper.

2. Related Works

Recently, advances in camera array technology have attracted great interest in various fields such as machine vision, robotics, and virtual reality. In the field of 3D reconstruction, camera arrays demonstrate the potential to capture depth information of the scene and provide powerful support for generating 3D models [29,30]. In addition, camera arrays have a variety of applications in target measurement [31], target tracking [32,33], light field refocusing [34], and super-resolution reconstruction [35].

2.1. Camera Arrays Development

Depending on the purpose and structure, camera arrays can be classified into different types, including stereoscopic, multi-view, or LF systems. These systems consist of multiple cameras arranged in specific configurations to capture multi-dimensional images or video of the same scene from various viewpoints. Regarding different shapes of camera arrays, researchers have proposed various types, such as planar camera arrays [14,36,37], circular camera arrays [38–40], and spherical camera arrays [41–43], to meet different application needs. Planar camera arrays were originally designed to capture LF information [44,45] and contributed to the development of LF cameras [46]. While LF cameras can capture the intensity and direction of light through microlenses to obtain angular information about the scene, they are limited by the resolution of the sensor, which requires the operator to choose between dense spatial sampling or angular sampling [12]. In contrast, planar camera arrays provide better image resolution by mounting multiple independent cameras on a regularly shaped structure. However, planar camera arrays currently lack the capability to detect, track, and identify distant, small, and dim targets effectively. Therefore, there is an urgent need for the development of equipment to promote the advancement of related technology.

2.2. LF Relative Pose Estimation

As a type of LF imaging device, camera arrays exhibit strong similarities in terms of imaging effects and calibration methods with LF cameras. Moreover, research on the calibration of LF cameras is extensive and relatively mature. Therefore, studying the current state of LF camera calibration is very helpful for our research.

In recent years, many pose estimation algorithms for LF cameras have been developed to leverage their unique characteristics, resulting in improved performance. Pless [47] defined a generalized camera model as a set of rays and established correspondences between rays intersecting the same 3D point, known as the generalized epipolar constraint (GEC). Considering all degenerate camera configurations, Li et al. [48] employed a linear algorithm to solve the relative pose from the GEC efficiently without ambiguities, which is also applicable to LF cameras. Johannsen et al. [49] were the first to introduce the GEC into LF cameras and considered the geometric constraints between projections within a single LF, effectively utilizing the characteristics of LF cameras. Zhang et al. [50] investigated how the ray manifolds associated with geometric features, including points, lines, and planes, transform when the relative pose changes and exploited these transformations to recover the pose. Nousias et al. [51] proposed a complete pipeline of LF SfM where the approach to pose estimation is the same as [48]. Zhang et al. [52] established the homography in ray space between two LF cameras and used rays captured by the two LF cameras to solve the relative pose.

However, the rays extracted from LF images using a small baseline are highly dense and susceptible to noise corruption in the aforementioned methods. To mitigate the drawbacks associated with rays, Nousias et al. [53] introduced an LF projection matrix to encapsulate the correspondence between LF features and 3D points, employing direct linear transformation (DLT) to estimate the absolute pose of LF cameras. The LF features used are directly derived from LF images. Nevertheless, the precision of the recovered 3D points may still be compromised by the limited baseline. In the latest research [54], a model of LF-point-LF-point correspondence was established, detailing the relationship between LF features (LF-points) extracted from a pair of LF raw images. This model eliminates the need for 3D points. During the resolution of this model, the rotation and translation were separated and addressed individually. The relative pose is represented by the LF-point components and the intrinsic parameters of the LF camera. In this article, we propose a comprehensive calibration model for the infrared camera array. This model establishes linear constraints on the ray space correspondence within the camera array at two different spatial positions. By doing so, it maintains the minimum degrees of freedom for the model.

3. Design and Construction of Infrared Camera Array System

3.1. System Composition and Principle

As shown in Figure 2, the 3 × 3 infrared camera array system measures 270 × 251.3 × 50 mm externally. It is made up of nine non-cooled LWIR cameras, four five-port gigabit switches, three GPUs, and a power supply module.

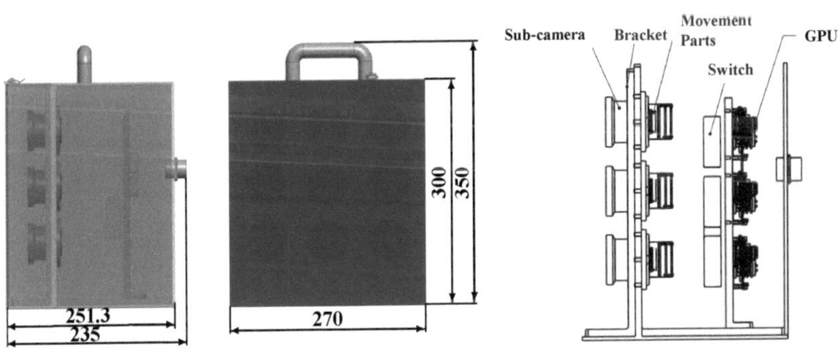

Figure 2. System outline and composition diagram.

Each LWIR camera consists of a lens with adjustable back focus and a detector. They are mounted as a unit on a bracket. A set of three cameras share a five-port gigabit switch and GPU, which are combined in one unit and powered by a power module.

The power supply module receives 12 V of external power and converts it to the voltage required by each component to ensure proper operation. Eight LWIR cameras operate simultaneously, triggered by the exposure signal from one camera. All nine LWIR cameras capture images of the external scene at a consistent exposure time and frame rate.

The image processing module processes three streams of image data, compresses and streams the data, and sends it to the host computer through the switch. When a store command is received, it transfers the raw camera data to the storage module. The storage module then writes these data to a storage card. The continuous image storage time is greater than 30 min at 50 Hz. The data can be exported and deleted through the PC management system.

The 3 × 3 infrared camera array system's main specifications are listed in Table 1.

Table 1. Proposed infrared camera array system.

Item	Infrared Camera Array System
Camera Array Scale	3 × 3
Response Waveband	8–14 μm
Optical Focal Length	35 mm (±5%)
Lens *F* Number	*F*/0.8
Single Camera Pixel Size	640 × 512
Pixel Pitch	12 μm
Frame Rate	50 Hz
Dynamic Range	0–200 °C
Quantization Bit Depth	14 bit
Frame Synchronization Precision	5 ms
Light Axis Consistency	2 pixels
Continuous Image Storage Duration	>30 min@50 Hz
Continuous Stable Working Time	>12 h
Operating Environment Temperature	−20–50 °C
Power Consumption	150 W
Volume	<350 mm × 300 mm × 300 mm
Weight	<10 kg
Power Supply Type	12 V

3.2. Non-Cooled LWIR Camera Design Proposal

The camera features a non-cooled LWIR sensor and a lens. Figure 3 shows the hardware design of this camera, and its main specifications are listed in Table 2.

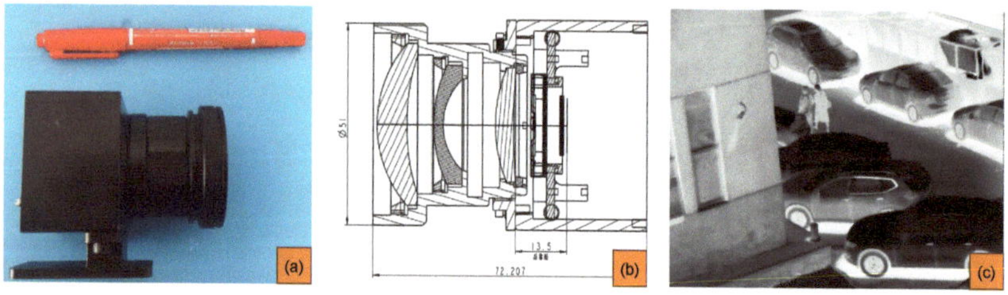

Figure 3. (**a**) Image captured by a LWIR camera. (**b**) Optical structure diagram. (**c**) LWIR camera imaging effect.

Table 2. Proposed non-cooled LWIR camera.

Item	Non-Cooled LWIR Camera
Detector Material	Alum Oxide
Response Waveband	8–14 μm
Detector Elements	640 × 512
Pixel Pitch	12 μm
Detector Origin	China
Optical Focal Length	35 mm (±5%), *F*/0.8
Frame Rate	50 Hz
Effective Field of View	Azimuth: 8.80°, Elevation: 7.04°
External Synchronization	TTL Signal

The optical system is a passive non-thermalized optical structure. It consists of three lenses made of sulfur glass and zinc selenide (ZnSe). The total length of the optical system is 50 mm, the total weight is about 63.2 g, and the back focal length is greater than 14 mm. The detector's pixel size is 12 μm × 12 μm, with a Nyquist spatial frequency of about $1/(2 \times 12\ \mu m) \approx 40$ lp/mm. The modulation transfer function (MTF) value at 40 lp/mm is over 0.4, and the energy concentration exceeds 0.3.

Due to the fact that a temperature change will cause refractive index changes, optical materials and lens barrel materials undergo thermal expansion. This causes the optical system to defocus, which affects the image quality of the system. Therefore, the system uses aluminium alloy as the lens barrel material (the coefficient of thermal expansion is about 23.6×10^{-6}/K). At the same time, we consider the mechanical structure and the effect of the optical components on the image quality when subjected to temperature changes. Through the reasonable combination of different materials, the system can meet the image quality requirements from low temperatures (−20 °C) to high temperatures (50 °C).

The optical system consists of five lenses. The surface of each lens is coated with a transmittance enhancement film. The transmittance of a single piece can reach 97%, and the total transmittance of the optical system is $\tau = 0.973$. The rear flange is connected to the detector, and the flange is connected to the lens by threads so that the rear leaning distance can be fine-tuned and locked by three set screws around the circumference.

3.3. Function and Specification of the System

The infrared camera array and image acquisition system is primarily used for high-frame-rate imaging of specific application scenarios. The system can control camera imaging parameters, synchronize image data acquisition, provide real-time display previews, and store and retrieve data. Image acquisition between sub-cameras requires high time synchronization accuracy and a consistent output frame rate. The control and image data readout interfaces for the camera array are designed with versatility in mind and have the ability to store and replay raw data. Its main technical indicators are shown in Table 1

4. Infrared Camera Array Self-Calibration Method

Our goal is to improve the detection capability of spatial dim targets by synthesizing multi-view array images. The infrared camera array's calibration accuracy directly impacts the multi-view synthesis's effectiveness. Therefore, we propose a comprehensive self-calibration model for the infrared camera array (as shown in Figure 4). First, this model extends the spatial resolution of the LF by capturing a scene at two spatial locations (i.e., dual-array frames). Then, we derive a set of linear constraints regarding the correspondence of ray space between the dual-array frames and maintain the minimum necessary degrees of freedom for the calibration model. Additionally, we also introduce an initial pair adaptive selection strategy to provide a set of precise initial 3D points for our self-calibration model.

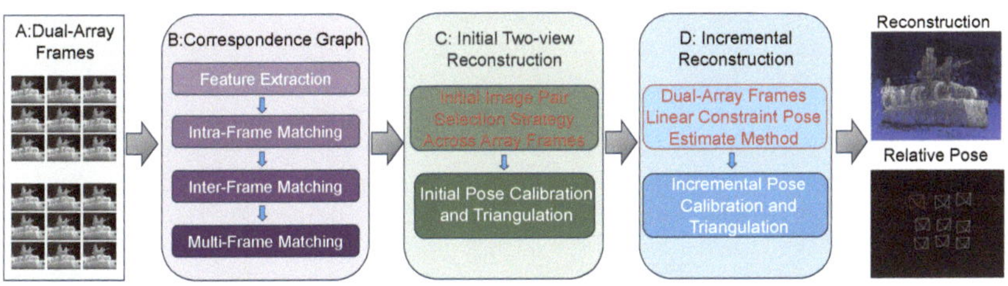

Figure 4. The pipeline of our proposed method is as follows: First, dual-array frame images are obtained. Then, a correspondence search is constructed. Next, an initial image pair selection strategy across array frames is used to select the optimal image pair, and the initial two views are reconstructed. Finally, linear constraints between dual-array frames are introduced to calibrate the remaining SAIs.

4.1. Dual-Array Frames Acquisition

Our method utilizes dual-array frames to calibrate the infrared camera array. Specifically, the initial array frame is acquired by recording a scene from the first perspective. Then, the second array frame is acquired through the parallax–baseline relationship:

$$d = \frac{b \cdot f}{Z}, \tag{1}$$

where f is the focal length, d represents the difference between images (parallax), and Z represents the depth of the object from the camera array. The baseline b required for the dual array frames can be calculated through these three easily obtainable parameters. Subsequently, the infrared camera array is moved to the second perspective for recording.

4.2. Correspondence Graph Build

The second stage of our pipeline focuses on constructing a correspondence graph, which involves four steps [55].

4.2.1. Feature Extraction and Intra-Frame Matching

For dual-array frames, sparse feature points are detected using the difference of Gaussian (DoG) [56] corner measurement method, and their RootSIFT descriptors [57] are computed. Standard distance ratio testing [56] is used to match the RootSIFT descriptors of the central SAI with those of other SAIs. Only matching feature sets that appear in at least four SAIs are retained.

4.2.2. Inter-Frame Matching

For efficiency, only the descriptors of the central SAI are used for matching. Combining ratio testing with left–right consistency checks, correspondences between frame pairs are determined. Since descriptors between different LF viewpoints are expected to have low similarity, a more lenient threshold is used for the ratio test. The five-point algorithm [58] is used within the RANSAC framework to fit an essential matrix, and outliers are discarded.

4.2.3. Establishing Multi-Frame Matches

Starting from the correspondences determined in the previous step, a directed graph is constructed, where each vertex represents a matching feature, and each pair of vertices represents a feature pair. This process ultimately identifies a list of feature matches for the same 3D points across multiple array frames.

4.3. Initial Two-View Reconstruction

To compute the initial reconstruction using two images, there needs to be enough parallax between them. Fachada et al. [59] propose to use the central image of a multi-array image to determine the initial image pair. However, since the shooting position of the multi-array image is human-controlled, the use of the central image as the initial image pair results in a non-optimal parallax between the two views, which leads to poor triangulation results. In *OpenSfM* [60], an adaptive initial image pairs selection method is proposed. It starts by trying to fit a rotation-only camera model to the two images. It only considers image pairs that have a significant portion of the correspondences that cannot be explained by the rotation model. It computes the number of outliers of the model and accepts it only if the portion of outliers is larger than 30%. However, this method does not consider the characteristics of multi-array images, so it is necessary to iterate through all the images to determine the rotation model. As shown on the left side of Figure 5, every image needs to fit the rotation-only camera model 17 times, so the computation time is long.

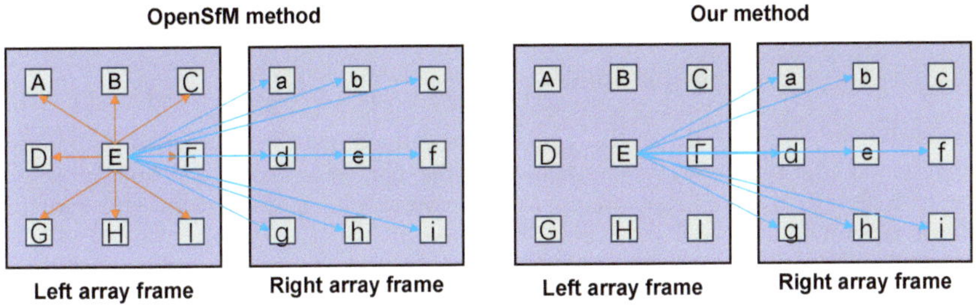

Figure 5. Two initial image pair adaptive selection methods. *A–I* are images from the left array frame, and *a–i* are images from the right array frame. The initial image pair selection of our strategy only needs to be considered from the across array frames.

In our calibration model, a dual-array frame is used to expand the parallax. Therefore, the selection of the initial image pair only needs to be considered from the across array frames. As shown on the right side of Figure 5, every image needs to fit the rotation-only camera model eight times. Specifically, we select two images from different array frames, and their matched point pairs are $\mathbf{p}_{Array1} = \{\mathbf{p}_{Array1,1}, \cdots, \mathbf{p}_{Array1,i}, \cdots, \mathbf{p}_{Array1,N}\}$ and $\mathbf{p}_{Array2} = \{\mathbf{p}_{Array2,1}, \cdots, \mathbf{p}_{Array2,i}, \cdots, \mathbf{p}_{Array2,N}\}$. The disparity is calculated by fitting a rotation matrix to the two SAIs using the five-point algorithm [58]. Then, the feature points of the right SAI $\mathbf{p}_{Array2}$ are transformed to the coordinate system of the left SAI:

$$\mathbf{p}_{rect,i} = \mathbf{R} \cdot \mathbf{p}_{Array2,i}{}^{T} \qquad \forall i \in N, \tag{2}$$

where $\mathbf{p}_{Array2,i}{}^{T}$ represents the transpose matrix of $\mathbf{p}_{Array2,i}$, $\mathbf{R}$ denotes the rotation matrix, and $\mathbf{p}_{rect,i}$ denotes the rectified feature point. Then, the disparity between the two SAIs is calculated according to:

$$disp_i = \sqrt{\sum_{k=1}^{d} (\mathbf{p}_{rect,i}(k) - \mathbf{p}_{Array1,i}(k))^2} \qquad \forall i \in N, \tag{3}$$

where $disp_i$ represents the disparity between $\mathbf{p}_{rect,i}$ and $\mathbf{p}_{Array1,i}$, and d is the dimension of the points. Subsequently, feature points with a disparity larger than the threshold value are labeled as outliers. Then, an index list of outliers is obtained. We compute the number of outliers of the model and accept it only if the portion of outliers is larger than 30% (the same as *OpenSfM* [60]). The accepted image pairs are sorted by the number of outliers of the rotation-only model. Finally, two-view pose calibration and bootstrap reconstruction are performed based on the optimal initial pair.

4.4. Incremental Reconstruction

After achieving good initial reconstruction and pose calibration, we integrate the remaining SAIs incrementally. The camera array consists of multiple pinhole lenses, and there are no specific geometric constraints between the projections within a single LF. However, since our calibration model adopts a dual-array frame, we studied the geometric constraints between the projections of the camera array within the LF composed of dual-array frames. Ultimately, a dual-array frame linear constraint estimator is proposed. This linear constraint estimator reduces reliance on the 2D–3D pose solution of the perspective-n-point (PnP) method [61], thereby enhancing the robustness of pose calibration.

Specifically, as shown in Figure 6, assume that images A and a comprise the initial image pair, and their absolute poses $\mathbf{P}_A = [\mathbf{R}_A|\mathbf{t}_A]$ and $\mathbf{P}_a = [\mathbf{R}_a|\mathbf{t}_a]$ have been calibrated in the previous section using the 2D-2D method. Next, assume that the absolute pose $\mathbf{P}_B = [\mathbf{R}_B|\mathbf{t}_B]$ of image B is calibrated based on the 2D–3D efficient perspective-n-point

(EPnP) method [62]. Then, we can calculate the relative pose between image A and image B:

$$[\mathbf{R}_1|\mathbf{t}_1] = \mathbf{P}_B^{-1} \cdot \mathbf{P}_A = \left[\mathbf{R}_B^T \mathbf{R}_A \middle| \mathbf{R}_B^T(\mathbf{t}_B - \mathbf{t}_A)\right]. \tag{4}$$

According to the principle that the relative pose of the camera array remains unchanged before and after two shots, we can know that the relative pose obtained from images a and b is also $[\mathbf{R}_1|\mathbf{t}_1]$. Ultimately, we can calculate the absolute pose of image b through the linear constraint between the dual-array frames as:

$$\mathbf{P}_b = [\mathbf{R}_a|\mathbf{t}_a] \cdot [\mathbf{R}_1|\mathbf{t}_1] = [\mathbf{R}_a\mathbf{R}_1|\mathbf{R}_a\mathbf{t}_1 + \mathbf{t}_a] \tag{5}$$

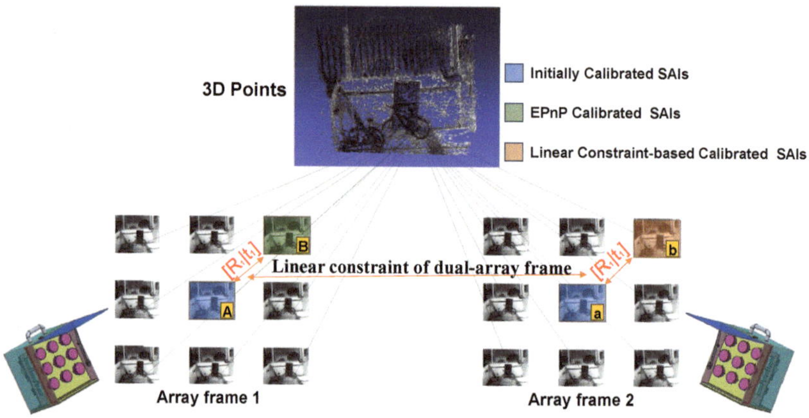

Figure 6. The linear constraint in dual-array frames. The relative poses between A–B and a–b in the dual-array frames are identical.

The algorithm proposed by us is shown in Algorithm 1.

Algorithm 1 Dual-array frame linear constraint pose estimate method

Input: SAIs feature points $\mathbf{p}$; The initially reconstructed 3D points; Initial image pair absolute poses $\mathbf{P}_A$ and $\mathbf{P}_a$
Output: SAIs absolute poses

for Uncalibrated images remain **do**
  EPnP [62] calibrate one SAI absolute pose $\mathbf{P}_i$
  Perform local Bundle Adjustment (BA) optimization and re-triangulation
  if The corresponding image of $\mathbf{P}_i$ in another array frame is uncalibrated **then**
    Compute the relative pose between $\mathbf{P}_i$ and $\mathbf{P}_A$ (or $\mathbf{P}_a$) by Equation (4)
    Calibrate the pose of the corresponding image in another array frame based on linear constraints by Equation (5)
    Perform local BA optimization and re-triangulation
  else
    Continue
  end if
end for
return SAIs absolute poses

A calibration result of our proposed method is shown in Figure 7. We analyzed the effectiveness of our method in addressing cumulative reprojection errors. While the EPnP algorithm [62] is a commonly used and efficient method for camera pose estimation, its limitations and sensitivity to various factors may lead to errors. However, by integrating our proposed method, we can obtain the pose of an unknown view from the known view's pose through the matrix's linear operation. Therefore, this approach reduces the model's degrees of freedom and thus suppresses the cumulative reprojection errors.

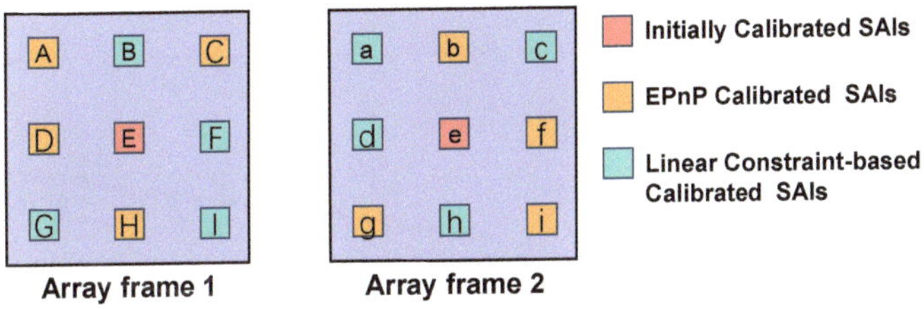

Figure 7. A calibration result with the introduction of linear constraints in dual-array frames.

5. Experiments

To comprehensively study the proposed method, we conducted extensive experiments for evaluation. In Section 5.1, we obtain the experimental parameters and construct a dual-array frame dataset using our infrared camera array. In Section 5.2, we comprehensively evaluate our calibration model compared to state-of-the-art approaches. In Section 5.3, we evaluate the execution time of our calibrated model. In Section 5.4, we conduct experiments on dim target SNR enhancement using the poses obtained from our calibration model. By comparing the SNR of refocused dim targets, we finally validate the effectiveness of our proposed infrared camera array system and self-calibration method.

5.1. Acquisition of Experimental Parameters and Construction of Dataset

To evaluate the performance of the proposed pose calibration method, it is first necessary to obtain the intrinsic parameters of each sub-camera of the infrared camera array. Secondly, it is also essential to construct a suitable dataset for its evaluation.

5.1.1. Acquisition of Intrinsic Parameters

Since the infrared camera array consists of nine long-wave LWIR sub-cameras, we designed a thermostatically controlled heating chessboard to obtain high-precision intrinsic parameters. As shown in Figure 8, the thermostatically controlled heating checkerboard measures 1.5×2 m and features a 14×10 grid pattern and a heating module to maintain a constant temperature. It can present high-quality patterns in the view of the infrared camera array.

Figure 8. An illustration of constant temperature heating checkerboard and the patterns in the infrared camera view.

Specifically, through rotation and translation, we obtain 15 images of the checkerboard grid with different poses. These images were then fed into the Camera Calibrator toolbox in matlab, and the intrinsic parameters (as shown in Table 3) of the infrared camera array were derived according to Zhang's method [27]. Figure 9 shows the accuracy of the calibrated intrinsic parameters, with the mean reprojection error (MRE) for all nine lenses being less than 0.12 pixels.

Table 3. The intrinsic parameters of the infrared camera array (pixel).

Num	f_x	f_y	c_x	c_x	K_1	K_2
1	2765.33	2765.21	264.67	132.64	−0.15	7.18
2	2736.96	2736.24	234.54	110.62	0.19	0.38
3	2696.24	2696.41	306.88	166.90	−0.31	24.91
4	2716.73	2717.15	256.03	94.12	−0.01	−5.87
5	2703.07	2704.70	238.08	104.41	−0.10	−16.03
6	2725.08	2724.96	293.52	120.73	−0.03	−7.36
7	2736.01	2736.14	278.08	93.17	−0.04	−5.98
8	2705.35	2706.76	284.03	91.35	−0.17	10.22
9	2684.96	2688.74	301.01	88.30	−0.10	−6.36

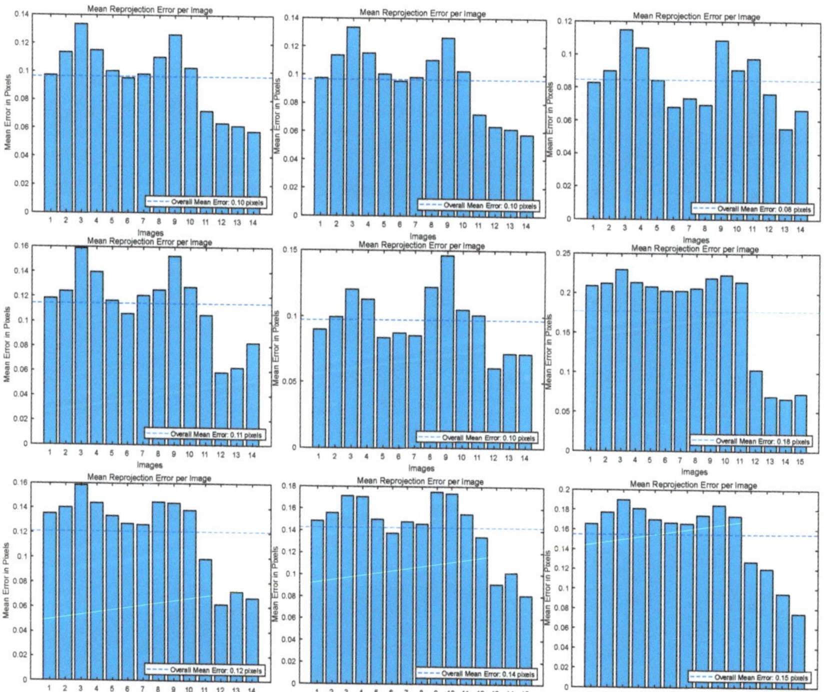

Figure 9. Mean reprojection error of infrared camera array.

5.1.2. Construction of Self-Developed Dataset

We constructed an infrared dual-array frame dataset to evaluate the proposed method. Specifically, we first selected four scenes with abundant details, where the distance between the objects and the camera array is 15–30 m. Calibrating at this distance ensures that the sub-cameras are clear at infinity. Then, we captured four scenes at two different spatial locations according to Section 4.1. Finally, we constructed a dual-array frame dataset, as shown in the first row of Figure 10.

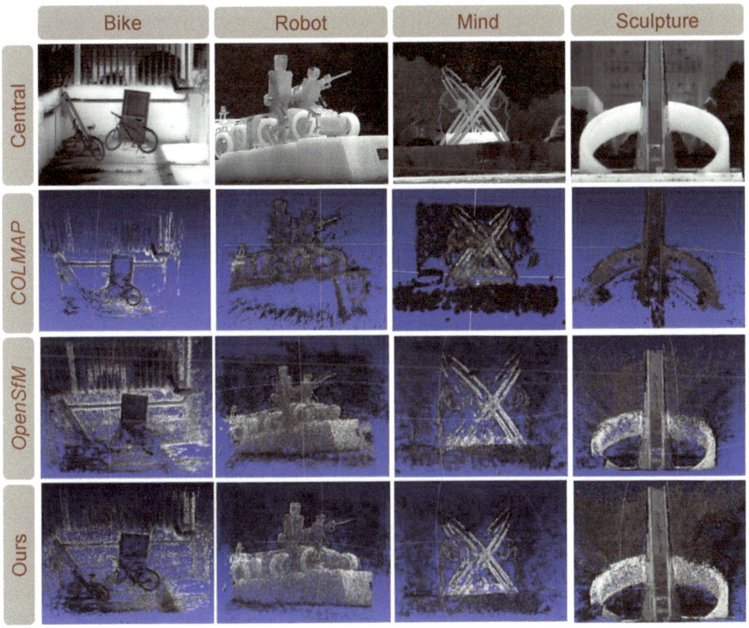

Figure 10. Estimated structures of four scenes. Frames from "Bike", "Robot", "Mind", and "Sculpture" (**top**); structure obtained from *COLMAP* [28] (**second row**); structure obtained from *OpenSfM* [60] (**third row**); structure obtained with our method (**last row**).

5.2. Calibration Model Evaluation

In this section, we comprehensively evaluate our calibration model in comparison with state-of-the-art approaches. Firstly, in the initial two-view reconstruction stage, we evaluate the effect of the adaptive initial image pair selection strategy proposed. Then, in the incremental reconstruction stage, we compare the final calibration results of the proposed self-calibration model with *COLMAP* [28] and *OpenSfM* [60] and thereby validate the gain brought by introducing linear constraints between the dual-array frames. Finally, we perform a scene reconstruction comparison with *COLMAP* and *OpenSfM* on real images.

5.2.1. Initial Two-View Pose Calibration and Reconstruction Comparison

In this stage, we demonstrate the results of two-view reconstruction using three initial image pair selection methods (i.e., Fachada et al.'s [59] method, the *OpenSfM* [60] method, and our method) in four scenes. In Table 4, we first present the selection of the initial image pair under different methods (refer to the numbers in Figure 8). Then, we evaluate the methods based on the number of inliers of the two views, the number of triangulated points, and the algorithm runtime. The boldfaced entries indicate the best results, while the underlined entries represent the second-best results. Specifically, Fachada et al.'s [59] method triangulates near the maximum number of points in the Bike scene, but fails to triangulate in the Mind scene. This is because the Fachada et al. [59] method fixes the use of the central SAI of the dual-array frame as the initial image pair (***E–e***). This method is limited by the quality of the acquired dual-array frames. In addition, this method determines the initial image pair by specifying the image, so its running time is 0. The *OpenSfM* [60] method determines the initial image pair for a rotation-only camera model. Therefore, it performs best in terms of inliers. However, its selection of initial image pairs are all concentrated in the same array frame, which still suffers from insufficient parallax. It leads to a smaller number of final triangulated points than our method. This method also has the longest running time compared to our method. When acquiring dual-array frame images, we artificially control the images to have good parallax between the dual-array frames. Therefore, our method is set to determine the initial image pairs across array frames. The initial image pair selected by our method from the across array frames matches a small number of inliers, but its final triangulation has the largest number of points. This is because our method balances between inliers and parallax. Moreover, our method has the shortest running time.

Table 4. Initial two-viewrt pose calibration and reconstruction comparison.

Scenes	Initial Image Pair			Two-View Inliers			Triangulated			Time		
	[59]	[60]	Ours	[59]	[60]	Ours	[59]	[60]	Ours	[59]	[60]	Ours
Bike	*E–e*	*a–i*	*F–e*	254/317	**446/479**	200/271	210	57	**215**	-	0.094 s	**0.056** s
Robot	*E–e*	*F–H*	*C–h*	78/107	**208/239**	205/232	68	219	**228**	-	0.269 s	**0.185** s
Mind	*E–e*	*C–I*	*D–g*	71/131	**722/752**	588/607	0	335	**533**	-	0.261 s	**0.175** s
Sculpture	*E–e*	*F–g*	*C–c*	201/318	**765/797**	141/264	46	31	**122**	-	0.115 s	**0.086** s

5.2.2. Incremental Pose Calibration and Reconstruction Comparison

In this stage, we compare the performance of the proposed method with *COLMAP* [28] and *OpenSfM* [60]. Specifically, during the experimental process, we first ensure that we use the same feature extraction and matching methods. Then, we conduct high-quality auto-reconstruction experiments in four scenes using *COLMAP* [28]. Next, since the *OpenSfM* [60] parameters are adjustable, we conduct three sets of experiments under different min-track lengths (MTL), where MTL is an important parameter in SfM, affecting the reconstruction quality and processing speed. Finally, experiments with our method are conducted based on linear constraints in dual-array frames.

In terms of performance evaluation, due to the absence of ground truth for the relative poses of the camera array, we reproject the reconstructed 3D points onto all SAIs to calculate the mean reprojection error (*MRE*) according to:

$$MRE=\frac{1}{N}\sum_{i=1}^{N}\sqrt{(x_{proj,i}-x_{GT,i})^2+(y_{proj,i}-y_{GT,i})^2}, \tag{6}$$

where $(x_{GT,i}, y_{GT,i})$ represents a real image point, $(x_{proj,i}, y_{proj,i})$ represents a reprojected point, and N represents the number of reconstructed 3D points.

Our experimental results are presented in Table 5. Experimental results show that our proposed method consistently maintains the minimum *MRE* in all four scenes. Compared to the previous improvements, it has been increased by up to 35.2%. This demonstrates the effectiveness of introducing linear constraints in the dual-array frame.

Table 5. Comparison of *MRE* ↓ (pixel).

Methods	Bike	Robot	Mind	Sculpture
COLMAP [28]	0.66	0.68	0.52	0.71
OpenSfM-MTL(2) [60]	0.79	0.62	0.53	0.66
OpenSfM-MTL(4) [60]	0.88	0.71	0.57	0.69
OpenSfM-MTL(6) [60]	0.86	0.74	0.69	0.72
Ours	**0.57**	**0.65**	**0.46**	**0.62**

5.2.3. Structure Estimation Quality Comparison

In addition, we conducted quality comparison experiments for structure estimation. We visually evaluate the quality of pose calibration by displaying the structures recovered in four scenes using three methods. Specifically, as shown in Figure 10, the overall structure estimated by *COLMAP* [28] in the four scenes is not good. *OpenSfM* [60] estimates structure details less effectively than our method. These visual results are consistent with the conclusions in Table 5.

5.3. Execution Time Comparison

The introduction of linear constraints on the dual-array frames allows the calibration model to maintain a minimum number of degrees of freedom and improves computational efficiency. Therefore, in this stage, we perform a comparison of execution times. Specifically, we perform pose calibration in each of the four scenes. Table 6 demonstrates the comparison of the three methods in execution time. Our method achieves the shortest running time since the introduction of linear constraints, with an average improvement of 31.8%. This validates the industrial practicality of our proposed calibration model.

Table 6. Execution time comparison.

Scenes	COLMAP [28]	OpenSfM [60]	Ours
Bike	37.86 s	25.52 s	**21.56** s
Robot	35.93 s	28.70 s	**21.58** s
Mind	23.41 s	23.91 s	**19.86** s
Sculpture	29.31 s	27.58 s	**21.25** s

5.4. Enhanced Dim Target SNR Experiment

To validate the effectiveness of the infrared camera array system and its self-calibration methods for dim target perception, the experiment of enhancing the SNR of dim targets is conducted in this stage using the poses provided by the three calibration methods. The metric calibration method [27] is the most commonly used checkerboard-grid-based calibration method. The *OpenSfM* [60] method and our method are the self-calibration methods that were verified well in the previous section.

5.4.1. Evaluation Metrics

Wu et al. [2] introduced an infrared image noise theory and analyzed the advantages of multi-view synthetic aperture imaging in improving the *SNR* of dim targets. Specifically, the *SNR* after multi-view synthesis is calculated according to:

$$SNR = E_t / E_n, \tag{7}$$

$$E_t = \sum_{i \in I} |I_{res}(i)| / |T|, \tag{8}$$

$$E_n = \sum_{i \in U \setminus T} |I_{res}(i)| / (|U| - |T|), \tag{9}$$

E_t is the energy of the target, E_n is the noise energy, I_{res} is the residual image after background suppression, T is the set of image elements occupied by the target, and U is the set of full image elements. U and T denote the number of image elements in the set, respectively. Wu et al. [2] stated that the noise in the synthetic aperture image obeys the distribution $\bar{n} \sim N\left(0, \frac{\sigma^2}{K}\right)$, so calculating the spatial average of the noise energy can be equated to calculating the expectation of the noise energy, i.e.,

$$\begin{aligned} E_n &= \sum_{i \in U \setminus T} |I_{res}(i)| / (|U| - |T|) = E(|n|) \\ &= \int_{-\infty}^{+\infty} |x| \frac{\sqrt{K}}{\sqrt{2\pi}\sigma} e^{-\frac{Kx^2}{2\sigma^2}} dx = \frac{\sqrt{2K}}{\sqrt{\pi}\sigma} \int_{0}^{+\infty} x e^{-\frac{Kx^2}{2\sigma^2}} dx \\ &= \sqrt{\frac{2}{K\pi}} \sigma = \frac{C_1}{\sqrt{K}}, \end{aligned} \tag{10}$$

where C_1 is a constant. It is known from the above equation that the noise energy is inversely proportional to the 0.5 power of the number of cameras. The target energy remains constant in the synthetic imaging process, so

$$SNR = \frac{E_t}{E_n} = \frac{E_t}{\sqrt{C_1}}\sqrt{K}. \tag{11}$$

By theoretical derivation, the SNR of the image obtained by synthetic aperture imaging is proportional to the 0.5 power of the number of cameras.

5.4.2. Experiment

Yang et al. [63] proposed a one interpolation refocusing method. This method can change the focus region to bring an object on a specified plane into focus. Therefore, we use this method to refocus on dim targets based on the relative pose obtained from the three calibration methods mentioned. Specifically, we use our infrared camera array to capture three groups of unmanned aerial vehicles (UAVs) at distances between 100 and 200 m. As shown in Figure 11, the UAV is a dim target in all nine sub-aperture images.

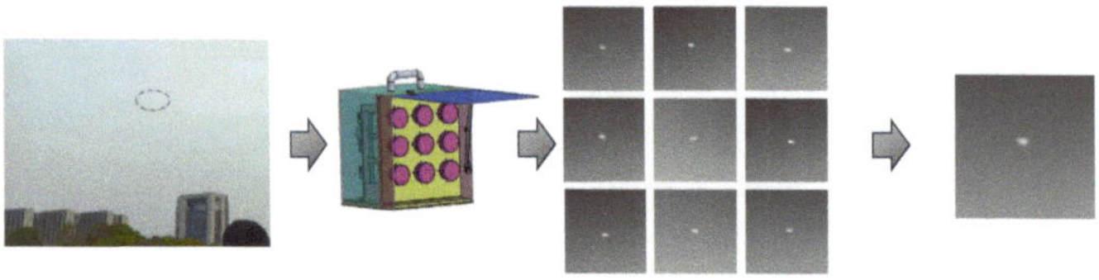

Figure 11. Schematic diagram of the SNR enhancement method for dim target.

We conducted a visual analysis of the experimental results. Figure 12 shows that the relative poses from the three methods successfully focused dim targets in each sub-aperture view into a single target. The image noise in the refocused images is significantly smoother than that in the central sub-view. The metric calibration method [27] performed the worst, as indicated by the diffusion of the refocused dim targets. Our method achieved the best results in refocusing dim targets, especially at 100 m. Although the center sub-camera's depth of field (DoA) was not at the optimal imaging plane, missing the structural features of the UAV, the structural information was restored through the synthesis of the nine views. The *OpenSfM* [60] method did not focus well on dim targets at greater distances, leading to a decrease in target brightness. We also conducted a visual analysis in the energy domain. Figure 13 reveals that the relative poses from the three methods all successfully refocused the dim targets. The noise in the central sub-image was smoothed, and the target energy was slightly reduced, which is negligible for detection tasks.

Then, we conducted quantitative experimental comparisons. In Table 7, we first compare the SNR. The results reveal that the SNR of the refocused target significantly improved compared to that of the central sub-image. Our method contributed the most to the SNR enhancement, averaging a 2.71 times improvement. This is close to the theoretical 3 times SNR increase. Next, we compared the size changes of dim targets before and after refocusing at the pixel level. The metric calibration method [27] resulted in refocused targets becoming noticeably smaller. We analyzed that this was due to the inadequate refocusing effect causing energy loss at the target location, which led to the pixels at the original target position being classified as background. In the first scene, our method also led to smaller refocused targets because the central sub-image's imaging was poor, eventually recovering the structural features of the UAV through multi-view refocusing. In the other two scenes, our method resulted in refocused targets becoming larger, but this difference is negligible when compared to the actual size of the dim targets. The *OpenSfM* [60] method's refocused target size was similar to ours. Finally, we analyzed the target energy before and after refocusing. The results show that our method incurs the lowest energy loss for the refocused targets. Compared to the energy of the target in the central sub-view image, our method on average loses 3.8% of the energy, while the metric calibration method [27] and *OpenSfM* [60] method incur energy losses of 6.0% and 4.5%, respectively.

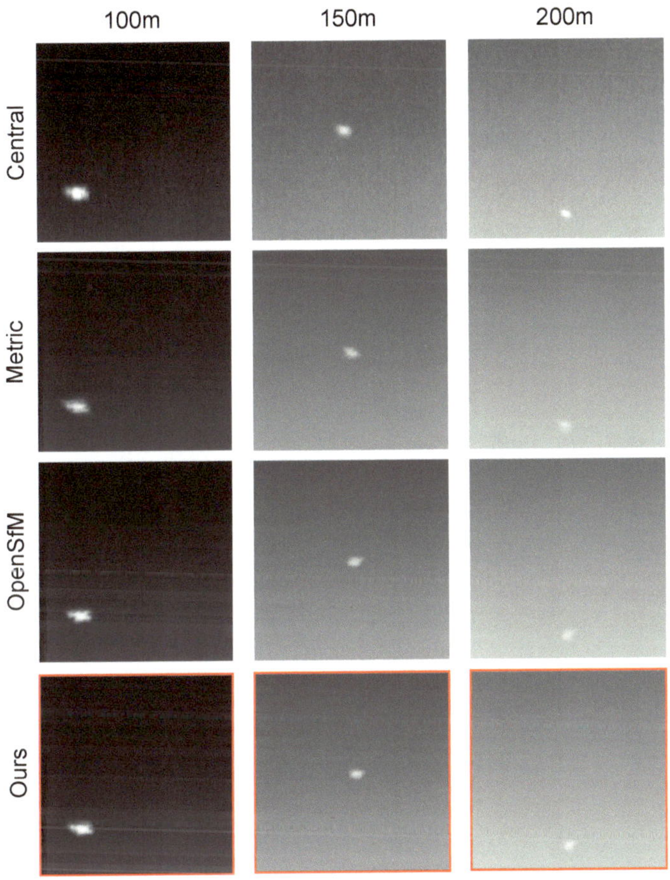

Figure 12. The dim target in the central sub-view (**top**); the dim target refocused in the metric calibration method [27] (**second row**); the dim target refocused in *OpenSfM* [60] method (**third row**); the dim target refocused in our method (**bottom row**).

Table 7. Multi-view synthetic comparison of dim targets.

Distance	SNR				Target Size (pixel)				Signal Intensity			
	Central	[27]	[60]	Ours	Central	[27]	[60]	Ours	Central	[27]	[60]	Ours
100	51	112	129	**137**	59	**30**	45	43	**243.271**	230.286	236.867	238.256
150	38	79	68	**100**	**40**	**40**	51	50	**200.400**	188.255	190.745	192.74
200	30	70	75	**86**	37	**30**	35	39	**233.919**	218.167	219.714	221.231

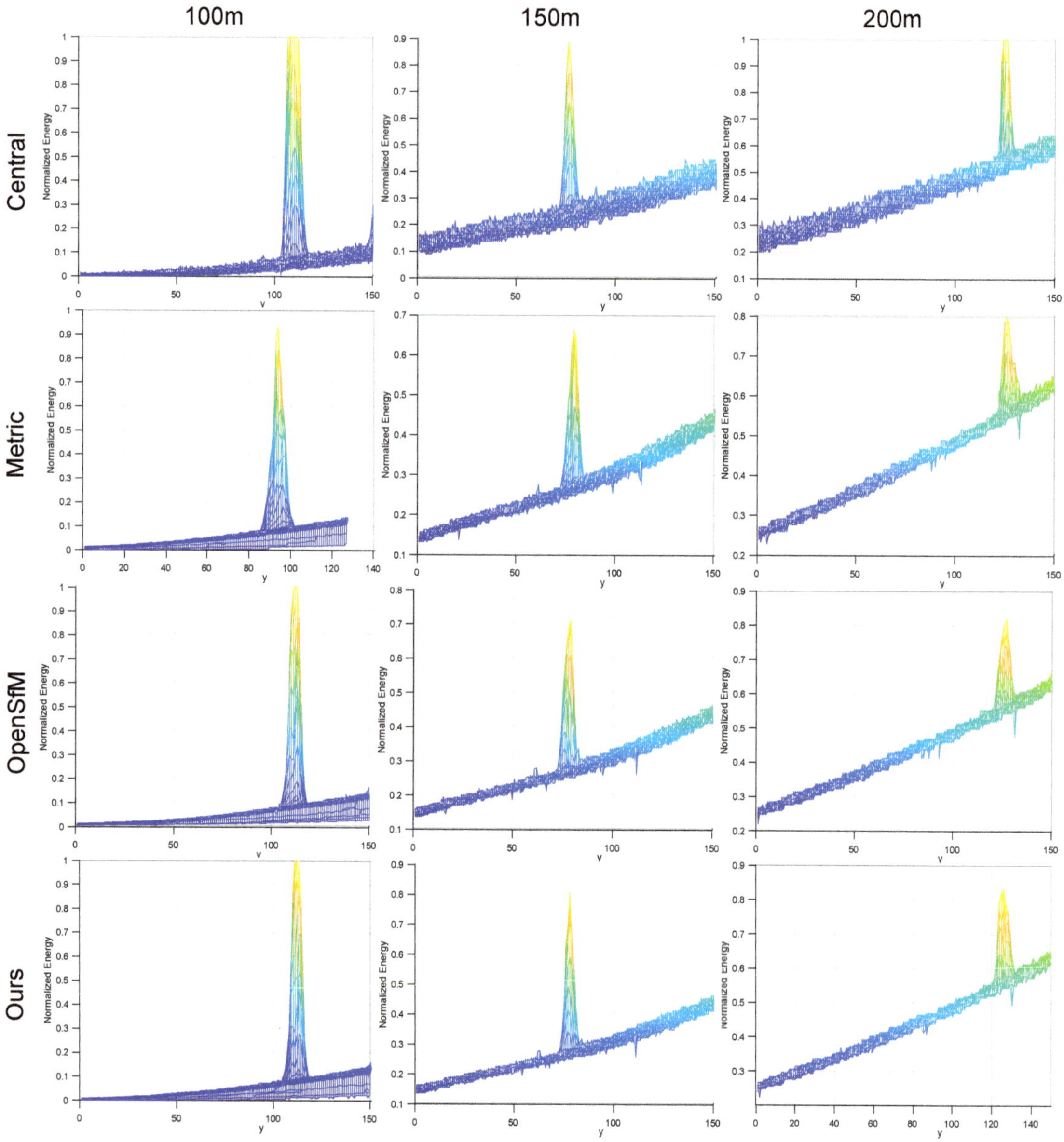

Figure 13. The normalized energy in the central sub-view (**top**); the normalized energy refocused in the metric calibration method [27] (**second row**); the normalized energy refocused in *OpenSfM* [60] method (**third row**); the normalized energy refocused in our method (**bottom row**).

6. Discussion

The infrared camera array we designed aims to provide a solution for enhancing the detection capability of dim targets at long distances. With this solution, the detection distance of the target is directly related to the baseline length between the sub-cameras. In this study, the baseline between adjacent sub-cameras of the infrared camera array is 10 cm, which can preliminarily satisfy multi-view synthesis for dim targets within a range of 300 m. Therefore, this camera array is merely an experimental verification device.

In practical tasks, to meet the detection of targets at different distances, the baseline between sub-cameras can be expanded or reduced as needed. To further reduce the impact of noise on the target, an appropriate increase in the number of sub-cameras can also be considered.

The self-calibration model we have designed is intended to provide high-precision camera poses for the infrared camera array without the constraints of calibration objects and scenes. The method draws on the advanced techniques of light-field camera calibration. It explores the structural priors of the camera array before and after multiple captures and organizes the disordered images for pose calibration. Although the calibration effect has been improved compared to methods based on structure from motion (SfM), the selection of calibration scenes still presents a challenging issue. This includes the distance of scene objects from the images, the area occupied by scene objects in the images, and other problems. Therefore, further experimental validation is required to address these issues.

In the future, we plan to conduct more extensive experiments to validate the performance of our infrared camera array and self-calibration model under different conditions and environments. We also aim to optimize the calibration process and reduce the computational complexity of the algorithm. Moreover, we will explore the integration of our method with other advanced technologies, such as deep learning and computer vision, to further enhance the detection capability of dim targets.

We believe that these future work directions will not only address the limitations of our current research but also contribute to the development of new applications and technologies in the field of infrared imaging and detection.

7. Conclusions

In this paper, we design an infrared camera array and propose a comprehensive self-calibration model, which aims to achieve good perception of spatial dim targets. The design of the infrared camera array meets the requirements of excellent mobility, highly synchronized image acquisition accuracy, impressive spatial resolution, and the capability to detect dim targets in the LWIR spectrum. The proposed self-calibration model introduces a linear constraint between the dual-array frames. This method reduces the degrees of freedom of the model and ensures that the calibration of the relative pose has a minimum cumulative reprojection error. Experimental validation on self-developed datasets confirms the superiority of our method. Finally, using our infrared camera array and its self-calibration model, we improve the SNR of spatial dim targets and demonstrate the value of its application in industrial environments. Looking ahead, we plan to conduct more extensive experiments under various conditions and environments to further validate the performance of our infrared camera array and self-calibration model. We aim to optimize the calibration process and reduce the computational complexity of the algorithm to enhance its practical applicability. Additionally, we are exploring the integration of our method with cutting-edge technologies such as deep learning and computer vision to further enhance the detection capability of dim targets.

Author Contributions: Methodology, Y.Z.; Investigation, T.W.; Resources, W.A.; Writing—original draft, Y.Z.; Writing—review & editing, T.W. and J.Y. All authors have read and agreed to the published version of the manuscript.

Funding: This work was supported in part by the Foundation for Innovative Research Groups of the National Natural Science Foundation of China under Grant 61921001.

Data Availability Statement: The original contributions presented in the study are included in the article, further inquiries can be directed to the corresponding author.

Conflicts of Interest: The authors declare no conflict of interest.

References

1. Levoy, M. Light fields and computational imaging. *Computer* **2006**, *39*, 46–55. [CrossRef]
2. Wu, T.; Zhang, Y.; Yang, J. Refocusing-based signal-to-noise ratio enhancement method for dim targets in infrared array cameras. In Proceedings of the Third International Symposium on Computer Engineering and Intelligent Communications (ISCEIC 2022), Xi'an, China, 16–18 September 2023; Volume 12462, pp. 249–254.
3. Zhu, J.; Xie, Z.; Jiang, N.; Song, Y.; Han, S.; Liu, W.; Huang, X. Delay-Doppler Map Shaping through Oversampled Complementary Sets for High-Speed Target Detection. *Remote Sens.* **2024**, *16*, 2898. [CrossRef]
4. Zhu, H.; Liu, S.; Deng, L.; Li, Y.; Xiao, F. Infrared small target detection via low-rank tensor completion with top-hat regularization. *IEEE Trans. Geosci. Remote Sens.* **2019**, *58*, 1004–1016. [CrossRef]
5. Liu, T.; Yang, J.; Li, B.; Wang, Y.; An, W. Infrared Small Target Detection via Nonconvex Tensor Tucker Decomposition with Factor Prior. *IEEE Trans. Geosci. Remote Sens.* **2023**, *61*, 1–17. [CrossRef]
6. Hao, Y.; Liu, Y.; Zhao, J.; Yu, C. Dual-Domain Prior-Driven Deep Network for Infrared Small-Target Detection. *Remote Sens.* **2023**, *15*, 3827. [CrossRef]
7. Kim, C.; Zimmer, H.; Pritch, Y.; Sorkine-Hornung, A.; Gross, M. Scene reconstruction from high spatio-angular resolution light fields. *ACM Trans. Graph.* **2013**, *32*, 73. [CrossRef]
8. Dansereau, D.G.; Schuster, G.; Ford, J.; Wetzstein, G. A wide-field-of-view monocentric light field camera. In Proceedings of the IEEE Conference on Computer Vision and Pattern Recognition (CVPR), Honolulu, HI, USA, 21–26 July 2017; pp. 5048–5057.
9. Taguchi, Y.; Agrawal, A.; Ramalingam, S.; Veeraraghavan, A. Axial light field for curved mirrors: Reflect your perspective, widen your view. In Proceedings of the IEEE Computer Society Conference on Computer Vision and Pattern Recognition (CVPR), San Francisco, CA, USA, 13–18 June 2010; pp. 499–506.
10. Lumsdaine, A.; Georgiev, T. The focused plenoptic camera. In Proceedings of the IEEE International Conference on Computational Photography (ICCP), San Francisco, CA, USA, 16–17 April 2009; pp. 1–8.
11. Wei, L.Y.; Liang, C.K.; Myhre, G.; Pitts, C.; Akeley, K. Improving light field camera sample design with irregularity and aberration. *ACM Trans. Graph.* **2015**, *34*, 1–11. [CrossRef]
12. Ng, R.; Levoy, M.; Brédif, M.; Duval, G.; Horowitz, M.; Hanrahan, P. Light Field Photography with a Hand-Held Plenoptic Camera. Ph.D. Thesis, Stanford University, Stanford, CA, USA, 2005.
13. Yang, J.C.; Everett, M.; Buehler, C.; McMillan, L. A real-time distributed light field camera. *Render. Tech.* **2002**, *2002*, 2.
14. Wilburn, B.; Joshi, N.; Vaish, V.; Talvala, E.V.; Antunez, E.; Barth, A.; Adams, A.; Horowitz, M.; Levoy, M. High performance imaging using large camera arrays. *ACM Trans. Graph.* **2005**, *24*, 765–776. [CrossRef]
15. Zhang, C.; Chen, T. A self-reconfigurable camera array. In *ACM SIGGRAPH 2004 Sketches*; Springer: Cham, Switzerland, 2004; p. 151.
16. Zhang, M.; Vogelbacher, M.; Hagenmeyer, V.; Aleksandrov, K.; Gehrmann, H.J.; Matthes, J. 3-D refuse-derived fuel particle tracking-by-detection using a plenoptic camera system. *IEEE Trans. Instrum. Meas.* **2022**, *71*, 1–15. [CrossRef]
17. Pu, X.; Wang, X.; Gao, X.; Wei, C.; Gao, J. Polarizing Camera Array System Equipment and Calibration Method. *IEEE Trans. Instrum. Meas.* **2023**, *73*, 1–15. [CrossRef]
18. Lins, R.G.; Givigi, S.N.; Kurka, P.R.G. Vision-based measurement for localization of objects in 3-D for robotic applications. *IEEE Trans. Instrum. Meas.* **2015**, *64*, 2950–2958. [CrossRef]
19. Heinze, C.; Spyropoulos, S.; Hussmann, S.; Perwaß, C. Automated robust metric calibration algorithm for multifocus plenoptic cameras. *IEEE Trans. Instrum. Meas.* **2016**, *65*, 1197–1205. [CrossRef]
20. Gao, Y.; Cui, H.; Wang, X.; Huang, Z. Novel precision vision measurement method between area-array imaging and linear-array imaging especially for dynamic objects. *IEEE Trans. Instrum. Meas.* **2022**, *71*, 1–9. [CrossRef]
21. Peng, J.; Xu, W.; Liang, B.; Wu, A.G. Virtual stereovision pose measurement of noncooperative space targets for a dual-arm space robot. *IEEE Trans. Instrum. Meas.* **2019**, *69*, 76–88. [CrossRef]
22. Li, X.; Li, W.; Yin, X.; Ma, X.; Zhao, J. Camera-mirror binocular vision-based method for evaluating the performance of industrial robots. *IEEE Trans. Instrum. Meas.* **2021**, *70*, 1–14. [CrossRef]
23. Kaczmarek, A.L.; Blaschitz, B. Equal baseline camera array—Calibration, testbed and applications. *Appl. Sci.* **2021**, *11*, 8464. [CrossRef]
24. Perez, A.J.; Perez-Cortes, J.C.; Guardiola, J.L. Simple and precise multi-view camera calibration for 3D reconstruction. *Comput. Ind.* **2020**, *123*, 103256. [CrossRef]
25. Vaish, V.; Wilburn, B.; Joshi, N.; Levoy, M. Using plane+ parallax for calibrating dense camera arrays. In Proceedings of the PIEEE Computer Society Conference on Computer Vision and Pattern Recognition (CVPR), Washington, DC, USA, 27 June–2 July 2004; Volume 1.
26. Hamzah, R.A.; Ibrahim, H. Literature survey on stereo vision disparity map algorithms. *J. Sensors* **2016**, *1*, 8742920. [CrossRef]
27. Zhang, Z. A flexible new technique for camera calibration. *IEEE Trans. Pattern Anal. Mach. Intell.* **2000**, *22*, 1330–1334. [CrossRef]
28. Schonberger, J.L.; Frahm, J.M. Structure-from-motion revisited. In Proceedings of the IEEE Conference on Computer Vision and Pattern Recognition (CVPR), Las Vegas, NV, USA, 27–30 June 2016; pp. 4104–4113.
29. Pei, Z.; Li, Y.; Ma, M.; Li, J.; Leng, C.; Zhang, X.; Zhang, Y. Occluded-object 3D reconstruction using camera array synthetic aperture imaging. *Sensors* **2019**, *19*, 607. [CrossRef] [PubMed]

30. Ke, J.; Watras, A.J.; Kim, J.J.; Liu, H.; Jiang, H.; Hu, Y.H. Towards real-time 3D visualization with multiview RGB camera array. *J. Signal Process. Syst.* **2022**, *94*, 329–343. [CrossRef] [PubMed]
31. Yang, Y.; Tang, D.; Wang, D.; Song, W.; Wang, J.; Fu, M. Multi-camera visual SLAM for off-road navigation. *Robot. Auton. Syst.* **2020**, *128*, 103505. [CrossRef]
32. Ali, I.; Suominen, O.J.; Morales, E.R.; Gotchev, A. Multi-view camera pose estimation for robotic arm manipulation. *IEEE Access* **2020**, *8*, 174305–174316. [CrossRef]
33. Chi, J.; Liu, J.; Wang, F.; Chi, Y.; Hou, Z.G. 3-D gaze-estimation method using a multi-camera-multi-light-source system. *IEEE Trans. Instrum. Meas.* **2020**, *69*, 9695–9708. [CrossRef]
34. Liu, P.; Li, X.; Wang, Y.; Fu, Z. Multiple object tracking for dense pedestrians by Markov random field model with improvement on potentials. *Sensors* **2020**, *20*, 628. [CrossRef] [PubMed]
35. Wang, Y.; Yang, J.; Guo, Y.; Xiao, C.; An, W. Selective Light Field Refocusing for Camera Arrays Using Bokeh Rendering and Superresolution. *IEEE Signal Process. Lett.* **2019**, *26*, 204–208. [CrossRef]
36. Wang, T.C.; Efros, A.A.; Ramamoorthi, R. Occlusion-aware depth estimation using light-field cameras. In Proceedings of the IEEE International Conference on Computer Vision (ICCV), Santiago, Chile, 7–13 December 2015; pp. 3487–3495.
37. Schilling, H.; Diebold, M.; Rother, C.; Jähne, B. Trust your model: Light field depth estimation with inline occlusion handling. In Proceedings of the IEEE Conference on Computer Vision and Pattern Recognition (CVPR), Salt Lake City, UT, USA, 18–23 June 2018; pp. 4530–4538.
38. Solav, D.; Moerman, K.M.; Jaeger, A.M.; Genovese, K.; Herr, H.M. MultiDIC: An open-source toolbox for multi-view 3D digital image correlation. *IEEE Access* **2018**, *6*, 30520–30535. [CrossRef]
39. Abedi, F.; Yang, Y.; Liu, Q. Group geometric calibration and rectification for circular multi-camera imaging system. *Opt. Express* **2018**, *26*, 30596–30613. [CrossRef] [PubMed]
40. Ge, P.; Wang, Y.; Wang, B. Universal calibration for a ring camera array based on a rotational target. *Opt. Express* **2022**, *30*, 14538–14552. [CrossRef] [PubMed]
41. Brady, D.J.; Gehm, M.E.; Stack, R.A.; Marks, D.L.; Kittle, D.S.; Golish, D.R.; Vera, E.; Feller, S.D. Multiscale gigapixel photography. *Nature* **2012**, *486*, 386–389. [CrossRef] [PubMed]
42. Lin, X.; Wu, J.; Zheng, G.; Dai, Q. Camera array based light field microscopy. *Biomed. Opt. Express* **2015**, *6*, 3179–3189. [CrossRef]
43. Thomson, E.E.; Harfouche, M.; Kim, K.; Konda, P.C.; Seitz, C.W.; Cooke, C.; Xu, S.; Jacobs, W.S.; Blazing, R.; Chen, Y.; et al. Gigapixel imaging with a novel multi-camera array microscope. *eLife* **2022**, *11*, e74988. [CrossRef]
44. Venkataraman, K.; Lelescu, D.; Duparré, J.; McMahon, A.; Molina, G.; Chatterjee, P.; Mullis, R.; Nayar, S. Picam: An ultra-thin high performance monolithic camera array. *ACM Trans. Graph.* **2013**, *32*, 1–13. [CrossRef]
45. Lin, J.; Lin, X.; Ji, X.; Dai, Q. Separable coded aperture for depth from a single image. *IEEE Signal Process. Lett.* **2014**, *21*, 1471–1475. [CrossRef]
46. Georgiev, T.; Lumsdaine, A. Focused plenoptic camera and rendering. *J. Electron. Imaging* **2010**, *19*, 021106.
47. Pless, R. Using many cameras as one. In Proceedings of the IEEE Computer Society Conference on Computer Vision and Pattern Recognition (CVPR), Madison, WI, USA, 18–20 June 2003; Volume 2.
48. Li, H.; Hartley, R.; Kim, J.h. A linear approach to motion estimation using generalized camera models. In Proceedings of the IEEE Conference on Computer Vision and Pattern Recognition (CVPR), Anchorage, AK, USA, 23–28 June 2008; pp. 1–8.
49. Johannsen, O.; Sulc, A.; Goldluecke, B. On linear structure from motion for light field cameras. In Proceedings of the IEEE International Conference on Computer Vision (ICCV), Santiago, Chile, 7–13 December 2015; pp. 720–728.
50. Zhang, Y.; Yu, P.; Yang, W.; Ma, Y.; Yu, J. Ray space features for plenoptic structure-from-motion. In Proceedings of the IEEE International Conference on Computer Vision (ICCV), Venice, Italy, 22–29 October 2017; pp. 4631–4639.
51. Nousias, S.; Lourakis, M.; Bergeles, C. Large-scale, metric structure from motion for unordered light fields. In Proceedings of the IEEE Conference on Computer Vision and Pattern Recognition (CVPR), Long Beach, CA, USA, 15–20 June 2019; pp. 3292–3301.
52. Zhang, Q.; Li, H.; Wang, X.; Wang, Q. 3D scene reconstruction with an un-calibrated light field camera. *Int. J. Comput. Vis.* **2021**, *129*, 3006–3026. [CrossRef]
53. Nousias, S.; Lourakis, M.; Keane, P.; Ourselin, S.; Bergeles, C. A linear approach to absolute pose estimation for light fields. In Proceedings of the International Conference on 3D Vision (3DV), Fukuoka, Japan, 25–28 November 2020; pp. 672–681.
54. Zhang, S.; Jin, D.; Dai, Y.; Yang, F. Relative pose estimation for light field cameras based on LF-point-LF-point correspondence model. *IEEE Trans. Image Process.* **2022**, *31*, 1641–1656. [CrossRef]
55. Schöps, T.; Sattler, T.; Häne, C.; Pollefeys, M. Large-scale outdoor 3D reconstruction on a mobile device. *Comput. Vis. Image Underst.* **2017**, *157*, 151–166. [CrossRef]
56. Lowe, D.G. Distinctive image features from scale-invariant keypoints. *Int. J. Comput. Vis.* **2004**, *60*, 91–110. [CrossRef]
57. Bellavia, F.; Colombo, C. RootsGLOH2: Embedding RootSIFT 'square rooting' in sGLOH2. *IET Comput. Vis.* **2020**, *14*, 138–143. [CrossRef]
58. Nistér, D. An efficient solution to the five-point relative pose problem. *IEEE Trans. Pattern Anal. Mach. Intell.* **2004**, *26*, 756–770. [CrossRef] [PubMed]
59. Fachada, S.; Losfeld, A.; Senoh, T.; Lafruit, G.; Teratani, M. A calibration method for subaperture views of plenoptic 2.0 camera arrays. In Proceedings of the IEEE 23rd International Workshop on Multimedia Signal Processing (MMSP), Tampere, Finland, 6–8 October 2021; pp. 1–6.

60. Adorjan, M. Opensfm: A Collaborative Structure-from-Motion System. Ph.D. Thesis, Vienna University of Technology, Wien, Austria, 2016.
61. Lourakis, M.; Terzakis, G. A globally optimal method for the PnP problem with MRP rotation parameterization. In Proceedings of the IEEE International Conference on Pattern Recognition (ICPR), Milan, Italy, 10–15 January 2021; pp. 3058–3063.
62. Lepetit, V.; Moreno-Noguer, F.; Fua, P. EPnP: Efficient perspective-n-point camera pose estimation. *Int. J. Comput. Vis.* **2009**, *81*, 155–166. [CrossRef]
63. Yang, J.; Xiao, C.; Wang, Y.; An, C.; An, W. High-precision refocusing method with one interpolation for camera array images. *IET Image Process.* **2020**, *14*, 3899–3908. [CrossRef]

remote sensing MDPI

Article

Thinned and Sparse Beamforming for Semicircular FDAs in the Transmit–Receive Domain

Anyi Wang [1], Xiao Huang [1], Yanhong Xu [1,*], Xiao Meng [2] and Yumeng Lu [1]

1 Xi'an Key Laboratory of Network Convergence Communication, College of Communication and Information Engineering, Xi'an University of Science and Technology, Xi'an 710054, China; wanganyi@xust.edu.cn (A.W.); 21207040036@stu.xust.edu.cn (X.H.); 23207040040@stu.xust.edu.cn (Y.L.)
2 School of Physics, Xidian University, Xi'an 710071, China; xmeng@xidian.edu.cn
* Correspondence: yhxu@xust.edu.cn

Abstract: The thinned and sparse beamforming for semicircular FDAs were investigated, where the excitation amplitudes were also considered in thinned semicircular FDAs, and only the elements' positions were incorporated into the sparse semicircular FDA. Firstly, the transmit–receive model was introduced to handle the inherent time-varying issue of FDA, followed by the thinned and sparse implementations successively. Note that three types of non-linearly varying frequency offsets (FO), i.e., log-FO, sin-FO, and tanh-FO, were adopted during the investigations. Under the same assumption that 50% of the elements should be saved, the sidelobe levels (SLLs) of the thinned semicircular -FDAs were reduced by 5.8 dB, 4.4 dB, and 4.4 dB, and the widths of the mainlobes were all widened by 3° in their angle dimension. Compared with the thinned semicircular FDAs, the phenomenon of mainlobe widening was alleviated in the sparse semicircular FDAs where the SLLs were reduced by 2.2 dB, 3.7 dB and 3.5 dB, and the mainlobes' widths in the angle dimension were widened by 1°, 0° and 1°, respectively. It should be highlighted that the sparse semicircular FDA with sin-FO did not broaden the mainlobe in the angle dimension. Therefore, it can be concluded that a sparse semicircular FDA is superior over a thinned semicircular FDA, since it can reduce the same cost with a higher array resolution.

Keywords: semicircular array configuration; frequency diverse array; thinned and sparse array antenna; transmit–receive beamforming

1. Introduction

Unlike conventional phased arrays (PAs), frequency diverse arrays (FDAs) exhibit a more flexible ability in beampattern control and signal processing by applying the frequency diverse technique across the array's aperture [1,2]. Therefore, it can provide potential applications in detecting targets in severely cluttered environment [3–5], jointly estimating the range, angle, and velocity parameters of targets [6], suppressing swarm interference [7] and suppressing deception jamming while interrupting and interfering with the enemy's radar signal [8], providing high resolution and wide bandwidth imaging [9,10], detecting moving targets in partially homogeneous environments [11], secure communications [12–15], wide-swath synthetic aperture radar (SAR) imaging [16,17], SAR ground moving target indication (SAR GMTI) [18,19], spaceborne SAR scattered-wave deception jamming [20], etc. It can be concluded that FDA is superior over conventional PA in future multi-task radars, satellites, navigation systems, radio astronomy, wireless security communications, etc., and has become one of the hot spots in relevant research fields.

Just as its name implies, the frequency offset (FO) applied between the adjacent elements plays a significant role in the beampattern control of an FDA. Theoretically, an S-shaped transmission beampattern is achieved for a uniformly spaced FDA with linearly increasing/decreasing FO, which is not suitable for accurate target positioning. Therefore, extensive research has been conducted to investigate the effects of different types of non-linearly varying FOs on FDAs' performance. In specific, logarithmically increasing frequency offsets (log-FOs) [21,22], Cantor multi-stage FOs [23], random permutation power increasing FOs [24], sinusoidally increasing frequency offsets (sin-FOs) [25,26], tangent hyperbolic function FOs [27], segmented triangular FOs [28], window function-based FOs [29–33], and other non-linearly varying FOs obtained with intelligent optimization algorithms [34–37] have been explored in detail. Meanwhile, the array's configuration is also an important factor that greatly influences the performance of an FDA, such as non-uniform linear array configurations [38,39], circular array configurations [40], semicircular array configurations [41,42], arc array configurations [43,44], and concentric circular array configurations [45]. Among them, the semicircular configuration is a typical conformal structure which can enable the array to achieve the largest scanning angle with the least number of elements.

Another important issue of FDAs is the time-varying characteristic when only considering the transmission mode, which means that the phase term of each element in the FDA is accumulated with respect to time during the signal's duration. This property enables an FDA to provide continuous electric beam steering during the signal's duration, on one hand, and increases the difficulty in controlling the mainbeam on the other hand. To alleviate this problem, some researchers have devoted themselves to time-modulated FOs [46,47]. However, it has been pointed out that the time-varying issue inherently exists when only considering the transmission mode [48,49]. Under this circumstance, researchers have started to consider equivalent transmit beamforming at the receiver [50,51]. In the following, two transmit–receive signal processing chains are proposed along with the detailed derivation of the corresponding signal models [41,52]. In [41], a series of mixers and low-pass filters (LPFs) were utilized to handle the time-varying terms where a series of analog-to-digital converters (ADCs) were adopted to facilitate the subsequent digital beamforming procedure, while only a mixer and an ADC were adopted in [52] to greatly reduce the system's cost.

In many scenarios, an array with large elements is prevalent, since it can provide a higher angle resolution. Nevertheless, the huge number of elements in the array would always mean the extremely high complexity of signal processing and a huge cost. In order to maintain the advantages of high resolution and high sensitivity of the array system while greatly reducing the consumption cost, sparse technology is considered to reduce the number of elements by percentages and consequently lower the complexity of the feed channel and network. Note that the generalized sparse arrays refer to non-uniform arrays, which can be divided into thinned arrays and sparse arrays according to whether the interelement spacing is an integer multiple of the initial interelement spacing. Thinned technology has been widely used in PA [53,54]. At present, some researchers have conducted studies on the sparse FDAs. In [55], the modified genetic algorithm (GA) was used to create a sparse multi-carrier FDA with a linear configuration. In [56], the artificial bee colony algorithm was used to create a sparse linear FDA. In [57], the FO and element position of the linear FDA were optimized simultaneously to synthesize the thinned FDA. An effective non-iterative beamforming method for sparse FDAs was proposed in [58]. Nevertheless, the works above mostly focused on the sparse linear FDAs, and corresponding investigations on conformal FDAs are quite limited. Furthermore, the works above only considered the transmission mode, and the time-varying issue was not resolved. Under these circumstances, we intended to conduct investigations on thinned and sparse semicircular FDAs in the transmit–receive dimensions. Note that the GA was adopted during the optimization, since it is one of the most efficient and effective algorithms that has been widely used in optimization problems of arrays' configuration [59–62]. Our main contributions are briefly summarized as follows.

(1) A thinned transmit–receive beamforming for semicircular FDAs was investigated, where three types of non-linearly varying FOs, i.e., log-FO, sin-FO, and tanh-FO, were adopted to further disturb the range-angle-coupled characteristic of the FDA's beampattern. Note that the amplitudes of excitation were also incorporated into the

optimization to achieve low sidelobe level (SLL) during thinning, since a conformal array always exhibits a high SLL.

(2) A sparse transmit–receive beamform for semicircular FDAs was investigated successfully, the principle of which is to rearrange the elements at a certain angle interval under the premise of a fixed array aperture. In particular, the sparse elements of the redistributed arrangement were 50% of the full array's elements, and the improved GA was used to carry out the sparse layout array. The differences in the angle between the two array elements were the individual genes, and the search space was reduced.

(3) The performance of the thinned and sparse forms of semicircular FDAs' transmit–receive beamforming were examined. Note that the first and the last elements were fixed to guarantee the array's resolution in the angle dimension. The resulting beampatterns are presented from various aspects. Comparisons and analysis of thinned and sparse semicircular arrays with the above three FOs are provided in detail to provide suggestions for other researchers.

The rest of the article is arranged as follows. In Section 2, the transmit–receive beamforming signal model of semicircular FDAs is established. In Section 3, thinned and sparse beamforming for semicircular FDAs in the transmit–receive domain using GA are proposed, which achieved a low SLL while reducing the number of elements. Section 4 presents the simulation's results and analyzes them in detail to demonstrate the effectiveness of the proposed method. Finally, Section 5 summarizes the work and draws the conclusion of our work.

2. The Transmit–Receive Beamforming Signal Model of Semicircular FDAs

As depicted in Figure 1, a semicircular array consisting of N elements was considered, where the elements were uniformly distributed along a semicircular platform with a radius of r.

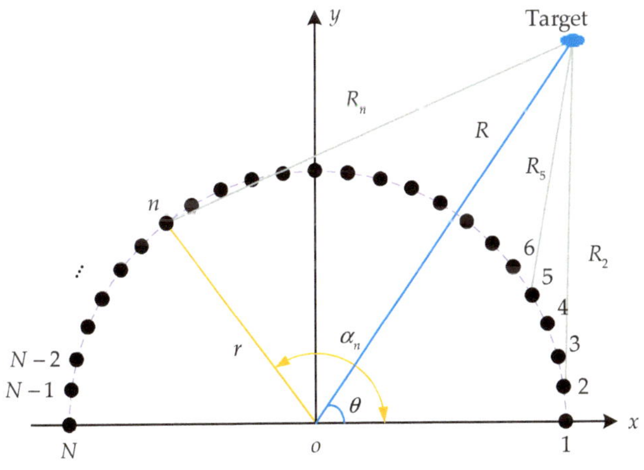

Figure 1. Configuration of the uniform semicircular FDA.

The operating frequency of the nth element can be written as

$$f_n = f_0 + \Delta f_n \tag{1}$$

where f_0 represents the reference frequency and Δf_n is the FO corresponding to the nth element. To achieve the range-angle-decoupled beampattern, two types of non-linearly varying FOs, i.e., log-FO [41] and sin-FO [35], were adopted a linear FDAs in [52]. For comparison, the log-FO and sin-FO of the same expressions were also adopted here, and an another non-linearly varying FO [27] was also introduced. Secifically, the FO of the nth element can be expressed as

$$\Delta f_n = -\Delta f(\ln(n))^{1.5}, n = 1,2,\ldots,N \tag{2}$$

$$\Delta f_n = -9\Delta f \sin((n)/38.8),\ n = 1,2,\ldots,N \tag{3}$$

$$\Delta f_n = 10\tanh(0.03(n-1)),\ n = 1,2,\ldots,N \tag{4}$$

The distance from the nth element to the target can be expressed as

$$R_n = R - r\cos(\theta - \alpha_n) \tag{5}$$

where $\alpha_n = (n-1)\pi/(N-1)$, $n = 1,2,\ldots,N$ represents the angle of the nth element deviating from the positive x-axis, and R and θ, respectively, denote the range and angle parameters of the target from the coordinates' origin. The signal radiated by the nth element of the array has the form of

$$s_n(t) = \exp(j2\pi f_n t). \tag{6}$$

The signal transmitted by the nth element, reflected by the target in the far field, and received by the mth receiving element can be expressed as

$$s_{m,n}(R,\theta,t) = \exp\left(j\left(2\pi f_n\left(t - \frac{R_n + R_m}{c}\right)\right)\right) \tag{7}$$

where $R_m = R - r\cos(\theta - \alpha_m)$, and c is the speed of light.

Then the output of the mth receiving element is the summation of all the transmitted and reflected signals, i.e.,

$$x_m(R,\theta,t) = \sum_{n=1}^{N} s_{m,n}(R,\theta,t). \tag{8}$$

As described in the following, the inherent time-varying problem of the semicircular FDA was solved by combining the signal processing in the joint transmission and receiving domains. As shown in Figure 2, after being amplified in a low-noise amplifier (LNA), the output of the mth receiving element was mixed with a mixer of frequency f_0. Then the mixed signals were sampled with an analog-to-digital converter (ADC) before being further mixed with a set of digital mixers (DMs). In the sequel, the signal in each channel was filtered by a low-pass filter (LPF), weighted, and summed for digital beamforming.

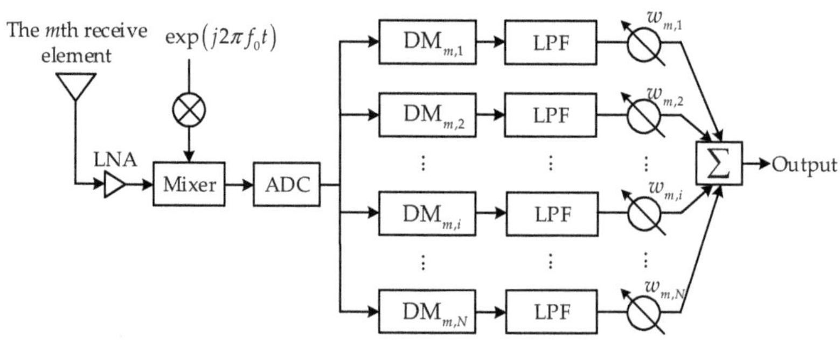

Figure 2. Receiving processing chain of the mth receiving element [52].

The output signal of the ith channel after filtered by LPF can be expressed as

$$\begin{aligned}\hat{a}_{m,i}(R,\theta) &= ((x_m(R,\theta,t)\times\exp(-j2\pi f_0 t))\times\exp(-j2\pi\Delta f_i t)) * h(t)\\ &= \exp\left(-j\frac{2\pi f_i(R_m+R_i)}{c}\right)\end{aligned} \tag{9}$$

where $\Delta f_i, i = 1,2,\ldots,N$ is the bandwidth of the ith DM used to compensate for the time-varying term caused by the FO. $h(t)$ represents the temporary response of a series of identical LPFs, and $*$ denotes the convolution operator. At this time, the time-varying terms are eliminated.

Therefore, the output of the mth receiving processing chain can be written as

$$\tilde{a}_m(R,\theta) = \sum_{i=1}^{N} w_{m,i}\hat{a}_{m,i}(R,\theta) = \sum_{i=1}^{N} w_{m,i}\exp\left(-j\frac{2\pi f_i(R_m+R_i)}{c}\right) \tag{10}$$

where $w_{m,i}$ represents the weight of the ith channel in the mth processing chain.

Consequently, the transmit–receive array factor of the semicircular FDA can be expressed as

$$\mathrm{AF}_{\mathrm{Semicir}-T-R}(R,\theta) = \sum_{m=1}^{N} u_m\tilde{a}_m(R,\theta) = \sum_{m=1}^{N}\sum_{i=1}^{N} v_{m,i}\hat{a}_{m,i}(R,\theta) \tag{11}$$

where $v_{m,i} = u_m w_{m,i}$, with u_m denoting the weight of the mth transmit–receive processing chain.

3. Thinned and Sparse Beamforming for Semicircular FDAs in Transmit–Receive Domain

In this section, thinned and sparse beamforming for semicircular FDAs was investigated based on the signal model established in Section 2. Specifically, the optimization model of thinned beamforming for semicircular FDAs was constructed as described in Section 3.1, where the elements' positions as well as the excitation amplitudes were incorporated into the problem. To realize the thinned semicircular FDA, some of the elements were deactivated and the positions of the remaining elements were not changed. In Section 3.2, the optimization model of the sparse semicircular FDA is derived, where only the elements' positions are considered. It should be highlighted that the interelements' spacing in sparse semicircular FDAs is not an integer multiple of the initial interelements' spacing. Note that the two elements at both ends of the array were retained to ensure that the array's aperture remained unchanged after the optimization.

3.1. Thinned Transmit–Receive Beamforming of Semicircular FDAs

As depicted in Figure 3, an illustration of the configuration of the thinned semicircular FDA is provided, where a series of black solid points is used to represent the full uniform semicircular array consisting of N array elements. The red '×' means that the elements are deactivated/removed from the uniform full array. Therefore, the remaining elements formed the non-uniform semicircular FDA with the spacing of the array's elements being an integer multiple of a basic quantity. The aim in this subsection was to seek for an array configuration and its corresponding excitations simultaneously to make the generated beampattern approximate the desired beampattern as closely as possible.

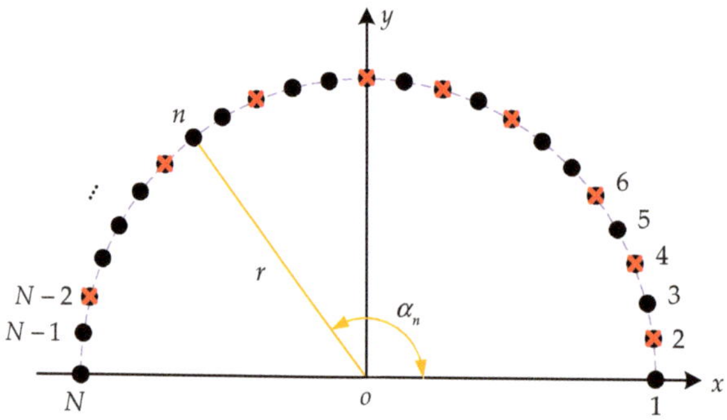

Figure 3. Configuration of a thinned semicircular FDA.

The transmit–receive power pattern of the semicircular FDA can be expressed as

$$\mathrm{P}_{\text{Semicir}-T-R}(R,\theta) = |\mathrm{AF}_{\text{Semicir}-T-R}(R,\theta)|^2 = \mathbf{v}^H\mathbf{Y}(R,\theta)\mathbf{v} \tag{12}$$

where $\mathbf{v} = \left[\mathbf{v}_1^T,\ldots,\mathbf{v}_m^T,\ldots,\mathbf{v}_N^T\right]^T$ is the transmit–receive weight vector of the semicircular FDA and $\mathbf{v}_m = [v_{m,1},\ldots,v_{m,i},\ldots,v_{m,N}]^{\mathrm{H}}$, $m = 1,\ldots,N$. The superscripts ^{T} and ^{H} represent the transpose operator and the conjugate transpose operator, respectively. $\mathbf{Y}(R,\theta) = \mathbf{a}(R,\theta)\mathbf{a}^{\mathrm{H}}(R,\theta) \in \mathbb{C}^{N^2\times N^2}$ represents the transmit–receive covariance matrix of the semicircular FDA, and $\mathbf{a}(R,\theta) = \left[\mathbf{a}_1^{\mathrm{T}}(R,\theta),\ldots,\mathbf{a}_m^{\mathrm{T}}(R,\theta),\ldots,\mathbf{a}_N^{\mathrm{T}}(R,\theta)\right]^{\mathrm{T}}$ is the transmit–receive steering vector with $\mathbf{a}_m(R,\theta) = [\hat{a}_{m,1}(R,\theta),\ldots,\hat{a}_{m,i}(R,\theta),\ldots,\hat{a}_{m,N}(R,\theta)]^T$, m = 1,…,N.

Note that the transmit–receive weight vector **v** plays an important role in the beamforming of thinned semicircular FDAs, rewritten the basic entry $v_{m,n}$ of $\mathbf{v}_m$ as

$$v_{m,n} = (q_m * \rho_m)w_{m,n} \tag{13}$$

where $0 < q_m \leq 1$ represents the excitation amplitude of the mth receiving element, and ρ_m is the selection variable. In particular, $\rho_m = 1$ indicates that the mth element works, and $\rho_m = 0$ means that the mth element does not work. To ensure that the array's aperture did not change after the thinned optimization, it was necessary to retain the two elements located on the left and right boundaries of the semicircular FDA, i.e., when $n = 1$ and $n = N$, $\rho_1 = \rho_N = 1$.

For the optimization problem of the thinned semicircular FDA, the transmit–receive weight vector **v** of the semicircular FDA was incorporated into the optimization problem. The goal was to find the best weight vector **v** such that the generated beampattern $\mathrm{P}_{\text{Semicir}-T-R}(R,\theta)$ was, in some sense, the best approximation of the desired pattern $\mathrm{P}_d(R,\theta)$. The error function was defined as the integral of the weighted error between the currently generated beampattern and the desired beampattern in the range and angle space, and could be constructed as

$$K(\mathbf{v}) = \iint_{(R,\theta)\in\Omega} W(R,\theta)|\mathrm{P}_{\text{Semicir}-T-R}(R,\theta) - \mathrm{P}_d(R,\theta)|\sin\theta dR d\theta \tag{14}$$

where $\Omega = \{(R,\theta)|R_{\min} \leq R \leq R_{\max},\ 0^\circ \leq \theta \leq 180^\circ\}$ represents the total observation area, $W(R,\theta)$ is the importance weight function, and $\sin(\theta)$ is generated by integration in the angle dimension.

The characteristics of the desired beampattern are described as

$$\mathrm{P}_d(R,\theta)\Big|_{(R,\theta)=(R_0,\theta_0)} = 1 \tag{15}$$

$$\mathrm{P}_d(R,\theta)\Big|_{(R,\theta)=(R_a,\theta_0),\ (R_b,\theta_0),\ (R_0,\theta_a)\text{ and }(R_0,\theta_b)} = 0 \tag{16}$$

$$\mathrm{P}_d(R,\theta)\Big|_{(R,\theta)\in\Omega\text{ and }(R,\theta)\notin\Omega_0} \leq \eta \tag{17}$$

where (R_0,θ_0) in (15) is the direction of the mainlobe; $(R,\theta) = (R_a,\theta_0)$, $(R,\theta) = (R_b,\theta_0)$, $(R,\theta) = (R_0,\theta_a)$, and $(R,\theta) = (R_0,\theta_b)$ are the four prescribed edge points of the mainlobe in the range and angle dimensions in (16). $\Omega_0 = \{(R,\theta)|R_a \leq R \leq R_b, \theta_a \leq \theta \leq \theta_b\}$ in (17) represents the expected position of the mainlobe, and η represents the acceptable peak SLL.

Therefore, the objective function of the transmit–receive beamforming optimization of the thinned semicircular FDAs can be established as

$$\min_{\mathbf{v}_{\text{thinned}}} K(\mathbf{v}_{\text{thinned}}) \tag{18}$$

where $\mathbf{v}_{\text{thinned}}$ is updated iteratively. Considering discrete sampling, Equation (18) can be expressed as

$$\min_{\mathbf{v}_{\text{thinned}}} \sum_{i=1}^{I}\sum_{j=1}^{J} W(R_i,\theta_j)\left|\mathrm{P}_{\text{Semicir}-T-R}(R_i,\theta_j) - \mathrm{P}_d(R_i,\theta_j)\right|\sin(\theta_j). \tag{19}$$

According to the analysis above, the fitness function of the problem of transmit–receive beampattern synthesis for the thinned semicircular FDA can be expressed as

$$f_1(\mathbf{v}_{\text{thinned}}) = \sum_{i=1}^{I}\sum_{j=1}^{J} W(R_i,\theta_j)\left|\mathrm{P}_{\text{Semicir}-T-R}(R_i,\theta_j) - \mathrm{P}_d(R_i,\theta_j)\right|\sin(\theta_j) \tag{20}$$

where $R_i, i = 1,\cdots,I$, and $\theta_j, j = 1,\cdots,J$ denote the ith and jth sample points in the range and angle dimensions, respectively.

It is known that the GA is widely used in the synthesis of thinned arrays due to its unique 0 and 1 characteristics. Therefore, the GA was adopted here, and the corresponding GA-based procedures used to solve the thinned optimization above is provided in Algorithm 1, where N represents the number of elements of the full semicircular FDA, N_{thinned} is the number of elements after synthesis of the thinned array, Q is the initial population, L is the number of iterations, P_c is the probability of crossover, and P_m is the probability of mutation.

Therefore, the transmit–receive beampattern of a semicircular FDA after synthesis of the thinned array can be written as

$$\hat{P}_{\text{thinned}-\text{Semicir}-T-R}(R,\theta) = \mathbf{v}_{\text{thinned}-\text{opti}}^{H}\mathbf{Y}(R,\theta)\mathbf{v}_{\text{thinned}-\text{opti}}. \tag{21}$$

Algorithm 1. GA-based optimization procedures of thinned semicircular FDAs

Input: $f_0, r, N, N_{\text{thinned}}, Q, L, P_c, P_m$
Step 1: Initialize Q populations, and each population has N individuals. Under the constraint of $\rho_1 = \rho_N = 1$, it is encoded by a decimal gene string. Set the generation number of l as $l = 1$.
Step 2: The fitness value of each individual is calculated according to (20).
Step 3: Select the optimal configuration parameters which can generate the minimum $f_1(\mathbf{v}_{\text{thinned}})$.
Step 4: Select a descendant from the elites of the last generation.
Step 5: New populations are generated according to the probability of crossover P_c and the probability of mutation P_m.
Step 6: Update l to $l = l + 1$, repeat Steps 2 to 5 until the maximum number of iterations L has been reached, and determine the optimal distribution of the elements' positions and the distribution of the amplitude values.
Step 7: Determine the optimal transmit–receive thinned weight vector $\mathbf{v}_{\text{thinned}-\text{opti}}$.
Output: The array's configuration and $\mathbf{v}_{\text{thinned}-\text{opti}}$

3.2. Sparse Transmit–Receive Beamforming of Semicircular FDAs

The model of an array with a non-uniform sparse semicircular FDA with N_S elements is shown in Figure 4. All the elements are randomly distributed on a semicircular platform with the radius r in the plane.

According to Equation (11), the transmit–receive array factor of sparsely distributed semicircular FDAs can be expressed as

$$\mathrm{AF}_{\text{sparse}-\text{Semicir}-T-R}(R,\theta) = \sum_{m=1}^{N_S}\sum_{i=1}^{N_S} v_{m,i}\exp\left(-j\frac{2\pi f_i(2R - r\cos(\theta-\beta_m) - r\cos(\theta-\beta_i))}{c}\right) \tag{22}$$

where $\beta_i = (i-1)\pi/(N_S-1)$, $i = 1,2,\ldots,N_S$ is the angle parameter of the ith element deviating from the positive x-axis, which is used to describe the position of the ith element. All the elements' positions are represented by a vector, i.e., $\boldsymbol{\beta}$, which is defined as

$$\boldsymbol{\beta} = [\beta_1, \beta_2, \ldots, \beta_{N_S-2}, \beta_{N_S}]^T. \tag{23}$$

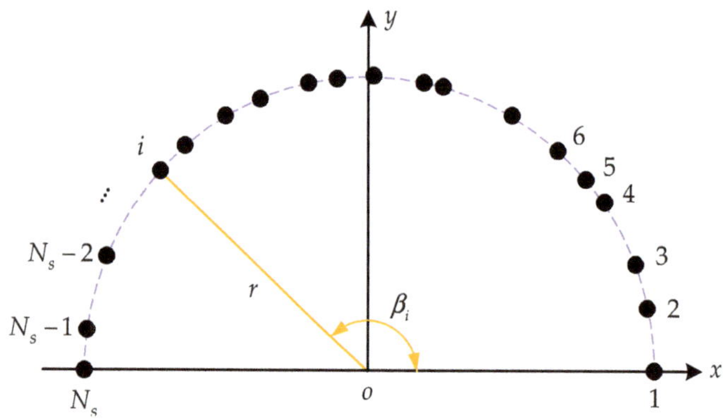

Figure 4. Configuration of a sparse semicircular FDA.

A new coding method is adopted. For a sparse array where the number of elements is N_S and the total length is D, the positions of the first element and the last element are already determined, i.e., $\beta_1 = 0$, $\beta_{N_S} = D$. Only the positions of the remaining $N_S - 2$ elements need to be determined. This method uses decimal coding to quantify the distributed parameters.

In order to reduce the mutual coupling between the elements, there are certain requirements for the positions of adjacent elements. The interval of the angles between two adjacent elements of a semicircular FDA can be expressed as

$$\min\{\beta_i - \beta_j\} \geq \beta_c,\ 1 \leq j < i \leq N_S \tag{24}$$

where β_c is the minimum angle interval between two adjacent elements. The minimum angle interval β_c can be calculated from the minimum distance interval β_0 and the semicircular FDA radius r by the cosine theorem, as follows

$$\beta_c = \arccos\left(\frac{2r^2 - \beta_0^2}{2r^2}\right) \tag{25}$$

where the minimum distance interval β_0 is $\lambda/2$.

Therefore, the azimuth angle of the nth element should satisfy the following constraint

$$\beta_n \leq 180^\circ - \beta_c - \beta_1. \tag{26}$$

If we split β_i into φ_i and $(i-1)\beta_c$, the following equation can be obtained

$$\beta=\begin{bmatrix}\beta_1\\ \beta_2\\ \beta_3\\ \vdots\\ \beta_{N_S}\end{bmatrix}=\begin{bmatrix}\varphi_1\\ \varphi_2+\beta_c\\ \varphi_3+2\beta_c\\ \vdots\\ \varphi_{N_S}+(N_S-1)\beta_c\end{bmatrix}=\begin{bmatrix}\varphi_1\\ \varphi_2\\ \varphi_3\\ \vdots\\ \varphi_{N_S}\end{bmatrix}+\begin{bmatrix}0^\circ\\ \beta_c\\ 2\beta_c\\ \vdots\\ (N_S-1)\beta_c\end{bmatrix}=\varphi+\begin{bmatrix}0^\circ\\ \beta_c\\ 2\beta_c\\ \vdots\\ (N_S-1)\beta_c\end{bmatrix} \quad (27)$$

In order to guarantee that Equations (24) and (27) hold, there must be

$$\varphi_1\le\varphi_2\le\varphi_3\cdots\le\varphi_{N_S}. \quad (28)$$

From Equations (26) and (27), it can be obtained that

$$\beta_{N_S}=\varphi_{N_S}+(N_S-1)\beta_c\le180^\circ-\beta_c. \quad (29)$$

From the two equations above, it can be seen that

$$\varphi_{N_S}\le180^\circ-N_S\cdot\beta_c, \quad (30)$$

In turn,

$$\varphi_1\le\varphi_2\le\varphi_3\cdots\le\varphi_{N_S}\in[0^\circ,\ 180^\circ-N_S\cdot\beta_c]. \quad (31)$$

In order to satisfy the constraint in (31), it is necessary to ensure that the generated random number δ_i satisfies $\delta_i\in[0,D-(N_S-1)\beta_c]$ and, after sorting it from small to large, β can be expressed as

$$\beta=[0,\beta_c,\ldots,(N_S-2)\beta_c,D]^T+[0,\delta_1,\ldots,\delta_{N_S-2},0]^T. \quad (32)$$

Through the operation above, the value of an individual gene's angle is indirectly converted into the difference in angle, which can reduce the searching space and consequently improve the efficiency of searching.

In the following, the element position vector β is incorporated into the sparse optimization problem of a semicircular FDA, i.e.,

$$F(\beta)=\iint_{(R,\theta)\in\Omega}W(R,\theta)|\mathrm{P}_{\mathrm{Semicir}-T-R}(R,\theta)-\mathrm{P}_d(R,\theta)|\sin(\theta)dRd\theta \quad (33)$$

where the expected beampattern is the same as that provided in Section 3.1.

The objective function in the optimization process is as follows

$$\min_{\beta}F(\beta). \quad (34)$$

The model optimizes the position of the element by optimizing the value of β, so that the generated beampattern can approximate the desired one.

Considering discrete sampling, Equation (34) can be equivalently expressed as

$$\min_{\beta}\sum_{p=1}^{P}\sum_{q=1}^{Q}W(R_p,\theta_q)|\mathrm{P}_{\mathrm{Semicir}-T-R}(R_p,\theta_q)-\mathrm{P}_d(R_p,\theta_q)|\sin(\theta_q). \quad (35)$$

According to the analysis above, the fitness function of the synthesis of the transmit–receive beampattern of the sparse semicircular FDA can be expressed as

$$f_2(\beta)=\sum_{p=1}^{P}\sum_{q=1}^{Q}W(R_p,\theta_q)|\mathrm{P}_{\mathrm{Semicir}-T-R}(R_p,\theta_q)-\mathrm{P}_d(R_p,\theta_q)|\sin(\theta_q). \quad (36)$$

The specific implementation of sparse transmit–receive beamforming of semicircular FDAs based on GA is provided in Algorithm 2.

Algorithm 2. GA-based sparse optimization procedures of semicircular FDAs

Input: $f_0, r, N_s, Q, L, P_c, P_m$
Step 1: Initialize Q populations, symbolizing the distance vector from the element to the reference element, where each population has N_S individuals.
Step 2: Generate an intermediate initial population with the difference in the adjacent elements' angles as the individuals.
Step 3: Arrange each individual's genes in an ascending order, transform the population of the differences in angle into the angle interval population.
Step 4: Let the generation $l = 1$ and calculate the fitness value of each individual according to (36).
Step 5: Select the optimal parameters which can generate the minimum $f_2(\boldsymbol{\beta})$.
Step 6: The individual is updated according to the probability of crossover P_c and the probability of mutation P_m.
Step 7: Arrange the genes in an ascending order, convert them into a population with angular intervals, and update l to $l = l + 1$.
Step 8: Repeat Step 4 to Step 7. The new populations are evaluated. Iterations continue until the maximum number of iterations has been reached.
Step 9: Determine the optimal element position vector $\boldsymbol{\beta}_{\text{sparse-opti}}$.
Output: $\boldsymbol{\beta}_{\text{sparse-opti}}$

4. Numerical Results and Discussion

As described in this section, simulation experiments were carried out to examine the effectiveness of the proposed thinned and sparse transmit–receive beamforming of semicircular FDAs. The whole observation area was set as Ω: 10 km $\leq R \leq$ 90 km, $0^\circ \leq \theta \leq 180^\circ$, and the space of interest was $(R_0, \theta_0) = (50\ \text{km}, 90^\circ)$. The sparse ratio was defined as ς. Other specific parameters of the simulation are listed in Table 1.

Table 1. Simulation parameters.

Parameters	Values	Parameters	Values	Parameters	Values
f_0	10 GHz	Δf	10 kHz	N	60
r	5λ	L	200	Q	50
P_c	0.8	P_m	0.05	ς	50%

Figure 5 provides the optimal positions of the elements and the distributions of amplitude of the thinned semicircular FDAs with log-FO, sin-FO, and tanh-FO, where a series of circles is used to represent the selected elements and the corresponding excitation amplitudes, while '$\times$' is used to represent the unselected elements. Figure 6 presents the distribution maps of the optimal positions of the elements of a semicircular FDA with log-FO, sin-FO, and tanh-FO after sparse distribution, where a series of circles is used to represent the distribution of elements in the full array, and a series of solid points is used to represent the distribution of the elements in the array of a sparse semicircular FDA. According to Equations (22)–(36), the synthesis of the sparse array took the elements' spacing as the constraint condition and rearranged the other elements' positions with the positions of the first and the last elements in the array being fixed.

Figure 7 provides the convergent curves of the thinned and sparse synthesis of the semicircular FDAs with log-FO, sin-FO, and tanh-FO. From Figure 7a, it can be seen that the fitness values decreased rapidly in the first 20 iterations and remained almost unchanged after 70 iterations. From Figure 7b, we can see that the fitness values decreased rapidly in the 15 iterations and remained unchanged after 20 iterations.

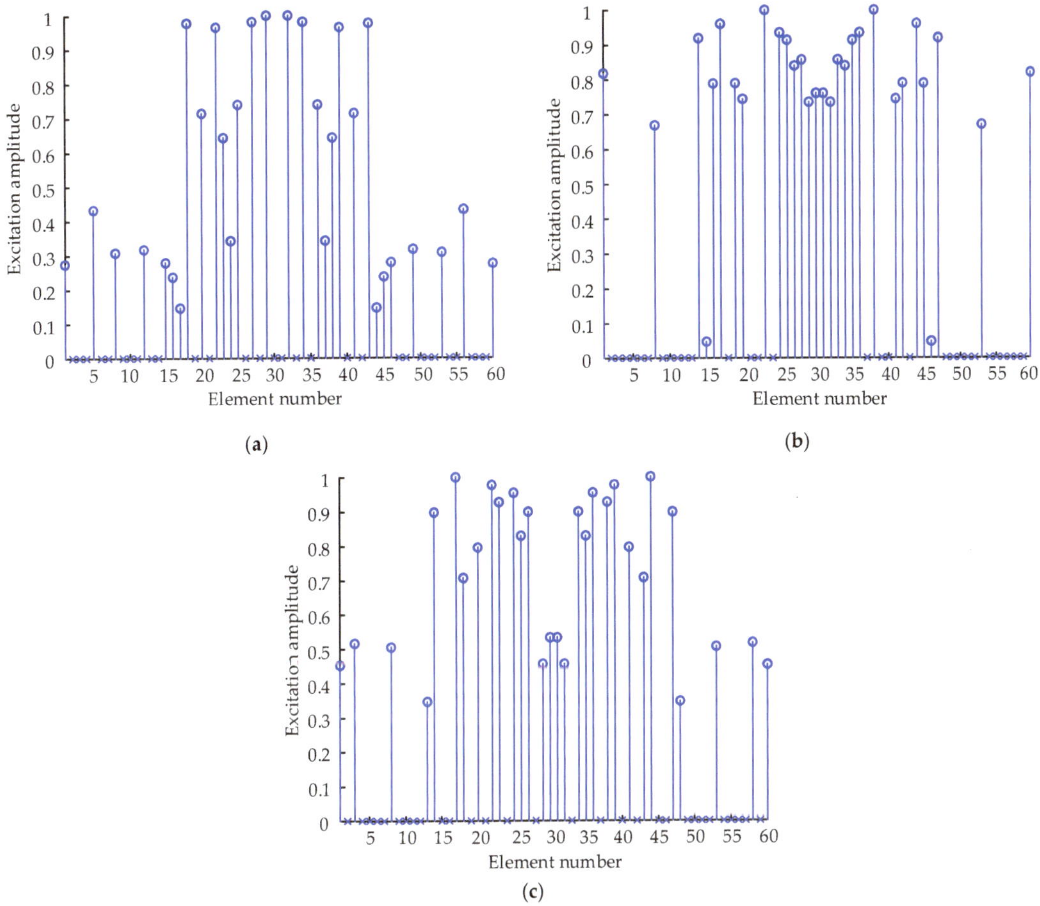

Figure 5. Positions of the elements in the array and distributions of amplitude of thinned semicircular FDAs with (**a**) log-FO, (**b**) sin-FO, and (**c**) tanh-FO.

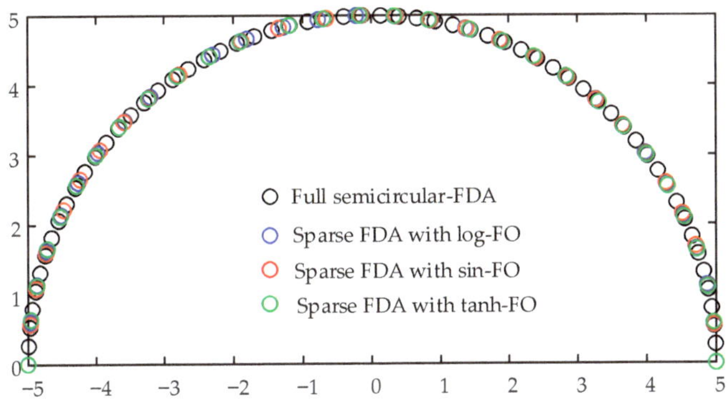

Figure 6. Optimal distribution of the elements' positions for sparse semicircular FDAs.

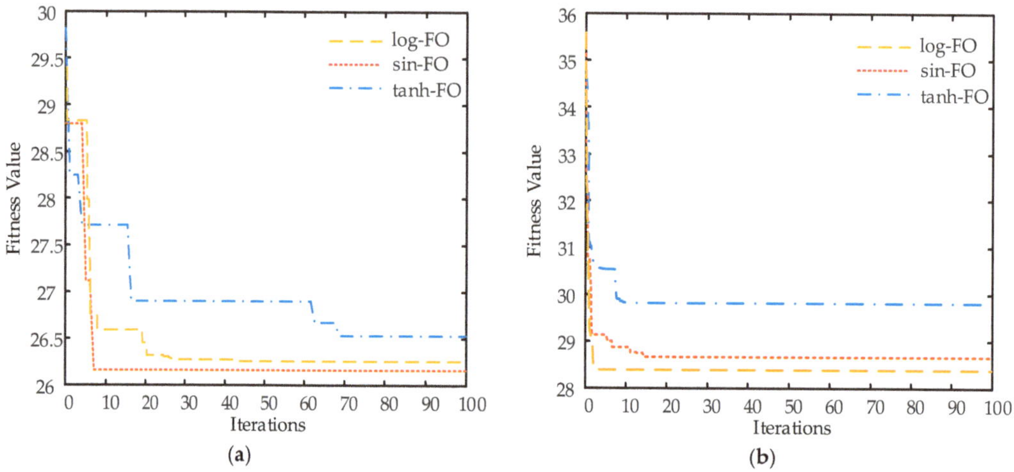

Figure 7. Convergent curves of the (**a**) thinned and (**b**) sparse semicircular FDAs.

Figure 8 provides the transmit–receive beampatterns of the full semicircular FDAs with log-FO, sin-FO, and tanh-FO. As can be seen from this figure, only one mainbeam was generated in the desired location of $(R_0, \theta_0) = (50\text{ km}, 90^\circ)$, and the peak SLLs were 10log10 (0.1) = −10 dB, 10log10 (0.1267) = −8.97 dB, and 10log10 (0.1566) = −8.05 dB, respectively, for these three full semicircular FDAs. The widths of the mainlobe of the transmit–receive beampattern of the full semicircular FDAs with log-FO, sin-FO, and tanh-FO were all 9° in the angle dimension, and the widths of the mainlobe in the range dimension wer 7.12 km, 3.56 km and 3.56 km, respectively, for log-FO, sin-FO, and tanh-FO.

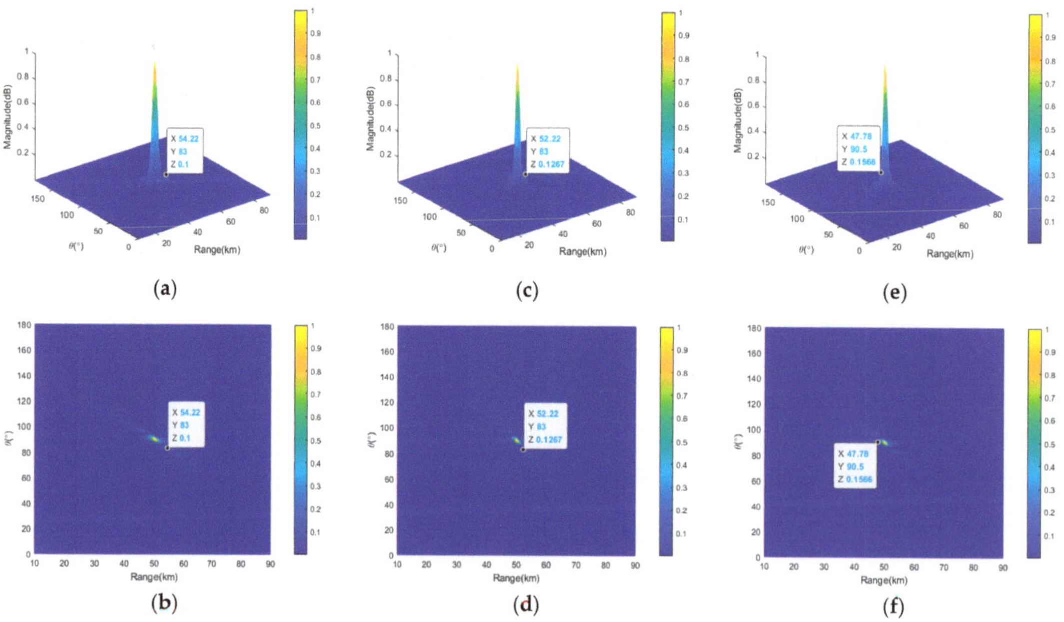

Figure 8. Transmit–receive beampatterns of full semicircular FDAs with (**a**,**b**) log-FO, (**c**,**d**) sin-FO, and (**e**,**f**) tanh-FO, where (**a**,**c**,**e**) are in 3D, and (**b**,**d**,**f**) are the projections of (**a**,**c**,**e**), respectively.

Figure 9 presents the transmit–receive beampatterns of the thinned semicircular FDAs with log-FO, sin-FO, and tanh-FO. As can be seen from this figure, similar phenomena can be seen, i.e., only one mainbeam was generated in the desired location of $(R_0, \theta_0) = (50\text{ km}, 90^\circ)$. The peak SLLs lowered by 5.78 dB (from 10log10 (0.1) = −10 dB to 10log10 (0.02645) = −15.78 dB), 4.39 dB (from 10log10 (0.1267) = −8.97 dB to 10log10 (0.04618) = −13.36 dB), and 4.41 dB (from 10log10 (0.1566) = −8.05 dB to 10log10 (0.05674) = −12.46 dB), respectively, for log-FO, sin-FO, and tanh-FO. The widths of the mainlobes of the three thinned semicircular FDAs with log-FO, sin-FO, and tanh-FO were all 12° in the angle dimension, and the widths of the mainlobe in the range dimension were 12 km, 7.56 km and 7.12 km, respectively, for log-FO, sin-FO, and tanh-FO. Correspondingly, the widths of the mainlobe are all widened by 3° (from 9° to 12°) in angle dimension. The widths of the mainlobes were widened by 4.88 km (from 7.12 km to 12 km), 4 km (from 3.56 km to 7.56 km), and 3.56 km (from 3.56 km to 7.12 km) in the range dimension.

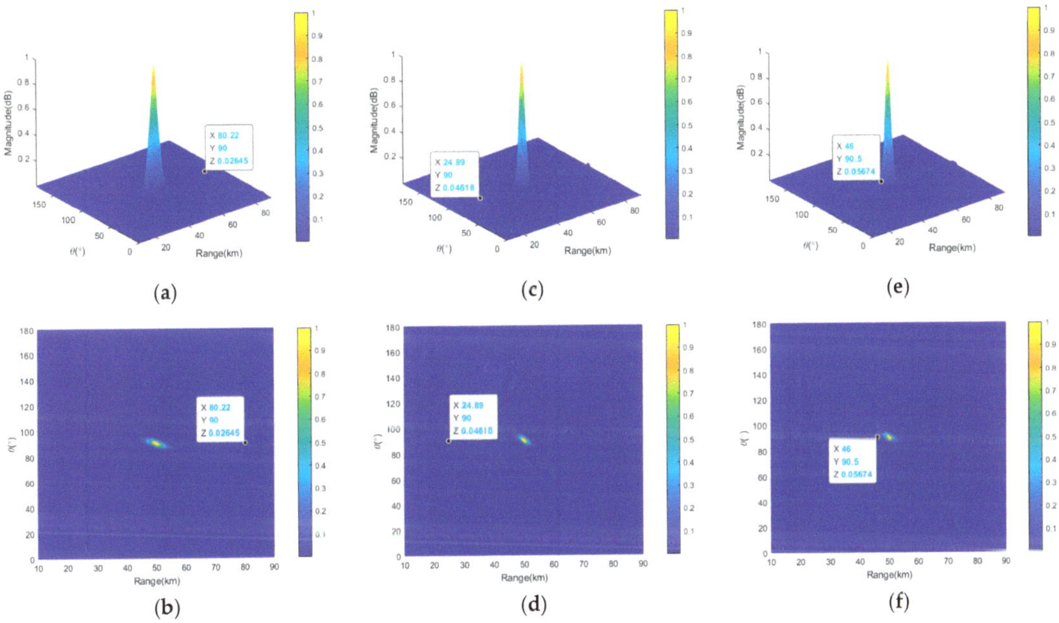

Figure 9. Transmit–receive beampatterns of thinned semicircular FDAs with (**a**,**b**) log-FO, (**c**,**d**) sin-FO, and (**e**,**f**) tanh-FO, where (**a**,**c**,**e**) are in 3D, and (**b**,**d**,**f**) are the projections of (**a**,**c**,**e**), respectively.

Figure 10 provides the transmit–receive beampatterns of the sparse semicircular FDAs with log-FO, sin-FO, and tanh-FO. It can also be clearly seen that only one mainbeam ws generated in the desired location of $(R_0, \theta_0) = (50\text{ km}, 90^\circ)$. The peak SLLs were 10log10 (0.06038) = −12.19 dB, 10log10 (0.05448) = −12.64 dB, and 10log10 (0.07042) = −11.52 dB, respectively, for these three sparse semicircular FDAs. The widths of the mainlobe of the transmit–receive beampattern of the sparse semicircle-FDAs with log-FO, sin-FO, and tanh-FO were 10°, 9° and 10° in the angle dimension, and the widths of the mainlobe in the range dimension were 8.66 km, 4.88 km and 4.00 km, respectively. Compared with Figure 8, the peak SLLs lowered by 2.19 dB (from 10log10 (0.1) = −10 dB to 10log10 (0.06038) = −12.19 dB), 3.67 dB (from 10log10 (0.1267) = −8.97 dB to 10log10 (0.05448) = −12.64 dB), and 3.47 dB (from 10log10 (0.1566) = −8.05 dB to 10log10 (0.07042) = −11.52 dB), respectively, for log-FO, sin-FO, and tanh-FO. The widths of the mainlobe were widened by 1° (from 9° to 10°), 0° (from 9° to 9°), and 1° (from 9° to 10°) in the angle dimension. The widths of the mainlobe were widened by 1.54 km (from 7.12 km to 8.66 km), 1.32 km (from 3.56 km to 4.88 km), and 0.44 km (from 3.56 km to 4 km) in the range dimension. Note that both in the thinned synthesis and in the sparse synthesis, the constraints imposed on the mainlobes' widths were set as $\theta_{\text{width}} = 9^{\circ}$ in the angle dimension, which is the same as the counterpart before the synthesis, and R_{width} = 7.5 km, 4.0 km, and 4.0 km for the semicircular FDAs with log-FO, sin-FO, and tanh-FO, respectively, which were slightly larger than their counterparts in the range dimension before the synthesis. Comparisons between the thinned and sparse syntheses are listed in Table 2. Moreover, it can be obviously seen from Figures 8–10 that the range-angle information decoupled beampattern of the semicircular FDAs with non-linearly varying FOs can be maintained after thinned and sparse synthesis.

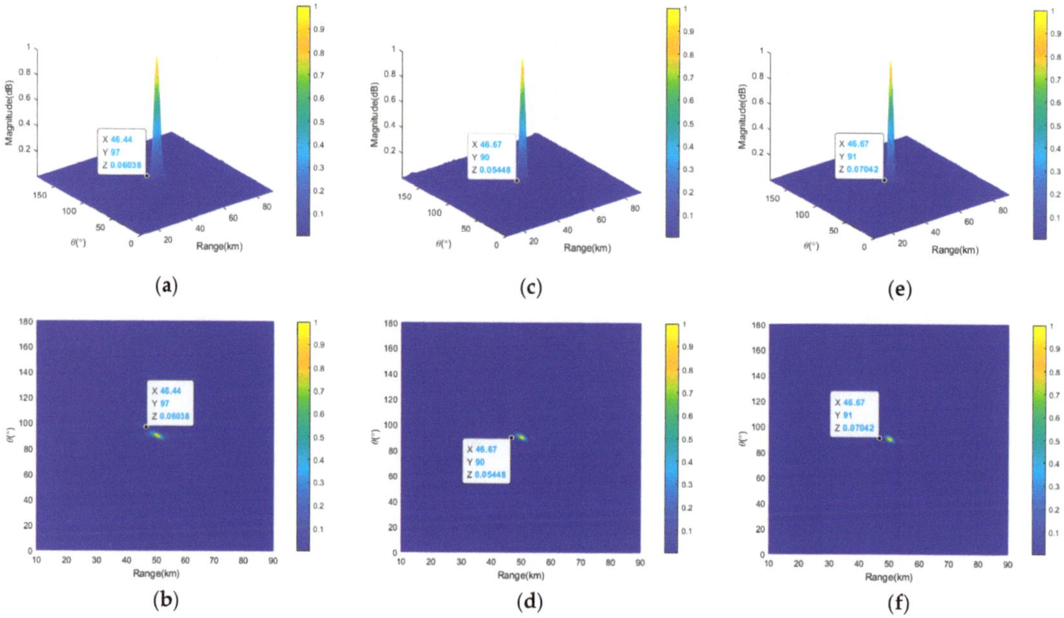

Figure 10. Transmit–receive beampatterns of sparse semicircular FDAs with (**a**,**b**) log-FO, (**c**,**d**) sin-FO, and (**e**,**f**) tanh-FO, where (**a**,**c**,**e**) are in 3D, and (**b**,**d**,**f**) are the projections of (**a**,**c**,**e**), respectively.

Table 2. Comparisons of the peak SLLs and widths of the mainlobe of transmit–receive beampatterns of the semicircular FDAs before and after thinned and sparse synthesis.

	Full FDA			Thinned FDA			Sparse FDA		
	Log-FO	**Sin-FO**	**Tanh-FO**	**Log-FO**	**Sin-FO**	**Tanh-FO**	**Log-FO**	**Sin-FO**	**Tanh-FO**
Peak SLL (dB)	−10.00	−8.97	−8.05	−15.78	−13.36	−12.46	−12.19	−12.64	−11.52
Angle (°)	9.0	9.0	9.0	12.0	12.0	12.0	10.0	9.0	10.0
Range (km)	7.12	3.56	3.56	12.0	7.56	7.12	8.66	4.88	4.00

To further clearly compare the performance of the three semicircular FDAs before and after the thinned and sparse synthesis, the projections and profiles in the range and angle dimensions of the transmit–receive beampatterns are provided below. Figure 11 shows the transmit–receive beampattern projections of the semicircular FDA with log-FO in the range and angle dimensions before and after thinned and sparse distribution. If we compare Figure 11b,c with Figure 11a, it can be seen that after the thinned and sparse synthesis, the peak SLL of the beampattern in the angle dimension was reduced by 5.97 dB and 2.19 dB, respectively (from 10log10 (0.1) = −10 dB to 10log10 (0.02532) = −15.97 dB and 10log10 (0.06038) = −12.19 dB). From Figure 11d–f, it can be seen that after thinned and sparse synthesis, the peak SLL in the range dimension was reduced by 5.78 dB and 2.59 dB (from 10log10 (0.1) = −10 dB to 10log10 (0.02645) = −15.78 dB and 10log10 (0.0551) = −12.59 dB). The details of the peak SLLs are listed in Table 3.

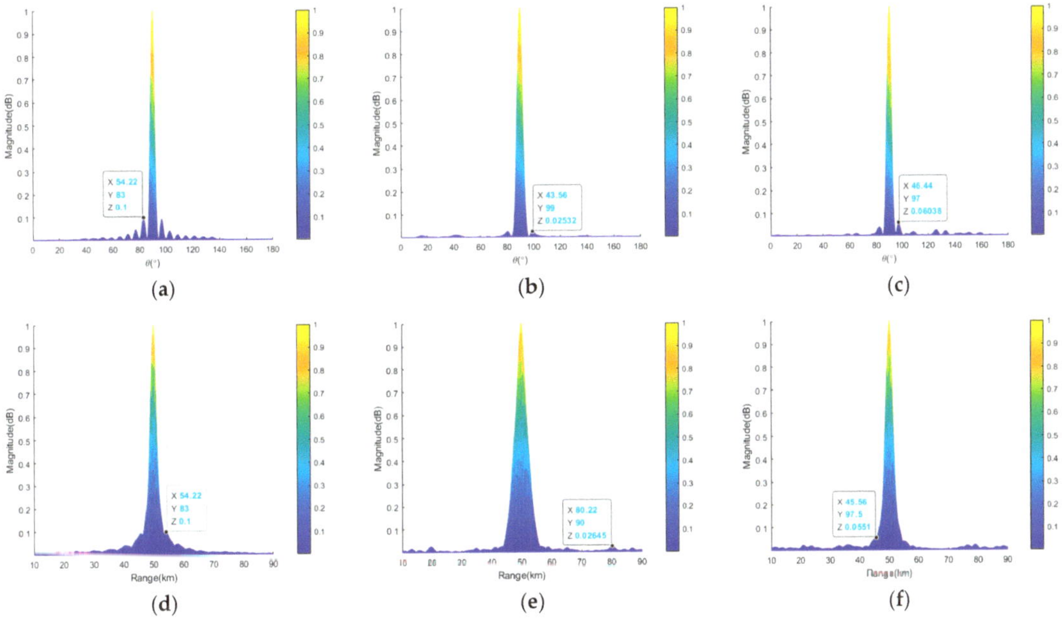

Figure 11. Projections of the transmit–receive beampatterns of semicircular FDAs with log-FO in the range-angle dimensions: (**a**) full array in the angle dimension; (**b**) thinned array in the angle dimension; (**c**) sparse array in the angle dimension; (**d**) full array in the range dimension; (**e**) thinned array in the range dimension; (**f**) sparse array in the range dimension.

Table 3. Comparison of the peak SLLs of semicircular FDAs with log-FO in the range and angle dimensions before and after thinned and sparse synthesis.

	Full FDA	Thinned FDA	Sparse FDA
Peak SLL in angle (dB)	−10.00	−15.97	−12.19
Peak SLL in range (dB)	−10.00	−15.78	−12.59

Figure 12 provides the transmit–receive beampattern profiles of the semicircular FDA with log-FO before and after thinned and sparse synthesis. It can be seen from Figure 12a that the peak SLL of the semicircular FDA with log-FO was reduced by 14.51 dB and 2.52 dB (from −15.50 dB to −30.01 dB and −18.02 dB) in the angle dimension before and after the application of thinned and sparse synthesis, respectively. It can be seen from Figure 12b that the corresponding part increased by 1.01 dB and 2.14 dB (from −16.79 dB to −15.78 dB and −14.65 dB) in the range dimension, respectively. The specific peak SLLs are provided in Table 4.

Figure 13 shows the projections of transmit–receive beampatterns of the semicircular FDA with sin-FO in the range and the angle dimensions before and after thinned and sparse distribution. If we compare Figure 13b,c with Figure 13a, it can be seen that after the thinned and sparse synthesis, the peak SLL of the beampattern in the angle dimension was reduced by 5.55 dB and 7.3 dB, respectively (from 10log10 (0.1267) = −8.97 dB to 10log10 (0.03529) = −14.52 dB and 10log10 (0.02362) = −16.27 dB). From Figure 13d–f, it can be seen that after thinned and sparse synthesis, the peak SLL in the range dimension was reduced by 4.39 dB and 3.67 dB (from 10log10 (0.1267) = −8.97 dB to 10log10 (0.04618) = −13.36 dB and 10log10 (0.05448) = −12.64 dB). The details of the peak SLLs are listed in Table 5.

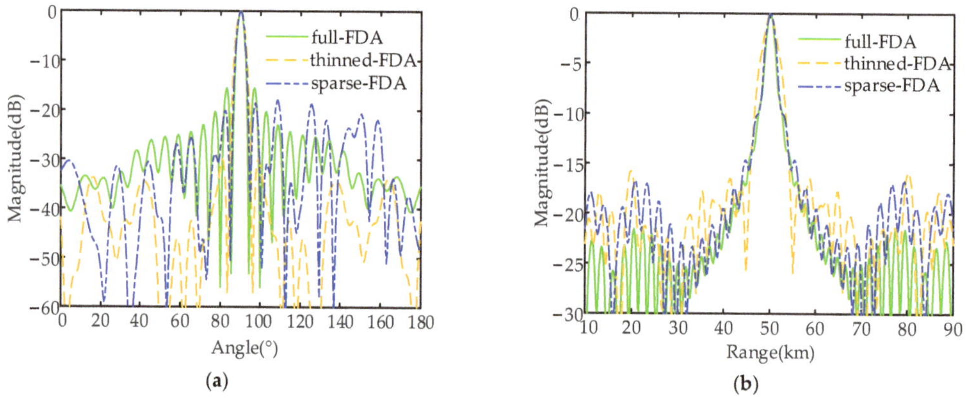

Figure 12. Normalized transmit–receive beampattern profiles of semicircular FDAs with log-FO (a) cut at a range of 50 km, and (b) cut at an angle of 90°.

Table 4. Peak SLLs of the transmit–receive beampattern profiles of the semicircular FDA with log-FO in Figure 11.

Peak SLLs	Full FDA	Thinned FDA	Sparse FDA
Angle (dB)	−15.50	−30.01	−18.02
Range (dB)	−16.79	−15.78	−14.65

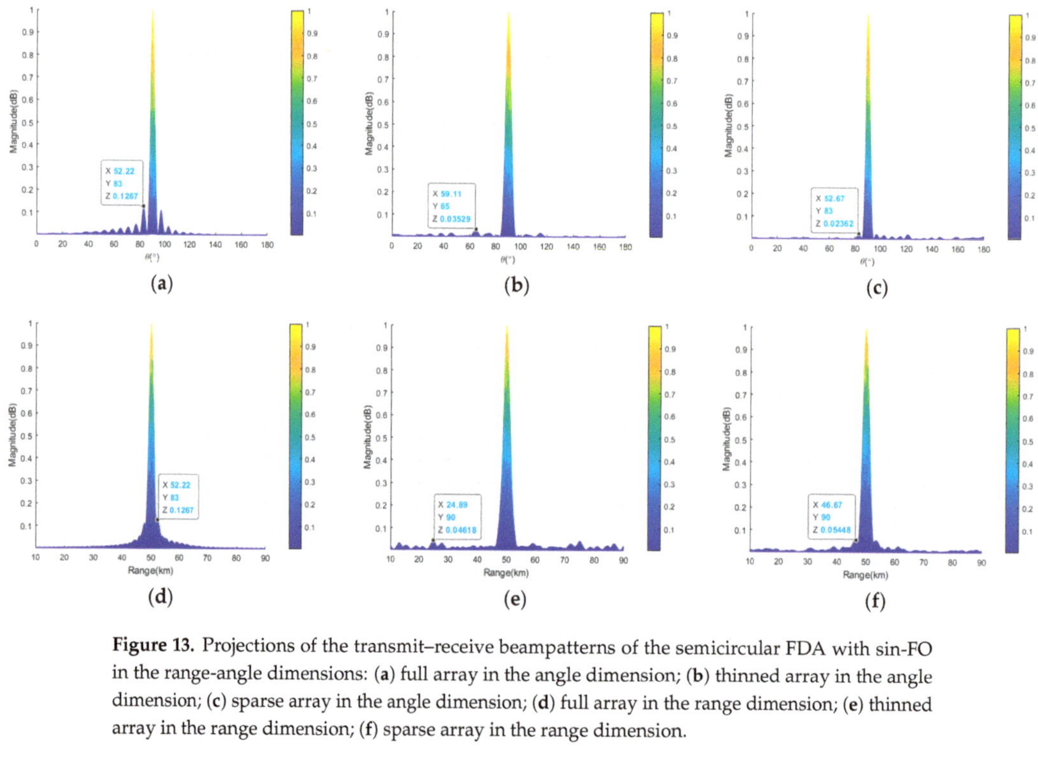

Figure 13. Projections of the transmit–receive beampatterns of the semicircular FDA with sin-FO in the range-angle dimensions: (a) full array in the angle dimension; (b) thinned array in the angle dimension; (c) sparse array in the angle dimension; (d) full array in the range dimension; (e) thinned array in the range dimension; (f) sparse array in the range dimension.

Table 5. Comparison of peak SLLs of the semicircular FDA with sin-FO in the range and angle dimensions before and after thinned and sparse synthesis.

	Full FDA	Thinned FDA	Sparse FDA
Peak SLL in Angle (dB)	−8.97	−14.52	−16.27
Peak SLL in Range (dB)	−8.97	−13.36	−12.64

Figure 14 provides the transmit–receive beampattern profiles of the semicircular FDA with sin-FO before and after thinned and sparse synthesis. It can be seen from Figure 14a that the peak SLL of the semicircular FDA with sin-FO was reduced by 10.58 dB and 3.15 dB (from −15.5 dB to −26.08 dB and −18.65 dB) in the angle dimension before and after the application of thinned and sparse synthesis, respectively. It can be seen from Figure 14b that the corresponding part increased by 3.945 dB and 3.225 dB (from −9.415 dB to −13.36 dB and −12.64 dB) in the range dimension, respectively. The specific peak SLLs are provided in Table 6.

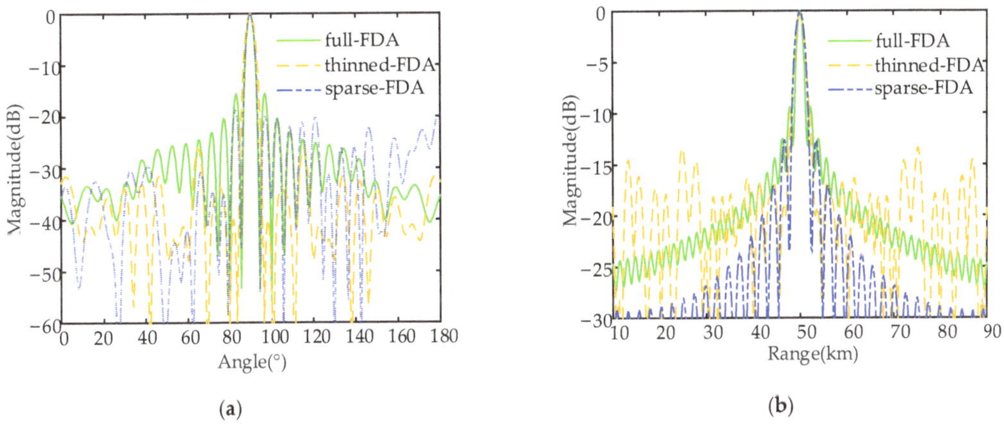

Figure 14. Normalized transmit–receive beampattern profiles of the semicircular FDA with sin-FO (**a**) cut at a range of 50 km, and (**b**) cut at an angle of 90°.

Table 6. Peak SLLs of the transmit–receive beampattern profiles of the semicircular FDA with sin-FO in Figure 13.

Peak SLLs	Full FDA	Thinned FDA	Sparse FDA
Angle (dB)	−15.50	−26.08	−18.65
Range (dB)	−9.415	−13.36	−12.64

Figure 15 shows the projections of the transmit–receive beampatterns of the semicircular FDA with tanh-FO in the range and the angle dimensions before and after thinned and sparse distribution. If we compare Figure 15b,c with Figure 15a, it can be seen that after the thinned and sparse synthesis, the peak SLL of the beampattern in the angle dimension was reduced by 4.38 dB and 8.06 dB, respectively (from 10log10 (0.09219) = −10.35 dB to 10log10 (0.03364) = −14.73 dB and 10log10 (0.01441) = −18.41 dB). From Figure 15d–f, it can be seen that after thinned and sparse synthesis, the peak SLL in the range dimension was reduced by 4.41 dB and 3.65 dB (from 10log10 (0.1566) = −8.05 dB to 10log10 (0.05674) = −12.46 dB and 10log10 (0.07042) = −11.52 dB), respectively. The details of the peak SLLs are listed in Table 7.

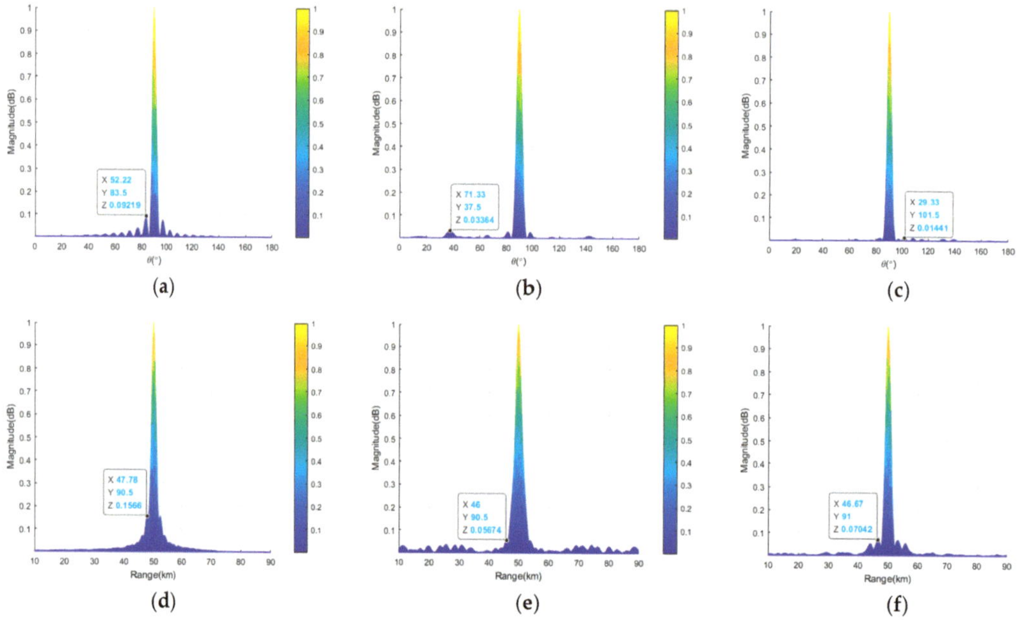

Figure 15. Projections of the transmit–receive beampatterns of the semicircular FDA with tanh-FO in the range and angle dimensions: (**a**) full array in the angle dimension; (**b**) thinned array in the angle dimension; (**c**) sparse array in the angle dimension; (**d**) full array in the range dimension; (**e**) thinned array in the range dimension; (**f**) sparse array in the range dimension.

Table 7. Comparison of the peak SLLs of the semicircular FDA with tanh-FO in the range and angle dimensions before and after thinned and sparse synthesis.

	Full FDA	Thinned FDA	Sparse FDA
Peak SLL in angle (dB)	−10.35	−14.73	−18.41
Peak SLL in range (dB)	−8.05	−12.46	−11.52

Figure 16 provides the transmit–receive beampattern profiles of the semicircular FDA with tanh-FO before and after thinned and sparse synthesis. It can be seen from Figure 16a that the peak SLL of the semicircular FDA with tanh-FO was reduced by 7.45 dB and 4.59 dB (from −15.50 dB to −22.95 dB and −20.09 dB) in the angle dimension before and after the application of thinned and sparse synthesis, respectively. It can be seen from Figure 16b that the corresponding part increased by 3.16 dB and 3.45 dB (from −8.35 dB to −11.51 dB and −11.80 dB) in the range dimension, respectively, for thinned and sparse synthesis. The specific peak SLLs are listed in Table 8.

Table 8. Peak SLLs of the transmit–receive beampattern profiles of the semicircular FDA with tanh-FO in Figure 15.

Peak SLLs	Full FDA	Thinned FDA	Sparse FDA
Angle (dB)	−15.50	−22.95	−20.09
Range (dB)	−8.35	−11.51	−11.80

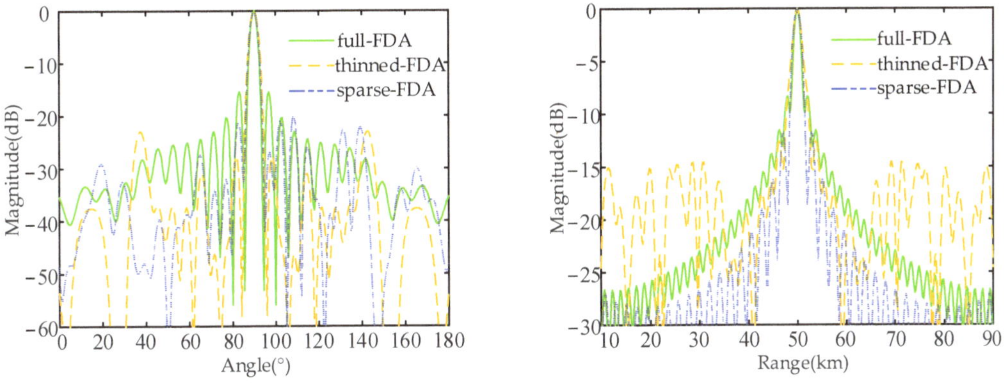

(a) (b)

Figure 16. Normalized transmit–receive beampattern profiles of the semicircular FDA with tanh-FO (**a**) cut at a range of 50 km and (**b**) cut at an angle of 90°.

5. Conclusions

FDA applies a tiny FO across the array's aperture, presenting a new choice for controlling the array's beampattern. Meanwhile, conformal array configurations have been widely explored due to their remarkable advantage of being easily conformed to the surface of the carrier platforms. When high resolution is required, a vast amount of elements should be included in an array, which would induce great cost. Under these circumstances, this study investigated the thinned and sparse beamforming of semicircular FDAs. In the thinned semicircular FDA, lower peak SLLs could be achieved by strategically turning off some antenna elements while optimizing the excitation amplitudes. In the sparse semicircular FDA, the array's position varied under the constraints. In particular, thinned and sparse synthesis were performed for semicircular FDAs with log-FO, sin-FO, and tanh-FO, respectively. Under the same assumption of saving 50% of the elements, the peak SLLs of the thinned semicircular FDAs were reduced by 5.8 dB, 4.4 dB, and 4.4 dB, respectively, for log-FO, sin-FO, and tanh-FO. Compared with the thinned semicircular FDA, the SLLs of sparse semicircular FDAs decreased by 2.2 dB, 3.7 dB, and 3.5 dB, respectively, for log-FO, sin-FO, and tanh-FO. Note that the sparse semicircular FDA with sin-FO did not widen the mainlobe in the angle dimension. It can be concluded that thinned optimization allows significant reduction of SLLs while sparse optimization allows nearly the same beamwidth in the angle dimension.

Author Contributions: Conceptualization, A.W. and Y.X.; methodology, A.W., X.H. and Y.X.; software, A.W., X.H., X.M. and Y.X.; validation, A.W., X.H., Y.X. and Y.L.; formal analysis, X.H. and Y.X.; data curation, X.H., X.M. and Y.X.; writing—original draft preparation, X.H. and Y.L.; writing—review and editing, A.W., X.M. and Y.X.; supervision, A.W. and Y.X.; funding acquisition, A.W. and Y.X. All authors have read and agreed to the published version of the manuscript.

Funding: This research was funded by the National Science Foundation of China (Grants 62471384, 62271386, and 61901357) and Shaanxi Science and Technology Association Youth Talent Lifting Program (Grant 20230149).

Data Availability Statement: The original contributions presented in the study are included in the article, further inquiries can be directed to the corresponding author.

Acknowledgments: The authors would like to thank the anonymous reviewers for their very competent comments and helpful suggestions.

Conflicts of Interest: The authors declare no conflicts of interest.

References

1. Antonik, P.; Wicks, M.C.; Griffiths, H.D.; Baker, C.J. Frequency Diverse Array Radars. In Proceedings of the 2006 IEEE Conference on Radar, Verona, NY, USA, 24–27 April 2006; IEEE: Syracuse, NY, USA, 2006; pp. 215–217.
2. Ding, Z.; Xie, J.; Xu, J. A Joint Array Parameters Design Method Based on FDA-MIMO Radar. *IEEE Trans. Aerosp. Electron. Syst.* **2023**, *59*, 2909–2919. [CrossRef]
3. Huang, L.; Li, X.; Wan, W.; Ye, H.; Zhang, S.; Wang, W.-Q. Adaptive FDA Radar Transmit Power Allocation for Target Detection Enhancement in Clutter Environment. *IEEE Trans. Veh. Technol.* **2023**, *72*, 11111–11121. [CrossRef]
4. Zhang, T.; Wang, Z.; Xing, M.; Zhang, S.; Wang, Y. Research on Multi-Domain Dimensionality Reduction Joint Adaptive Processing Method for Range Ambiguous Clutter of FDA-Phase-MIMO Space-Based Early Warning Radar. *Remote Sens.* **2022**, *14*, 5536. [CrossRef]
5. Liu, Z.; Zhu, S.; Xu, J.; Lan, L.; He, X.; Li, X. Cooperated Moving Target Detection Approach for PA-FDA Dual-Mode Radar in Range-Ambiguous Clutter. *Remote Sens.* **2023**, *15*, 692. [CrossRef]
6. Wang, K.; Liao, G.; Xu, J.; Zhang, Y.; Wang, W. Joint Range Angle and Velocity Estimation Method for FDA-MIMO Radar Under Clutter Background. *IEEE Trans. Aerosp. Electron. Syst.* **2023**, *59*, 7866–7877. [CrossRef]
7. Chen, G.; Wang, C.; Gong, J.; Tan, M.; Liu, Y. Data-Independent Phase-Only Beamforming of FDA-MIMO Radar for Swarm Interference Suppression. *Remote Sens.* **2023**, *15*, 1159. [CrossRef]
8. Zaman, F.; Akhtar, Z.; Mehmood, S.; Akbar, S.; Phanomchoeng, G. A Novel FDA-MIMO Deceptive Jamming Method for Neutralizing Phased Array Radar. *IEEE Access* **2023**, *11*, 90449–90458. [CrossRef]
9. Wen, Y.; Zhang, Z.; Chen, Z.; Qiu, J.; Ren, M.; Meng, X. A Novel Time-Domain Frequency Diverse Array HRWS Imaging Scheme for Spotlight SAR. *Remote Sens.* **2022**, *14*, 1085. [CrossRef]
10. Wang, Z.; Song, Y.; Li, Y. Ultra-Wideband Imaging via Frequency Diverse Array with Low Sampling Rate. *Remote Sens.* **2022**, *14*, 1271. [CrossRef]
11. He, C.; Zhang, R.; Huang, B.; Xu, M.; Wang, Z.; Liu, L.; Lu, Z.; Jin, Y. Moving-Target Detection for FDA-MIMO Radar in Partially Homogeneous Environments. *Electronics* **2024**, *13*, 851. [CrossRef]
12. Zhang, Y.; Zhang, Y.; Wang, J.; Xiao, S.; Tang, W. Distance-Angle Beamforming for Covert Communications via Frequency Diverse Array: Toward Two-Dimensional Covertness. *IEEE Trans. Wirel. Commun.* **2023**, *22*, 8559–8574. [CrossRef]
13. Xie, H.; Zhang, Y.; Zhang, Y.; Liu, H.; Guo, R. Beamforming for LEO Satellite Communication Based on FDA. *Electron. Lett.* **2023**, *59*, e12889. [CrossRef]
14. Hu, Y.-Q.; Chen, H.; Ji, S.-L.; Wang, W.-Q. Ambient Backscatter Communication with Frequency Diverse Array for Enhanced Channel Capacity and Detection Performance. *IEEE Sens. J.* **2020**, *20*, 10876–10885. [CrossRef]
15. Jian, J.; Wang, W.-Q.; Chen, H.; Huang, B. Physical-Layer Security for Multi-User Communications with Frequency Diverse Array-Based Directional Modulation. *IEEE Trans. Veh. Technol.* **2023**, *72*, 10133–10145. [CrossRef]
16. Chen, Z.; Zhang, Z.; Zhou, Y.; Zhao, Q.; Wang, W. Elevated Frequency Diversity Array: A Novel Approach to High Resolution and Wide Swath Imaging for Synthetic Aperture Radar. *IEEE Geosci. Remote Sens. Lett.* **2022**, *19*, 1–5. [CrossRef]
17. Zhang, M.; Liao, G.; Xu, J.; He, X.; Liu, Q.; Lan, L.; Li, S. High-Resolution and Wide-Swath SAR Imaging with Sub-Band Frequency Diverse Array. *IEEE Trans. Aerosp. Electron. Syst.* **2023**, *59*, 172–183. [CrossRef]
18. Huang, L.; Li, X.; Wan, W.; Zhang, S.; Wang, W.-Q. Frequency Diverse Array Introduced into SAR GMTI to Mitigate Blind Velocity and Doppler Ambiguity. *IEEE Geosci. Remote Sens. Lett.* **2022**, *19*, 1–5. [CrossRef]
19. Huang, L.; Zong, Z.; Zhang, S.; Wang, W.-Q. 2-D Moving Target Deception Against Multichannel SAR-GMTI Using Frequency Diverse Array. *IEEE Geosci. Remote Sens. Lett.* **2022**, *19*, 1–5. [CrossRef]
20. Huang, B.; Wang, W.-Q.; Zhang, S.; Wang, H.; Gui, R.; Lu, Z. A Novel Approach for Spaceborne SAR Scattered-Wave Deception Jamming Using Frequency Diverse Array. *IEEE Geosci. Remote Sens. Lett.* **2020**, *17*, 1568–1572. [CrossRef]
21. Khan, W.; Qureshi, I.M.; Saeed, S. Frequency Diverse Array Radar with Logarithmically Increasing Frequency Offset. *Antennas Wirel. Propag. Lett.* **2015**, *14*, 499–502. [CrossRef]
22. Huang, G.; Ding, Y.; Ouyang, S.; Fusco, V. Frequency Diverse Array with Random Logarithmically Increasing Frequency Offset. *Microw. Opt. Technol. Lett.* **2020**, *62*, 2554–2561. [CrossRef]
23. Liu, M.; Wang, C.; Gong, J.; Tan, M.; Bao, L.; Zhou, C. Analysis of Cantor Multistage Frequency Offset FDA-MIMO Beampattern Performance. *IEEE Sens. J.* **2023**, *23*, 23270–23281. [CrossRef]
24. Zhu, Z.; Chen, W.; Yang, Y.; Shu, Q. Frequency Diverse Array Beampattern Synthesis with Random Permutated Power Increasing Frequency Offset. *Antennas Wirel. Propag. Lett.* **2022**, *21*, 1975–1979. [CrossRef]
25. Wang, Y.; Huang, G.; Li, W. Transmit Beampattern Design in Range and Angle Domains for MIMO Frequency Diverse Array Radar. *Antennas Wirel. Propag. Lett.* **2017**, *16*, 1003–1006. [CrossRef]
26. Wang, B.; Xie, J.; Zhang, J.; Ge, J. Dot-Shaped Beamforming Analysis of Subarray-Based Sin-FDA. *Front. Inf. Technol. Electron. Eng.* **2019**, *20*, 1429–1444. [CrossRef]
27. Saeed, S.; Qureshi, I.M.; Khan, W.; Salman, A. Elliptical Frequency Diverse Array Radars: Uniform and Non-Uniform Frequency Offsets. *Remote Sens. Lett.* **2017**, *8*, 478–487. [CrossRef]
28. Mahmood, M.; Mir, H. FDA Transmit Beampattern Synthesis Using Piecewise Trigonometric Frequency Offset. *IET Radar Sonar Navig.* **2019**, *13*, 1149–1153. [CrossRef]

29. Basit, A.; Qureshi, I.M.; Khan, W.; Rehman, S.U.; Khan, M.M. Beam Pattern Synthesis for an FDA Radar with Hamming Window-Based Nonuniform Frequency Offset. *Antennas Wirel. Propag. Lett.* **2017**, *16*, 2283–2286. [CrossRef]
30. Liao, Y.; Tang, H.; Chen, X.; Wang, W.-Q. Frequency Diverse Array Beampattern Synthesis with Taylor Windowed Frequency Offsets. *Antennas Wirel. Propag. Lett.* **2020**, *19*, 1901–1905. [CrossRef]
31. Nusenu, S.Y.; Basit, A.; Asare, E. FDA Transmit Beamforming Synthesis Using Chebyshev Window Function Technique to Counteract Deceptive Electronic Countermeasures Signals. *PIER Lett.* **2020**, *90*, 53–60. [CrossRef]
32. Wen, H.; Gu, P.-F.; He, Z.; Lian, J.-W.; Cao, J.; Sun, Y.-Z.; Wang, J.-X.; Ding, D.-Z. Frequency Diverse Array Beampattern Synthesis with Sinc- and Weighted-Sinc-Function-Based Frequency Offsets. *IEEE Trans. Antennas Propag.* **2024**, *72*, 1694–1707. [CrossRef]
33. Liao, Y.; Tang, H.; Luo, Z.-B.; Wang, J.; Wang, W.-Q. Analysis of Frequency Offset Errors Through Interval Arithmetic in Frequency Diverse Arrays. *IEEE Trans. Aerosp. Electron. Syst.* **2023**, *59*, 3424–3431. [CrossRef]
34. Xiong, J.; Wang, W.-Q.; Shao, H.; Chen, H. Frequency Diverse Array Transmit Beampattern Optimization with Genetic Algorithm. *Antennas Wirel. Propag. Lett.* **2017**, *16*, 469–472. [CrossRef]
35. Shao, X.; Hu, T.; Xiao, Z.; Zhang, J. Frequency Diverse Array Beampattern Synthesis with Modified Sinusoidal Frequency Offset. *Antennas Wirel. Propag. Lett.* **2021**, *20*, 1784–1788. [CrossRef]
36. Xiao, M.; Shao, X.; Hu, T.; Zhang, J.; Xiao, Z. Frequency Diverse Array Beampattern Synthesis Method with Low Sidelobes by Integrating Mayfly Algorithm and Convex Programming. *Electron. Lett.* **2023**, *59*, 13012. [CrossRef]
37. Wen, H.; Gu, P.-F.; He, Z.; Lian, J.; Wang, J.; Song, D.; Ding, D. Optimal Function-Based Frequency Offset Design Based on Polynomial Fitting for Frequency Diverse Array Beampattern Synthesis. *Antennas Wirel. Propag. Lett.* **2024**, *23*, 149–153. [CrossRef]
38. Ahmad, Z.; Jaffri, Z.U.A.; Bao, S.-D.; Chen, M. Frequency Diverse Array Radar with Non-Uniform Array Spacing Based on Sigmoid Function. *J. Electr. Eng.* **2022**, *73*, 36–42. [CrossRef]
39. Wang, W.-Q.; So, H.C.; Shao, H. Nonuniform Frequency Diverse Array for Range-Angle Imaging of Targets. *IEEE Sens. J.* **2014**, *14*, 2469–2476. [CrossRef]
40. Xu, W.; Pei, C.; Huang, P.; Tan, W.; Gao, Z. Beampattern Synthesis and Optimization Method Based on Circular Frequency Diverse Array Engineering Model. *Electronics* **2024**, *13*, 1618. [CrossRef]
41. Xu, Y.; Wang, A.; Xu, J. Range–Angle Transceiver Beamforming Based on Semicircular-FDA Scheme. *IEEE Trans. Aerosp. Electron. Syst.* **2022**, *58*, 834–843. [CrossRef]
42. Hei, Y.Q.; Ju, X.Y.; Ma, L.Y.; Li, W.T.; Shi, X.W. ANN-Assisted Quasi-Time-Invariant Beamforming for Retrodirective Frequency Diverse Array. *IEEE Trans. Antennas Propag.* **2024**, *72*, 4271–4282. [CrossRef]
43. Xu, W.; Deng, Z.; Huang, P.; Tan, W.; Gao, Z. Beampattern Synthesis and Optimization for Frequency Diverse Arc Array Based on the Virtual Element. *Electronics* **2023**, *12*, 2231. [CrossRef]
44. Xu, W.; Tian, Y.; Huang, P.; Tan, W.; Qi, Y. Frequency Diversity Arc Array with Angle-Distance Two-Dimensional Broadening Null Steering for Sidelobe Suppression. *Electronics* **2024**, *13*, 1640. [CrossRef]
45. Yesilyurt, U.; Arican, G.O.; Aksoy, E. Dot-shaped Transmit Beamforming with Nonuniform Time-modulated Concentric Circular Frequency Diverse Array. *Microw. Opt. Technol. Lett.* **2024**, *66*, e33890. [CrossRef]
46. Yao, A.-M.; Wu, W.; Fang, D.-G. Frequency Diverse Array Antenna Using Time-Modulated Optimized Frequency Offset to Obtain Time-Invariant Spatial Fine Focusing Beampattern. *IEEE Trans. Antennas Propag.* **2016**, *64*, 4434–4446. [CrossRef]
47. Cheng, Q.; Zhu, J.; Xie, T.; Luo, J.; Xu, Z. Time-Invariant Angle-Range Dependent Directional Modulation Based on Time-Modulated Frequency Diverse Arrays. *IEEE Access* **2017**, *5*, 26279–26290. [CrossRef]
48. Xu, Y.; Xu, J. Corrections to "Range-Angle-Dependent Beamforming of Pulsed-Frequency Diverse Array". *IEEE Trans. Antennas Propag.* **2018**, *66*, 6466–6468. [CrossRef]
49. Tan, M.; Wang, C.; Li, Z. Correction Analysis of Frequency Diverse Array Radar About Time. *IEEE Trans. Antennas Propag.* **2021**, *69*, 834–847. [CrossRef]
50. Xu, Y.; Shi, X.; Li, W.; Xu, J.; Huang, L. Low-Sidelobe Range-Angle Beamforming with FDA Using Multiple Parameter Optimization. *IEEE Trans. Aerosp. Electron. Syst.* **2019**, *55*, 2214–2225. [CrossRef]
51. Xu, J.; Liao, G.; Huang, L.; So, H.C. Robust Adaptive Beamforming for Fast-Moving Target Detection with FDA-STAP Radar. *IEEE Trans. Signal Process.* **2017**, *65*, 973–984. [CrossRef]
52. Xu, Y.; Huang, X.; Wang, A. Transmit–Receive Sparse Synthesis of Linear Frequency Diverse Array in Range-Angle Space Using Genetic Algorithm. *Sensors* **2023**, *23*, 3107. [CrossRef]
53. Haupt, R.L. Adaptively Thinned Arrays. *IEEE Trans. Antennas Propag.* **2015**, *63*, 1626–1632. [CrossRef]
54. Agha, M.H.; Al-Adwany, M.A.S.; Bayat, O.; Hamdoon, H.T. Optimization of Antenna Array Pattern for Uniformly Excited Rectangular Array via a Thinning Method. *J. King Saud Univ. Eng. Sci.* **2022**, *34*, 544–561. [CrossRef]
55. Shao, X.L.; Hu, T.Y.; Li, L.; Xiao, Z.L.; Rong, Y.J. Sparse Multi-Carrier Frequency Diverse Array Transmit Beampattern Optimization. In Proceedings of the 2021 Photonics & Electromagnetics Research Symposium (PIERS), Hangzhou, China, 21–25 November; IEEE: Hangzhou, China, 2021; pp. 1620–1625.
56. Yang, Y.-Q.; Wang, H.; Wang, H.-Q.; Gu, S.-Q.; Xu, D.-L.; Quan, S.-L. Optimization of Sparse Frequency Diverse Array with Time-Invariant Spatial-Focusing Beampattern. *Antennas Wirel. Propag. Lett.* **2018**, *17*, 351–354. [CrossRef]
57. Zhai, W.; Wang, X.; Greco, M.S.; Gini, F. Joint Optimization of Sparse FDAs for Time Invariant Transmit Beampattern Synthesis. *IEEE Signal Process. Lett.* **2022**, *29*, 110–114. [CrossRef]

58. Shao, X.; Hu, T.; Zhang, J.; Li, L.; Xiao, M.; Xiao, Z. Efficient Beampattern Synthesis for Sparse Frequency Diverse Array via Matrix Pencil Method. *Sensors* **2022**, *22*, 1042. [CrossRef] [PubMed]
59. El-Khamy, S.E.; El-Sayed, H.F.; Eltrass, A.S. Synthesis of Wideband Thinned Eisenstein Fractile Antenna Arrays with Adaptive Beamforming Capability and Reduced Side-Lobes. *IEEE Access* **2022**, *10*, 122486–122500. [CrossRef]
60. Jung, S.-H.; Lee, K.-I.; Oh, H.-S.; Jung, H.-K.; Yang, H.; Chung, Y.-S. Optimal Design of Thinned Array Using a Hybrid Genetic Algorithm. *J. Electromagn. Eng. Sci.* **2021**, *21*, 261–269. [CrossRef]
61. Khan, A.; Roy, J.S. Thinned Smart Antenna of a Semi-Circular Dipole Array for Massive MIMO Systems. *Adv. Electromagn.* **2023**, *12*, 17–25. [CrossRef]
62. Nosier, A.H.; Elkhawaga, A.M.; Nasr, M.E.; Mahmoud, N.M.; Hussein, A.H. Side Lobe Level Reduction and Array Thinning of Concentric Circular Antenna Arrays. *Mansoura Eng. J.* **2024**, *49*, 8. [CrossRef]

remote sensing

Article

A Coffee Plant Counting Method Based on Dual-Channel NMS and YOLOv9 Leveraging UAV Multispectral Imaging

Xiaorui Wang [1], Chao Zhang [2,*], Zhenping Qiang [1], Chang Liu [2], Xiaojun Wei [1] and Fengyun Cheng [2]

1 College of Big Data and Intelligent Engineering, Southwest Forestry University, Kunming 650224, China; wangxiaorui@swfu.edu.cn (X.W.); qzp@swfu.edu.cn (Z.Q.); xgyywxj@gmail.com (X.W.)
2 College of Forestry, Southwest Forestry University, Kunming 650224, China; missliu@swfu.edu.cn (C.L.); fengyun@swfu.edu.cn (F.C.)
* Correspondence: zhchgis@swfu.edu.cn; Tel.: +86-138-8813-4614

Abstract: Accurate coffee plant counting is a crucial metric for yield estimation and a key component of precision agriculture. While multispectral UAV technology provides more accurate crop growth data, the varying spectral characteristics of coffee plants across different phenological stages complicate automatic plant counting. This study compared the performance of mainstream YOLO models for coffee detection and segmentation, identifying YOLOv9 as the best-performing model, with it achieving high precision in both detection (P = 89.3%, mAP50 = 94.6%) and segmentation performance (P = 88.9%, mAP50 = 94.8%). Furthermore, we studied various spectral combinations from UAV data and found that RGB was most effective during the flowering stage, while RGN (Red, Green, Near-infrared) was more suitable for non-flowering periods. Based on these findings, we proposed an innovative dual-channel non-maximum suppression method (dual-channel NMS), which merges YOLOv9 detection results from both RGB and RGN data, leveraging the strengths of each spectral combination to enhance detection accuracy and achieving a final counting accuracy of 98.4%. This study highlights the importance of integrating UAV multispectral technology with deep learning for coffee detection and offers new insights for the implementation of precision agriculture.

Keywords: coffee; plant counting; UAV; multispectral image; YOLOv9; dual-channel NMS

1. Introduction

Coffee, recognized as one of the three major beverage crops globally [1], is cultivated and consumed extensively around the world [2]. In recent years, China has experienced rapid economic growth and rising living standards, leading to a significant shift in coffee consumption from a niche market predominantly enjoyed by the elite to a more widespread demographic. This transition reflects a growing trend toward mass consumption and diversification within the coffee market [3]. According to statistics, the domestic coffee market in China is growing at an annual rate of 15%, and it is expected to reach a market size of CNY 217.1 billion (about USD 30.8 billion) by 2025 [4]. This rapidly growing market demand offers significant development potential for the coffee industry and unprecedented opportunities for production regions.

Yunnan Province, as China's most important coffee-producing region, is particularly renowned for its small-grain coffee from the Baoshan area. The coffee planting area in Baoshan has steadily expanded, reaching 8610 hectares in 2022, with an annual output of 20,000 tons and a total industry output value exceeding CNY 3.7 billion (about USD 524 million), making it an essential production area for high-quality coffee in China [5].

Despite Yunnan Province's significant role in China's coffee industry—accounting for more than 98% of the nation's coffee planting area and production [6]—the region still faces challenges such as insufficient technological support, limited variety of coffee plants, and low resilience to risks. These issues have, to some extent, hindered the further development of Yunnan's coffee industry. Therefore, enhancing the technological content of the coffee industry, optimizing planting and processing techniques, and increasing the added value at each stage of the industrial chain have become key to promoting the high-quality development of Yunnan's coffee industry.

One critical aspect of improving the coffee industry is estimating coffee yields. Due to the unique mountainous coffee-growing environment in the Baoshan area of Yunnan, coffee farmers find it challenging to accurately estimate coffee yields, which affects their income. Research teams have discovered that the biennial cycle of coffee production in this area can be easily identified, and within the same planting area and growth cycle there is little difference in coffee yields. Therefore, the key factor affecting coffee yield lies in the number of coffee trees per unit area. To estimate coffee yield more accurately, it is essential to precisely count the number of coffee trees in the planting area.

Traditional crop counting is typically conducted manually, which is a labor-intensive, time-consuming, and expensive task, especially in steeply sloped mountainous areas like Baoshan in Yunnan. Current technologies, such as the Internet of Things (IoT) and Artificial Intelligence (AI), and their applications, can be fully integrated into the agricultural sector to ensure long-term agricultural productivity [7]. These technologies have the potential to enhance global food security by reducing crop yield gaps, minimizing food waste, and addressing issues of inefficient resource use [7]. Additionally, computer vision offers a new, non-destructive method for estimating crop yields [8]. During crop monitoring, the ideal images are orthophotos [9], which can be obtained primarily through satellite remote sensing and UAV remote sensing. Low-altitude UAVs, due to their lower flight height, can capture higher spatial resolution images and are not affected by cloud cover [10]. As a result, UAVs are garnering increasing attention in the field of crop monitoring.

UAVs have the capability to capture images across various time intervals, and when equipped with different sensors they can also acquire spectral data that reflect various aspects of crop growth [11–14]. These spectral indices allow for the calculation of vegetation indices, with each index highlighting specific aspects of plant health [15], thereby enhancing the monitoring of crop development. Different spectral bands provide unique insights into crop conditions. For instance, the Near-infrared (NIR) band reveals photosynthetic capacity and water content, both of which are indicators of plant health [16], making it a valuable tool for assessing plant vitality. The Red edge (RE) band, which tracks changes in chlorophyll absorption, can indicate various phenological stages of crops [17]. By combining data from different bands, it is possible to gather more comprehensive information about crop growth than what is achievable with standard RGB imagery.

With the rapid advancement of Artificial Intelligence in image processing, estimating crop counts using imagery has become feasible, significantly improving the accuracy of crop count estimation [18]. Artificial Intelligence is already widely used in various agricultural monitoring tasks, such as pest and disease identification, and yield estimation [19]. Deep learning, which requires large datasets to train models for optimal monitoring performance [20], benefits greatly from the use of UAVs, as they can easily capture vast amounts of data anytime and anywhere [21]. As a result, the integration of computer vision and deep learning in agricultural monitoring has garnered increasing attention and is poised to become a primary method for agricultural monitoring in the future.

The key challenge in coffee monitoring is distinguishing coffee plants from other intercropped crops, such as macadamia nuts and mangoes. The fundamental issue in counting coffee plants lies in accurate detection. Target detection methods can be divided into two categories: anchor-based and anchor-free methods [22]. In multi-object detection tasks, anchor-based algorithms utilize a pre-defined set of anchor boxes to cover a wider range of possible target locations and sizes, thereby improving detection accuracy and recall rates [23,24]. Anchor-based methods, which are more common, can be further classified into single-stage and two-stage approaches [25]. Single-stage methods directly perform object detection without explicitly proposing regions of interest (RoIs). These methods divide the image into grids and predict the presence of objects and their bounding box coordinates in each grid cell. Representative works include the YOLO (You Only Look

Once) series, SSD (Single Shot MultiBox Detector), and RetinaNet, among others [26]. The YOLO framework stands out for its balance between accuracy and speed, making it a reliable choice for rapid object identification in images. YOLO employs a regression-based approach to directly predict the output for detected objects, resulting in probability values for the target [27].

The popular object detection algorithm YOLO, combined with multispectral UAV technology, offers new insights for modern agricultural monitoring and production.

2. Materials and Methods

2.1. Overview of the Research Area

The study area (Figure 1) is located at Xinzai Farm, situated in Xinzai Village, in the southwest of Longyang District, Baoshan City, Yunnan Province. The farm spans an area of 5.36 square kilometers, with an elevation of 700 m above sea level. The geographic coordinates of the area place it within a region that experiences an average annual temperature of 21.5 °C and an average annual rainfall of 850 mm. The primary coffee variety grown here is Catimor (the variety studied in this research), with a smaller proportion of Typica also cultivated.

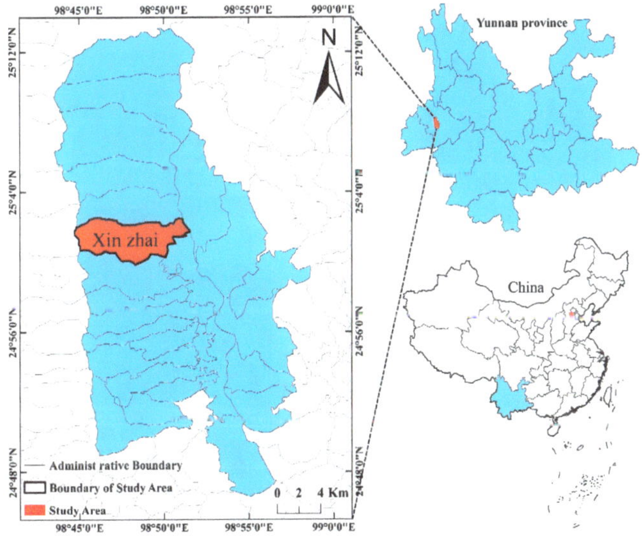

Figure 1. Administrative map of the study area.

2.2. Data and Preprocessing

2.2.1. Equipment

The research team utilized a DJI MAVIC 3M multispectral UAV, sourced from DJI Technology Co., Ltd., Shenzhen, China, capable of capturing data across several spectral bands: Green (G) at 560 nm ± 16 nm, Red (R) at 650 nm ± 16 nm, Red edge (Re) at 730 nm ± 16 nm, Near-infrared (Nir) at 860 nm ± 26 nm, and a full-color (RGB) band at 380–780 nm. The altitude of flight missions ranged between 30 and 100 m, with a 70% front overlap, using an orthographic imaging approach.

The hardware configuration for this study was as follows: the CPU used was an Intel(R) Core(TM) i9-14900KF at 3.20 GHz (Intel Corporation, Santa Clara, CA, USA), with 64 GB of memory, and the GPU was an NVIDIA GeForce RTX 4090 with 24 GB of computational capacity (NVIDIA Corporation, Santa Clara, CA, USA).

2.2.2. Data Collection

Data collection was conducted four times, in April 2023, December 2023, January 2024, and April 2024, during different phenological stages of coffee plants in the study area. This allowed us to capture images of the coffee plants during flowering, growth, fruit maturation, and post-harvest stages. For each time, 20 flight paths were set, covering a total area of 1.83 square kilometers.

2.2.3. Data Preprocessing

The UAV data collected were stitched using Pix4Dmapper software (version 4.5), resulting in 15 mosaic images of sample plots (Figure 2).

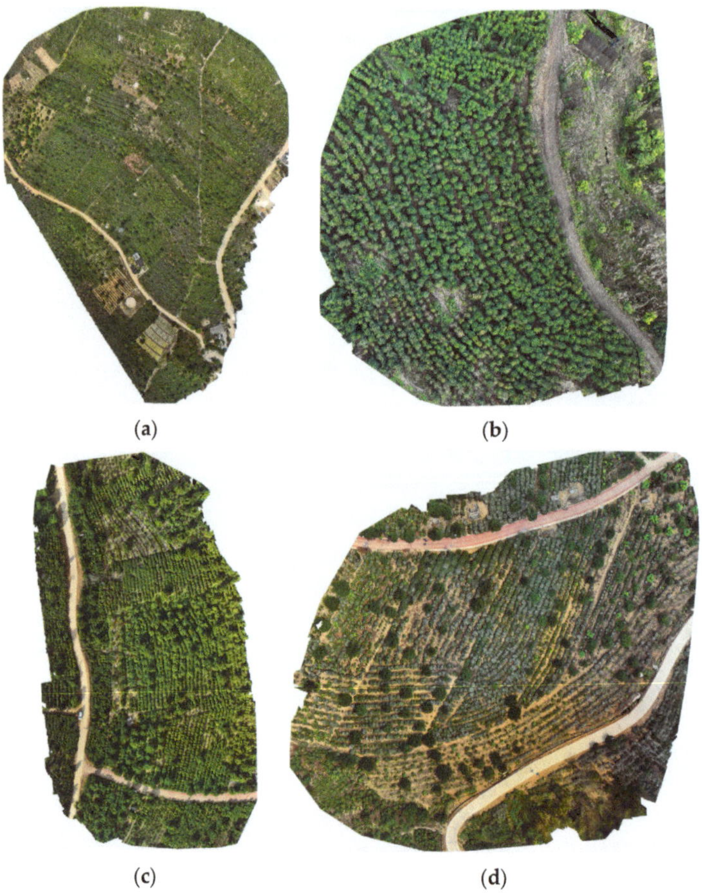

Figure 2. Stitched Images of Coffee at Different Phenological Stages: (**a**) Taken in April 2023, representing the late flowering stage; (**b**) Taken in December 2023, representing the coffee bean maturation stage; (**c**) Taken in January 2024, representing the post-harvest stage; (**d**) Taken in April 2024, representing the full flowering stage.

To enhance the robustness and generalizability of the model, the following criteria for data selection were applied:

1. Mosaic images from areas with a complex terrain and various land features were selected. These included roads, buildings, macadamia trees, mango trees, and coffee trees of different ages;

2. Different sample plot images for each phenological stage were selected, effectively avoiding the issue of having the same number of coffee plants across different phenological stages in the same experimental area.

Ultimately, six mosaic images as experimental data were selected for further processing. Since the YOLO model requires input images to be 640 × 640 pixels [28,29], we cropped the images accordingly. The cropped images were selected based on the following criteria:

1. Images that were clear and free from flight disturbances;
2. Images rich in terrain and land feature information, including various tree species such as mango, macadamia, and young coffee seedlings;
3. Images where coffee tree canopies exhibited overlap.

After filtering, 245 highly representative images were obtained, with a total of 4308 coffee trees. Of these, 1887 were non-flowering coffee trees, and 2421 were flowering coffee trees. Since the number and quality of coffee flowers significantly impact yield and quality, the selection of flowering coffee trees was intentionally increased for the subsequent analysis of coffee yield and quality.

For model training, the dataset was divided into an 80:20 ratio [30], with 80% allocated for training and 20% for validation. The training set contained 3492 coffee trees, including 1552 non-flowering and 1940 flowering trees. The validation set contained 816 coffee trees, with 335 non-flowering and 481 flowering trees labeled.

2.2.4. Spectral Combinations

Since the YOLO model requires a three-channel input, and the multispectral images obtained from the UAV contain five bands, different spectral bands need to be combined into sets of three to meet YOLO's input requirements. When selecting the YOLO model, standard RGB images typically employed in YOLO were used. For coffee plant counting, all spectral combinations were compared to determine the optimal spectral combination for accurately counting coffee plants (Table 1).

Table 1. UAV Band Combinations and Their Descriptions.

Band Combinations	UAV Bands	Wave Length
G3	Green, Green, Green	
N3	Nir, Nir, Nir	
R3	Red, Red, Red	Green: 560 nm ± 16 nm
Re3	Red edge, Red edge, Red edge	Nir: 860 nm ± 26 nm
RNRe	Red, Nir, Red edge	Red: 650 nm ± 16 nm
NGRe	Nir, Green, Red edge	Red edge: 730 nm ± 16 nm
RGN	Red, Green, Nir	
RGRe	Red, Green, Red edge	
RGB	A full-color (RGB) band	Red: 620–750 nm Green: 495–570 nm Blue: 450–495 nm

2.3. Methods

2.3.1. YOLOs

"You Only Look Once" (YOLO) is an object detection framework that uses a single convolutional neural network (CNN) to predict bounding boxes and class probabilities directly from full images in a single pass. It is widely known for its speed and accuracy. The architecture divides the image into a grid and assigns bounding boxes to detect objects within each grid cell.

Up to now, YOLO has evolved from YOLOv1 to YOLOv10 (with the source code for YOLOv10 yet to be released). Among the most widely used versions are YOLOv5 [31], YOLOv7 [32], and YOLOv8 [33,34]. This study compares the structural differences between these commonly used YOLO versions and the latest YOLOv9 [35], using RGB spectral data from UAV imagery to assess their performance in coffee plant detection.

Starting from YOLOv3 [36], the YOLO architecture is divided into three parts: backbone, neck, and head.

- Backbone: This is a convolutional neural network responsible for extracting key multi-scale features from the image. Shallow features, such as edges and textures, are extracted in the early stages of the network, while deeper layers capture high-level features like objects and semantics. YOLOv5 uses a variant of CSPNet (Channel-wise Spatial Pyramid Network) as its backbone [37], making the network lighter and optimizing inference speed. YOLOv7 employs an improved Efficient Layer Aggregation Network (ELAN) as its backbone [38], enhancing feature extraction through a more efficient layer aggregation mechanism. YOLOv8 utilizes a similar backbone to YOLOv5 but with an improved CSPLayer [39], thereby enhancing feature representation capabilities.
- Neck: Positioned between the backbone and head, the neck is responsible for feature fusion and enhancement. It extracts multi-scale features to detect objects of varying sizes, making it ideal for identifying coffee trees of different ages. YOLOv5's neck retains the Path Aggregation Network (PAN) structure, similar to YOLOv4 [40], but with lightweight improvements to optimize the speed of feature fusion. YOLOv7 enhances the model's ability to perceive multi-scale objects, maintaining a focus on small targets during feature aggregation. YOLOv7 strikes a better balance between efficiency and accuracy [41]. YOLOv8 employs the PANet structure, allowing for both top–down and bottom–up feature fusion [42], increasing the receptive field and enriching semantic information.
- Head: The head is the final part of the network, responsible for predicting the object's class, location, and confidence score based on the multi-scale features provided by the neck. YOLOv5 and YOLOv7 optimized the detection head to be more efficient while retaining high accuracy [43,44]. YOLOv8 introduced a new decoupled head structure that separates classification and regression tasks [45], improving the model's flexibility and stability.

YOLOv9 (Figure 3) focuses on solving the issue of information loss (the "information bottleneck") caused by various downscaling operations during the feedforward process in YOLOv7 [46]. YOLOv9 introduces two novel techniques: Programmable Gradient Information (PGI) and General Efficient Layer Aggregation Network (GELAN). These innovations not only resolve the information bottleneck but also push the boundaries of object detection accuracy and efficiency.

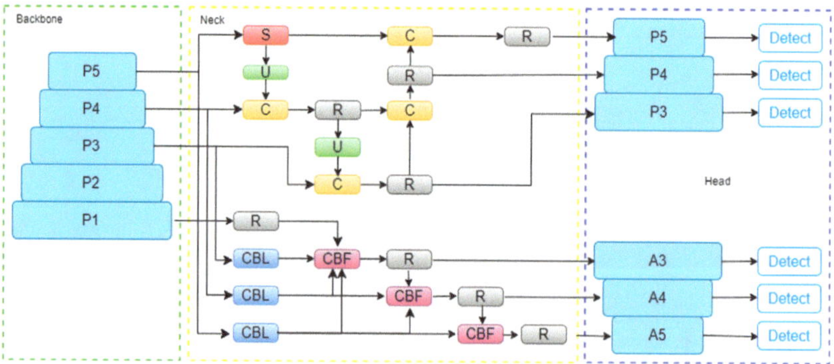

Figure 3. The framework of YOLOv9.

In Figure 3, the backbone layer operates as follows: P1 refers to the input image after two convolution operations; P2 indicates that P1 has been processed by the Rep-

NCSPELAN4 module. P3 is the result of applying a convolution operation and RepNCSPELAN4 to P2; similarly, P4 represents P3 processed with convolution and RepNCSPELAN4, and P5 indicates that P4 has undergone the same processing.

In the neck layer of Figure 3, R represents the RepNCSPELAN4 structure, CBL stands for CBLinear, CBF for CBFuse, C denotes concatenation (Concat), U for Upsample, and S represents the SPPELAN structure.

In the head layer of Figure 3, P3 refers to the output of the neck layer's P3 used as the primary input, while P4 and P5 represent the outputs of P4 and P5 in the neck layer, respectively. Likewise, A3 refers to the neck layer's P3 output used as the primary input, with A4 and A5 corresponding to P4 and P5. However, the processing of A3, A4, and A5 in the head layer differs from that of P3, P4, and P5 in the neck layer.

2.3.2. Dual-Channel NMS

Non-maximum suppression (NMS) [47] is an important post-processing step used in the YOLO object detection framework [48]. Its primary purpose is to filter out redundant or overlapping bounding boxes, ensuring that each detected object is represented by a single, high-quality bounding box.

When processing multispectral data, YOLO can only accept three-channel inputs. Therefore, different spectral bands were combined and fed separately into the YOLO model for detection. It was observed that different spectral combinations during various phenological stages had a significant impact on YOLO's detection performance. Specifically, RGB imagery was more effective for detecting growth stage images, while RGN imagery was better suited for detecting flowering stage images. Building on the strong performance of traditional NMS in YOLO, a dual-channel non-maximum suppression (dual-channel NMS) method is proposed to further enhance detection accuracy.

Dual-channel non-maximum suppression refers to the process of combining object detection results from YOLO across two different spectral combinations: RGB and RGN. By feeding the detection outputs from both channels into the NMS algorithm, redundant bounding boxes are removed, and overlapping detections are merged (Figure 4).

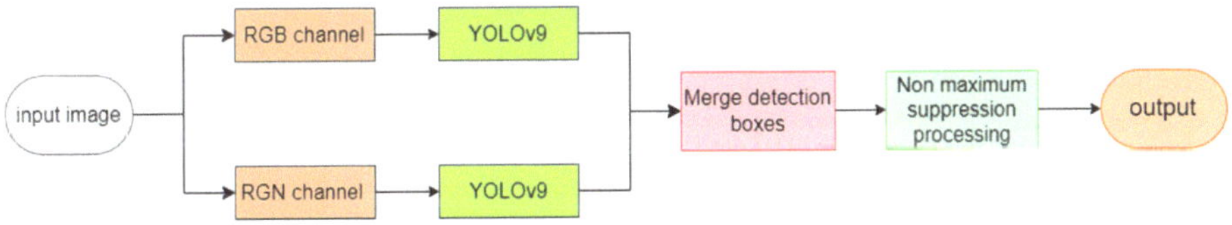

Figure 4. The framework of dual-channel NMS.

This method enhances the accuracy of coffee plant detection by ensuring more precise identification of individual coffee plants, leveraging the complementary information from both spectral combinations.

2.3.3. Workflow

The entire workflow of this study is illustrated in Figure 5. First, the RGB spectral composite images from the multispectral dataset captured by UAVs were fed into mainstream YOLO models for comparative analysis of coffee recognition performance, identifying the model that achieved the best results. Next, the multispectral data obtained from UAVs were combined to generate multiple sets of three-channel data, which were then input into the optimal YOLO model for the recognition and segmentation of coffee plants. A comparison of the recognition and segmentation performance across different spectral combinations revealed that RGB and RGN emerged as the two best combinations, each exhibiting distinct advantages. Finally, the segmentation results from both the RGB and RGN combinations were merged using the dual-channel non-maximum suppression method proposed in this study, leading to more accurate statistics of plant counts.

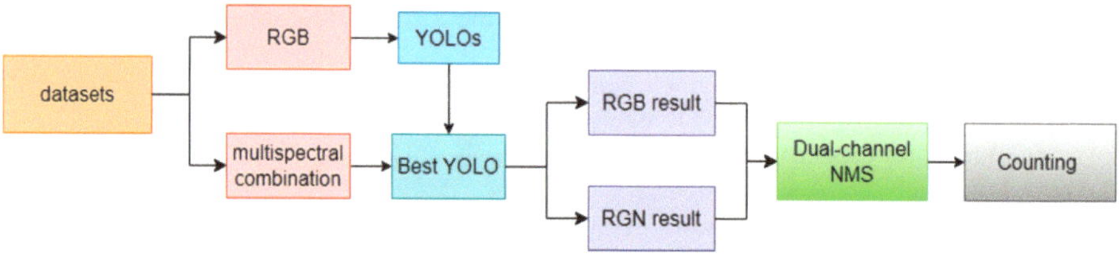

Figure 5. The workflow of this study.

2.4. Metrics

In this study, three common evaluation metrics were employed: recall (R), precision (P), and mean average precision (mAP). Additionally, for the evaluation of coffee plant counting, the crucial counting accuracy (CA) metric was used.

Recall refers to the proportion of true positive samples that the model successfully predicts as positive out of all actual positive samples, which is computed using Formula (1). It measures the model's ability to identify positive instances. A high recall value indicates that the model has a low rate of missing relevant objects and is effective at capturing most of the positive instances.

$$R(Recall) = \frac{TP}{TP + FN} \tag{1}$$

where TP represents true positives, and FN represents false negatives.

Precision refers to the proportion of true positive samples out of all samples predicted as positive by the model, as shown in Formula (2). It measures the accuracy of the model's positive predictions. A high precision value indicates a low probability of mistakenly identifying background or non-target objects as positive targets.

$$P(Precision) = \frac{TP}{TP + FP} \tag{2}$$

where TP represents true positives, and FP represents false positives.

When calculating the mean average precision (mAP), both the mAP50 and mAP50–95 were used.

mAP50 is the average precision calculated for each class at an IoU (Intersection over Union) threshold of 0.5 [49]. IoU measures the overlap between the predicted bounding box and the ground truth box, defined as the area of intersection divided by the area of union of both boxes. For a prediction to be considered correct, the predicted bounding box must overlap with the ground truth box by more than 50%. mAP50 is a relatively lenient metric and is often used for initial model evaluation.

mAP50–95 averages the mAP values at IoU thresholds ranging from 0.50 to 0.95 in increments of 0.05. This is a stricter and more comprehensive metric, assessing the model's performance across various IoU thresholds. It provides a more reliable evaluation, especially for detecting small objects, such as coffee seedlings.

Additionally, for evaluating the coffee plant counting results, the counting accuracy metric was used. This metric is straightforward, direct, and highly persuasive, which is computed using Formula (3):

$$CountAccuracy = \frac{Cnp}{Tnp} \tag{3}$$

where Cnp represents the correctly detected number of coffee plants, and Tnp represents the total number of detected plants.

3. Results

3.1. Models and Parameters

Currently, the mainstream YOLO models for object detection include YOLOv5, YOLOv7, YOLOv8, and YOLOv9, each with different strengths in detection tasks. For example, YOLOv8 is more lightweight, while YOLOv9 is better suited for detecting small objects. Unlike crops such as rice or tobacco, coffee tree cultivation is more complex. The environment includes non-agricultural features such as roads and buildings, as well as intercropped trees like mango and macadamia. Additionally, coffee plantations often contain trees of varying ages within the same plot.

To ensure the rigor of the experiments and considering that the creators of each YOLO version primarily used RGB images for model validation, which provide richer visual information, this study also selected high-resolution RGB images captured by UAV as its validation data. These images were input into different YOLO models for prediction, and their performance was analyzed and compared to determine the optimal model.

After multiple trials, it was observed that for the coffee dataset, the maximum number of training epochs was 811 for YOLOv5m, the minimum was 152 for YOLOv8m, and YOLOv9 required 500 epochs, with an average of 333.17 epochs. Since the models automatically stop training upon convergence and save the best results, the epoch value was set to 1000 for consistency. In addition to the epoch setting, key parameters such as batch size and workers were also consistent across all models (Table 2).

Table 2. Parameter for models.

Parameter	Value	Explain
epoch	1000	Number of complete passes through the training data
batch	16	Size of the data subset processed in one pass
workers	16	Number of sub-processes used for data loading

3.2. Optimal Model

When selecting the model, this study did not divide the coffee dataset by phenological stages; in other words, the dataset contained a mix of growth, flowering, and fruiting stages to make the model more generalizable. However, to facilitate future research, the coffee plants were categorized during the annotation process into flowering and non-flowering groups. Flowering coffee plants were labeled with red bounding boxes, while non-flowering coffee plants were labeled with blue bounding boxes. Experimental comparisons were conducted across different models in terms of both coffee detection and segmentation.

When comparing the coffee detection performance, varying results were observed across different models (Table 3).

Table 3. Coffee Detection Performance of Different YOLO Models.

Model	P (%)	R (%)	mAP50 (%)	mAP50–95 (%)
YOLOv5l	86.50%	90.40%	92.70%	60.80%
YOLOv5m	84.40%	91.60%	93.50%	63.80%
YOLOv5n	88.80%	89.90%	94.30%	59.70%
YOLOv5s	88.80%	88.30%	93.90%	60.90%
YOLOv5x	87.40%	90.50%	93.70%	61.00%
YOLOv7	85.20%	92.20%	93.70%	63.50%
YOLOv8l	86.70%	90.10%	94.50%	61.60%
YOLOv8m	85.90%	87.90%	93.90%	61.50%
YOLOv8n	88.10%	88.00%	93.90%	61.70%
YOLOv8s	88.00%	90.40%	94.10%	63.80%
YOLOv8x	86.10%	86.00%	93.00%	61.50%
YOLOv9	89.30%	87.80%	94.60%	64.60%

We selected 12 models from YOLOv5 to YOLOv9 for experimentation. The highest precision was achieved by YOLOv9 at 89.30%, while the lowest was YOLOv5m at 84.40%. The highest recall was attained by YOLOv7 at 92.20%, with the lowest being YOLOv8x at 86.00%. In terms of mAP50, YOLOv9 achieved the highest score of 94.60%, while YOLOv8x had the lowest at 93.00%. For mAP50–95, YOLOv9 again performed the best with a score of 64.60%, while YOLOv5m had the lowest at 63.80%. Overall, YOLOv9 demonstrated the best comprehensive performance.

Similarly, when comparing coffee segmentation results, it was found that different models exhibited varying performances. Additionally, there was no direct correspondence between coffee detection and segmentation performance (Table 4).

Table 4. Coffee Segmentation Performance of Different YOLO Models.

Model	P (%)	R (%)	mAP50 (%)	mAP50–95 (%)
YOLOv5l	86.20%	90.20%	92.30%	58.30%
YOLOv5m	84.60%	91.30%	93.40%	59.90%
YOLOv5n	87.90%	88.90%	93.00%	54.30%
YOLOv5s	88.30%	88.80%	93.80%	57.70%
YOLOv5x	87.50%	90.50%	93.80%	58.70%
YOLOv7	85.30%	92.30%	94.20%	60.80%
YOLOv8l	87.00%	90.40%	94.80%	57.80%
YOLOv8m	86.20%	87.40%	93.50%	58.60%
YOLOv8n	88.10%	88.40%	93.90%	58.00%
YOLOv8s	87.60%	90.80%	93.60%	61.10%
YOLOv8x	85.70%	85.50%	93.10%	58.70%
YOLOv9	88.90%	89.10%	94.80%	60.80%

Using the same approach as in coffee detection, we selected 12 models from YOLOv5 to YOLOv9 to perform single-plant segmentation on the detected coffee plants. The highest precision was achieved by YOLOv9 at 88.90%, while the lowest was YOLOv5m at 84.60%. The highest recall was achieved by YOLOv7 at 92.30%, with the lowest being YOLOv8x at 85.50%. For mAP50, both YOLOv9 and YOLOv8l achieved the highest score of 94.80%, while YOLOv5l had the lowest at 92.30%. For mAP50–95, YOLOv8s performed the best with 61.10%, and YOLOv5n the lowest at 54.30%. Thus, YOLOv9 demonstrated the best overall performance.

In conclusion, based on both detection and segmentation results, YOLOv9 exhibits the best and most stable overall performance. It is particularly effective in recognizing coffee plants at different growth stages (Figure 6) and segmenting individual coffee plants (Figure 7).

Figure 6 shows the identification of coffee plants at different phenological stages. The results demonstrate that the YOLOv9 model effectively detects coffee plants across various stages with minimal interference from other vegetation. However, misclassification can occur during the flowering stage, as seen in Figure 6d, where the bounding box for flowering coffee, which should be red, is mistakenly labeled as blue (indicating non-flowering coffee), and the model assigns it a low confidence score of only 0.28.

Figure 7 presents the segmentation results following the identification of coffee plants in Figure 6. It is clear that the YOLOv9 model performs well in segmenting individual coffee plants across various phenological stages, largely due to the accuracy of prior detection.

However, identification errors can occur during this process. For instance, in Figure 7d the coffee plant within the blue bounding box is misidentified as non-flowering, while manual observation confirms that it is actually in a flowering state. Therefore, during the coffee flowering phenological stage, if manual checks are not conducted and only the model is used for the automatic segmentation and counting of flowering coffee plants, missed detections may occur. This highlights the potential errors when using a single spectral combination for classifying coffee at different phenological stages and counting the plants.

Consequently, this study employs a dual-channel non-maximum suppression method in subsequent experiments to address this issue.

Figure 6. YOLOv9 Detection of Coffee Plants at Different Phenological Stages: (**a**) Coffee seedlings, roughly 12 months old, alongside weeds; (**b**) Flowering coffee, about 3 years old, with intercropped plants; (**c**) Fruiting coffee, around 4 years old, with intercropped plants; (**d**) Flowering coffee, approximately 3 years old, misclassified as non-flowering.

3.3. *Optimal Spectral Band Combination*

As discussed in Section 3.2, the YOLOv9 model is better suited for coffee detection and single-plant segmentation, with varying effectiveness across different phenological stages. To accurately identify coffee plants at different stages, various multispectral band combinations from UAV data were tested using the YOLOv9 model. The goal was to identify the optimal spectral combination that could accurately detect coffee plants in all phenological stages.

For this purpose, the coffee dataset was categorized into flowering, non-flowering, and mixed data, applied all spectral combinations, and then tested their performance in coffee detection and single-plant segmentation using the YOLOv9 model (Tables 5–7).

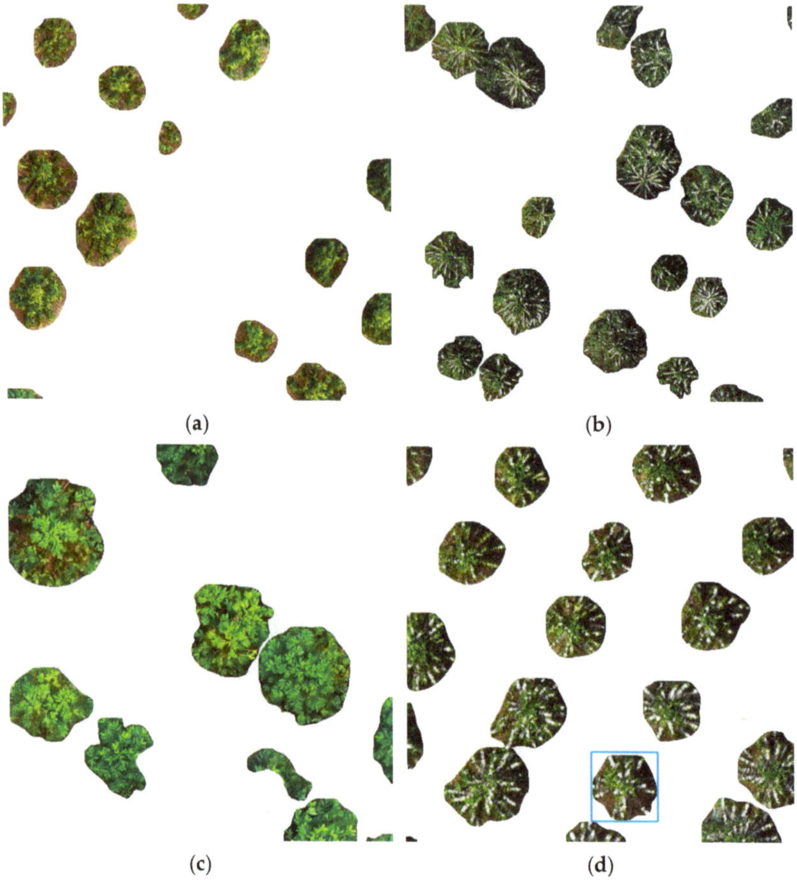

Figure 7. YOLOv9 Segmentation of Coffee Plants at Different Phenological Stages: (a) Segmentation results for the seedling stage; (b) Segmentation results for the flowering stage; (c) Segmentation results for the fruiting stage; (d) Segmentation results for the flowering stage.

Table 5. Performance Comparison of Different Band Combinations on Mixed Data Using YOLOv9.

Band Combinations	Category	Coffee Detection				Single-Plant Segmentation			
		P (%)	R (%)	mAP 50 (%)	mAP 50–95 (%)	P (%)	R (%)	mAP 50 (%)	mAP 50–95 (%)
G3	Mixed	87.20	89.00	92.10	60.60	87.60	89.20	92.60	57.60
N3	Mixed	87.60	85.00	90.60	59.80	88.10	85.50	91.20	57.20
R3	Mixed	88.10	82.80	90.70	59.40	88.40	83.30	91.30	56.70
Re3	Mixed	87.50	87.40	92.00	61.70	87.90	87.70	92.10	58.50
RNRe	Mixed	85.70	87.30	93.10	64.10	85.60	87.70	93.30	60.70
NGRe	Mixed	87.10	87.70	92.60	64.00	87.50	88.10	92.80	61.10
RGN	Mixed	89.10	88.20	94.70	65.30	89.20	88.30	94.40	61.40
RGRe	Mixed	87.20	87.60	93.40	64.40	86.70	88.90	92.90	60.50
RGB	Mixed	89.30	87.80	94.60	64.60	88.90	89.10	94.80	60.80

Table 6. Performance Comparison of Different Band Combinations on Flowering Data Using YOLOv9.

Band Combinations	Category	Coffee Detection				Single-Plant Segmentation			
		P (%)	R (%)	mAP 50 (%)	mAP 50–95 (%)	P (%)	R (%)	mAP 50 (%)	mAP 50–95 (%)
G3	Flower	86.60	90.30	93.70	61.20	87.10	90.40	93.70	57.60
N3	Flower	87.20	85.00	89.70	56.30	87.20	85.00	90.00	53.20
R3	Flower	88.90	87.70	93.50	63.10	88.50	87.60	93.20	59.20
Re3	Flower	86.70	87.10	91.80	58.90	87.20	87.50	91.90	55.30
RNRe	Flower	88.40	87.90	94.80	63.80	88.40	87.90	94.90	59.80
NGRe	Flower	86.40	90.10	93.80	63.00	86.60	90.40	93.40	59.90
RGN	Flower	89.30	88.20	95.50	65.60	89.30	88.20	94.90	61.40
RGRe	Flower	90.30	85.90	94.30	64.10	90.00	87.70	93.70	59.40
RGB	Flower	91.00	91.10	96.50	67.20	90.50	92.10	96.20	62.70

Table 7. Performance Comparison of Different Band Combinations on Non-flowering Data Using YOLOv9.

Band Combinations	Category	Coffee Detection				Single-Plant Segmentation			
		P (%)	R (%)	mAP 50 (%)	mAP 50–95 (%)	P (%)	R (%)	mAP 50 (%)	mAP 50–95 (%)
G3	Non-flowering	87.80	87.70	90.60	60.00	88.00	87.90	91.50	57.70
N3	Non-flowering	88.10	85.10	90.60	59.80	88.10	85.50	91.20	57.20
R3	Non-flowering	87.30	77.90	87.80	55.70	88.30	79.00	89.30	54.30
Re3	Non-flowering	88.30	87.60	92.20	64.40	88.60	87.90	92.30	61.60
RNRe	Non-flowering	82.90	86.70	91.40	64.40	82.80	87.40	91.60	61.50
NGRe	Non-flowering	87.70	85.20	91.50	65.00	88.30	85.90	92.10	62.40
RGN	Non-flowering	88.80	88.10	93.90	65.00	89.10	88.40	94.00	61.40
RGRe	Non-flowering	84.20	89.30	92.50	64.70	83.40	90.10	92.00	61.70
RGB	Non-flowering	87.70	84.40	92.70	62.00	87.40	86.10	93.30	58.90

When the mixed dataset was input into the YOLOv9 model (Table 5), the spectral combination with the highest detection precision was RGB, achieving 89.30%, while the lowest was RNRe at 85.70%. The highest recall was achieved by G3 at 89.00%, and the lowest by R3 at 82.80%. For mAP50, RGN performed the best at 94.70%, while N3 was the lowest at 90.60%. For mAP50–95, RGN was the highest at 65.30%, while R3 was the lowest at 59.40%. In terms of single-plant counting, the highest precision was achieved by RGN at 89.20%, while RNRe had the lowest at 85.60%. The highest recall was achieved by G3 at 89.20%, and the lowest by R3 at 83.30%. For mAP50, RGB had the highest score of 94.80%, while N3 was the lowest at 91.20%. For mAP50–95, RGN performed the best at 61.40%, while R3 had the lowest at 56.70%. Overall, RGN proved to be the slightly superior spectral combination.

Next, the flowering dataset was tested with the YOLOv9 model (Table 6). The results showed that for both coffee detection and single-plant segmentation in the flowering dataset, the RGB combination outperformed other spectral combinations, with N3 performing the worst. The RGN combination did not show any significant advantages.

Finally, the non-flowering dataset was tested with the YOLOv9 model (Table 7). For coffee detection, the highest precision was achieved by RGN at 88.80%, and the lowest by Re3 at 88.30%. The highest recall was achieved by RGRe at 89.30%, while the lowest was R3 at 77.90%. For mAP50, RGN achieved the highest score at 93.90%, while R3 had the lowest at 87.80%. Both RGN and NGRe achieved the highest mAP50–95 scores at 65.00%, with R3 having the lowest at 55.70%. In single-plant counting, RGN again had the highest precision at 89.10%, while RNRe had the lowest at 82.80%. The highest recall was achieved by RGRe at 90.10%, while R3 had the lowest at 79.00%. For mAP50, RGN had the highest score at

94.00%, while R3 had the lowest at 89.30%. NGRe had the highest mAP50–95 at 62.40%, while R3 had the lowest at 54.30%. Overall, RGN clearly outperformed the other spectral combinations, while R3 performed the worst. RGB did not show significant advantages or disadvantages.

In conclusion, different spectral combinations exhibit varying effectiveness for different phenological stages of coffee plants. The RGB combination is most effective for detecting flowering coffee, while the RGN combination is more effective for detecting non-flowering coffee.

To visually compare the RGB and RGN spectral combinations, the precision–recall (PR) curves were plotted to illustrate their performance in coffee detection (Figure 8). In the PR curve, the horizontal axis represents the recall, while the vertical axis represents the precision. Generally, as recall increases, precision tends to decrease, and vice versa. The PR curve reflects this trade-off. The closer the curve is to the top-right corner, the better the model is at maintaining both high precision and high recall, indicating more accurate predictions. Conversely, curves closer to the bottom-left corner suggest that the model struggles to balance precision and recall, resulting in less accurate predictions.

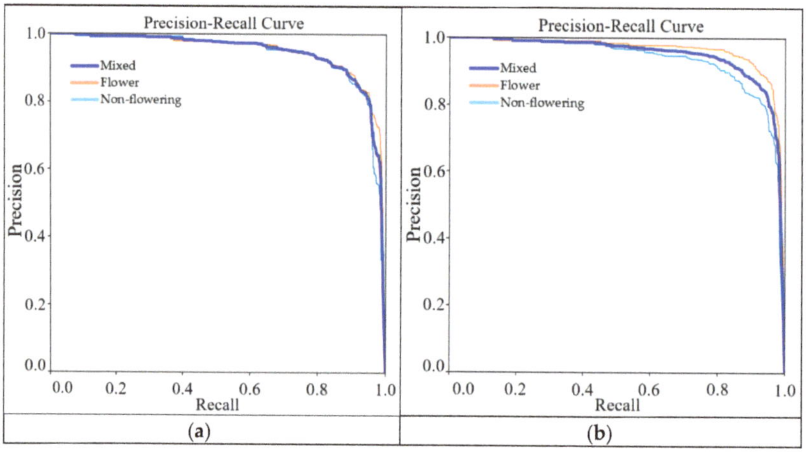

Figure 8. PR Curve: (**a**) RGN spectral combination; (**b**) RGB spectral combination.

It is clear from the curves that the RGN combination exhibits greater stability across different tasks, with minimal performance variation in coffee detection at various phenological stages. In contrast, the RGB combination shows less stability, with more significant performance fluctuations. The RGB combination performs best for detecting flowering coffee, even surpassing RGN in this specific task. Therefore, from a performance standpoint, the two spectral combinations demonstrate different strengths in different detection tasks.

3.4. *Coffee Counting*

From the experiments above, it was found that different spectral combinations produce varying performances in coffee detection and single-plant segmentation. There are several reasons for this. First, during detection coffee trees with low confidence scores are discarded. Second, large coffee canopies may lead to duplicated detections (Figure 9). Additionally, missed detections can occur, all of which directly affect the accuracy of coffee plant counting.

In Figure 9, blue boxes represent detection errors, with generally low confidence scores, meaning these would be discarded during segmentation. In Region A of Figure 9A, one coffee tree was detected as two, with a smaller box inside a larger one. In Region B of Figure 9B, an area was missed entirely by the RGB spectral combination. These issues contribute to inaccurate plant counting.

Figure 9. Detection Results of Different Spectral Combinations: (**A**) RGB spectral combination; (**B**) RGN spectral combination.

To accurately count coffee plants, the dual-channel non-maximum suppression method, described in Section 2.3.2, was proposed. This method merges the detection results from both the RGN and RGB spectral combinations and applies non-maximum suppression (NMS) to retain more accurate detections. However, after combining the two channels, overlapping bounding boxes may appear (Figure 10), similar to Region A in Figure 9A.

Figure 10. Overlapping Bounding Boxes After Dual-channel Merging.

To correctly handle overlapping boxes in various cases, multiple experiments were conducted, and the following rule was established: if the overlap area between two boxes exceeds 80% of one box's area or is less than 50% of the other box's area, it is considered a duplicate detection, and only the larger box is retained (Figure 11). Experimental results demonstrated that this method is effective.

(**a**) (**b**)

Figure 11. Results After Dual-channel NMS for Different Spectral Combinations: (**a**) RGB spectral combination; (**b**) RGN spectral combination.

As shown in Figure 11, after applying dual-channel non-maximum suppression, the results from RGB and RGN images are consistent, with a significant reduction in erroneous detections. However, two coffee trees were still missed.

Finally, coffee plant counting was performed using the results obtained after applying dual-channel non-maximum suppression, and these counts were compared with the plant counts from the unprocessed RGB and RGN combinations (Table 8).

Table 8. Coffee Plant Counting Across Different Phenological Stages.

Band Combinations	Category	Detected Plants	Correctly Detected Plants	Count Accuracy (%)
RGN	All	809	781	96.50
RGB		798	776	97.20
RGN + RGB		**826**	**813**	**98.40**
RGN	Flower	422	396	93.80
RGB		420	403	96.00
RGN + RGB		**432**	**422**	**97.70**
RGN	Non-flowering	387	385	99.40
RGB		378	373	98.70
RGN + RGB		**394**	**391**	99.20

As shown in Table 8, the dual-channel non-maximum suppression method proposed in this study achieved an accuracy of 97.70% on the flowering dataset. Although the accuracy on the non-flowering dataset was 99.20%, only 0.2% lower than the highest result of 99.40%, the accuracy on the mixed dataset reached 98.40%, surpassing the other spectral combinations. In fact, the model's performance on the mixed dataset is of greater interest as it enhances the model's generalizability.

4. Discussion

The primary objective of this study is to develop a robust model capable of accurately counting coffee plants at different phenological stages, which is essential for yield estimation. This model is particularly significant for Yunnan, where coffee is cultivated on steep, complex terrains and often intercropped with other plants. The integration of UAV multispectral data for plant counting presents a valuable approach for enhancing coffee detection efficiency, thereby providing a solid technical foundation for future automated harvesting.

Overall, the findings lead to several key conclusions that enrich the understanding of crop plant counting: (a) The experiments demonstrate the exceptional performance of the YOLOv9 model in coffee plant counting. Compared to other mainstream YOLO models, YOLOv9 consistently excels across all phenological stages and under complex intercropping conditions. This adaptability underscores the model's robustness and reliability, making it a valuable asset for coffee plant counting in diverse environments. (b) This research highlights the significant positive impact of multispectral features on model performance. By evaluating various spectral combinations, two critical combinations—RGB and RGN—were identified as essential for enhancing model accuracy in coffee plant counting. (c) The proposed dual-channel non-maximum suppression method allows for the effective utilization of the two most important spectral combinations, substantially improving counting accuracy.

In terms of model selection, while various YOLO variants excel in object detection, this study focused on the unique growth environments and phenological characteristics of coffee. YOLOv9 ultimately demonstrated the best overall performance, offering a balance between accuracy and efficiency. However, the authors recognize that YOLOv9 may not represent the pinnacle of performance. During the writing of this paper, newer versions like YOLOv10 and YOLOv11 were released, which offer improvements in both lightweight design and efficiency. Future research should consider evaluating these models for further optimization.

Regarding the choice of multispectral imagery, the results indicate that multi-band combinations significantly outperform single-band data (e.g., G3, N3, R3, and Re3) in distinguishing coffee from other crops. Traditional RGB bands remain vital, especially during the flowering stage, whereas RGN imagery proves more effective during non-flowering periods. This effectiveness is attributed to the distinctive white color of coffee flowers, facilitating their identification in RGB images. The Near-infrared (NIR) band in the RGN combination provides valuable insights into chlorophyll content and leaf moisture [50], assisting in differentiating non-flowering coffee plants from other crops. Previous studies have largely focused on single three-channel spectral combinations, but this research highlights that different spectral combinations are suited to different phenological stages, offering a new perspective for future crop detection studies.

The dual-channel non-maximum suppression method capitalizes on the strengths of different spectral combinations in coffee detection. Traditional YOLO models are limited to three-channel inputs, restricting the effective use of multispectral data that capture various growth characteristics across different bands. If only three spectral bands are employed, critical information related to crop growth may be overlooked. Moreover, fixed spectral combinations can only represent crop features at specific phenological stages. By merging the most informative spectral combinations—RGB and RGN—the model's performance in detecting, segmenting, and counting coffee plants is significantly enhanced.

To enable comparison with other studies, precision (P), recall (R), average precision (AP), and counting accuracy (CA) were used as performance metrics. These ratio-based metrics enable meaningful cross-study comparisons. The most relevant and recent studies were selected for comparison with the findings of this research (Table 9).

Table 9. Comparison with other studies.

	Lucas Santos Santana [51]	Hong Lin [26]	This Study
P (%)	84.70	**93.60**	89.30
R (%)	80.80	**90.70**	87.80
AP (%)	79.50	90.63	**94.60**
CA (%)	88.40	99.14	**99.20**

In terms of plant counting accuracy, the proposed method may lead to further improvements in the procedures used by other approaches.

For instance, the method proposed by Lucas Santos Santana (2023) [51] focused on counting coffee plants of various age groups across different sample plots. For consistency in comparison, the best results from different age groups were averaged. Since their study utilized only RGB images of 3-month-old and 6-month-old coffee plants, which do not flower, non-flowering coffee data were also employed for comparison. While the performance metrics in this study surpass those reported by Lucas Santos Santana (2023) [51], this research primarily addresses different phenological stages and has not yet focused specifically on seedlings. Coffee seedlings, due to their smaller height and diameter, are more challenging to distinguish from other shrubs. However, the method presented here represents a significant advancement over the metrics established in the earlier work.

Hong Lin's study (2023) [26] investigated tobacco plant counting using multispectral data. Although the crops differ, the methodologies share similarities. As with the previous comparison, the best results from their sample plots were averaged for evaluation. Since the tobacco plants in their study were in the non-flowering stage, non-flowering coffee data were also used for comparison. While Hong Lin's (2023) [26] study reported higher precision and recall, this approach achieved a higher average precision. It is important to recognize that the cultivation of tobacco and coffee occurs under vastly different conditions. Tobacco is typically grown without the complexity of intercropping, and strict standards for plant and row spacing in tobacco cultivation differ from the more flexible practices in coffee farming. These factors illustrate the greater difficulty in accurately counting coffee plants, indicating that the method proposed may contribute to improvements in the techniques described in their research.

5. Conclusions

This study presents a novel approach for coffee plant counting using multispectral data captured by UAVs. The key objectives and outcomes of this study are as follows: (a) We compared the performance of several mainstream YOLO models for coffee detection and segmentation tasks, and identified YOLOv9 as the most effective model, with it achieving high precision in both detection (P = 89.3%, mAP50 = 94.6%) and segmentation (P = 88.9%, mAP50 = 94.8%). YOLOv9 consistently outperformed other models in terms of accuracy and reliability. (b) We evaluated different combinations of multispectral data to determine their effectiveness for coffee detection across various phenological stages. We found that RGB performed best during the flowering stage, while RGN was more suitable for non-flowering periods, indicating that different spectral combinations cater to different growth stages. (c) We proposed and employed an innovative dual-channel non-maximum suppression (NMS) method, which merges detection results from both RGB and RGN spectral data. This method successfully leveraged the strengths of each spectral combination, leading to a significant improvement in detection accuracy, with the final coffee plant counting accuracy reaching 98.4%.

In conclusion, the integration of UAV multispectral data with advanced deep learning models such as YOLOv9 provides a powerful tool for enhancing coffee yield estimation. This method offers significant potential for improving not only coffee plant detection and counting but also a range of crop monitoring tasks, thereby contributing to sustainable and efficient agricultural practices. Future research will focus on further refining the technology to optimize yield estimation and ensure broader applications in precision agriculture.

Author Contributions: Conceptualization, C.Z.; methodology, X.W.(Xiaojun Wei); software, F.C.; validation, C.L.; formal analysis, Z.Q.; writing—original draft preparation, X.W. (Xiaorui Wang); funding acquisition, C.Z. All authors have read and agreed to the published version of the manuscript.

Funding: This research was funded by the National Natural Science Foundation of China (32160405); the Joint Special Project for Agriculture of Yunnan Province, China (202301BD070001-238,202301BD070001-008); and the Department of Education Scientific Research Fund of Yunnan Province, China (2023J0698).

Data Availability Statement: The data used in this research are collected and labeled by ourselves. We will publish all the codes and datasets in this study after the article is accepted https://github.com/legend2588/Coffee-plant-counting.git (accessed on 14 September 2024).

Conflicts of Interest: The authors declare no conflicts of interest.

References

1. Gaspar, S.; Ramos, F. Caffeine: Consumption and Health Effects. In Encyclopedia of Food and Health, 1st ed.; Caballero, B., Finglas, P.M., Toldrá, F., Eds.; Academic Press: 2016; pp. 573–578.
2. Chéron-Bessou, C.; Acosta-Alba, I.; Boissy, J.; Payen, S.; Rigal, C.; Setiawan, A.A.R.; Sevenster, M.; Tran, T.; Azapagic, A. Unravelling life cycle impacts of coffee: Why do results differ so much among studies? *Sustain. Prod. Consum.* **2024**, *47*, 251–266. [CrossRef]
3. Zhu, T. Research on the Current Situation and Development of China's Coffee Market. *Adv. Econ. Manag. Political Sci.* **2023**, *54*, 197–202.
4. China Industry Research Institute. *Annual Research and Consultation Report of Panorama Survey and Investment Strategy on China Industry*; China Industry Research Institute: Shenzhen, China, 2023; Report No. 1875749. (In Chinese)
5. Yunnan Statistics Bureau. *2023 Yunnan Statistical Yearbook*; Yunnan Statistics Bureau: Yunnan, China, 2023. (In Chinese)
6. Li, W.; Zhao, G.; Yan, H.; Wang, C. A Research Report on Yunnan Specialty Coffee Production. *Trop. Agric. Sci.* **2024**, *47*, 31–40. (In Chinese)
7. Alahmad, T.; Neményi, M.; Nyéki, A. Applying IoT Sensors and Big Data to Improve Precision Crop Production: A Review. *Agronomy* **2023**, *13*, 2603. [CrossRef]
8. Xu, D.; Chen, J.; Li, B.; Ma, J. Improving Lettuce Fresh Weight Estimation Accuracy through RGB-D Fusion. *Agronomy* **2023**, *13*, 2617. [CrossRef]
9. Zhang, Y.; Zhao, D.; Liu, H.; Huang, X.; Deng, J.; Jia, R.; He, X.; Tahir, M.N.; Lan, Y. Research hotspots and frontiers in agricultural multispectral technology: Bibliometrics and scientometrics analysis of the Web of Science. *Front. Plant Sci.* **2022**, *13*, 955340. [CrossRef]
10. Ivezić, A.; Trudić, B.; Stamenković, Z.; Kuzmanović, B.; Perić, S.; Ivošević, B.; Budēn, M.; Petrović, K. Drone-Related Agrotechnologies for Precise Plant Protection inWestern Balkans: Applications, Possibilities, and Legal Framework Limitations. *Agronomy* **2023**, *13*, 2615. [CrossRef]
11. Sishodia, R.P.; Ray, R.L.; Singh, S.K. Applications of Remote Sensing in Precision Agriculture: A Review. *Remote Sens.* **2020**, *12*, 3136. [CrossRef]
12. Jiménez-Brenes, F.M.; López-Granados, F.; Torres-Sánchez, J.; Peña, J.M.; Ramírez, P.; Castillejo-González, I.L.; de Castro, A.I. Automatic UAV-based detection of Cynodon dactylon for site-specific vineyard management. *PLoS ONE* **2019**, *14*, e0218132. [CrossRef]
13. Osco, L.P.; dos de Arruda, M.S.; Marcato Junior, J.; da Silva, N.B.; Ramos, A.P.M.; Moryia, É.A.S.; Imai, N.N.; Pereira, D.R.; Creste, J.E.; Matsubara, E.T.; et al. A convolutional neural network approach for counting and geolocating citrus-trees in UAV multispectral imagery. *ISPRS J. Photogramm. Remote Sens.* **2020**, *160*, 97–106. [CrossRef]
14. Bai, X.; Liu, P.; Cao, Z.; Lu, H.; Xiong, H.; Yang, A.; Yao, J. Rice plant counting, locating, and sizing method based on high-throughput UAV RGB images. *Plant Phenomics* **2023**, *5*, 0020. [CrossRef] [PubMed]
15. Barata, R.; Ferraz, G.; Bento, N.; Soares, D.; Santana, L.; Marin, D.; Mattos, D.; Schwerz, F.; Rossi, G.; Conti, L.; et al. Evaluation of Coffee Plants Transplanted to an Area with Surface and Deep Liming Based on multispectral Indices Acquired Using Unmanned Aerial Vehicles. *Agronomy* **2023**, *13*, 2623. [CrossRef]
16. Zeng, L.; Wardlow, B.D.; Xiang, D.; Hu, S.; Li, D. A review of vegetation phenological metrics extraction using time-series, multispectral satellite data. *Remote Sens. Environ.* **2020**, *237*, 111511. [CrossRef]
17. Boegh, E.; Soegaard, H.; Broge, N.; Hasager, C.; Jensen, N.; Schelde, K.; Thomsen, A. Airborne multispectral data for quantifying leaf area index, nitrogen concentration, and photosynthetic efficiency in agriculture. *Remote Sens. Environ.* **2002**, *81*, 179–193. [CrossRef]
18. Lin, H.; Tse, R.; Tang, S.K.; Qiang, Z.P.; Pau, G. The Positive Effect of Attention Module in Few-Shot Learning for Plant Disease Recognition. In Proceedings of the 2022 5th International Conference on Pattern Recognition and Artificial Intelligence (PRAI), Chengdu, China, 19–21 August 2022; IEEE: New York, NY, USA, 2022; pp. 114–120.
19. Wang, X.; Zhang, C.; Qiang, Z.; Xu, W.; Fan, J. A New Forest Growing Stock Volume Estimation Model Based on AdaBoost and Random Forest Model. *Forests* **2024**, *15*, 260. [CrossRef]
20. Alkhaldi, N.A.; Alabdulathim, R.E. Optimizing Glaucoma Diagnosis with Deep Learning-Based Segmentation and Classification of Retinal Images. *Appl. Sci.* **2024**, *14*, 7795. [CrossRef]
21. Bouachir, W.; Ihou, K.E.; Gueziri, H.E.; Bouguila, N.; Belanger, N. Computer vision system for automatic counting of planting microsites using UAV imagery. *IEEE Access* **2019**, *7*, 82491–82500. [CrossRef]
22. Buzzy, M.; Thesma, V.; Davoodi, M.; Mohammadpour Velni, J. Real-Time Plant Leaf Counting Using Deep Object Detection Networks. *Sensors* **2020**, *20*, 6896. [CrossRef]
23. Zhang, S.; Chi, C.; Yao, Y.; Lei, Z.; Li, S.Z. Bridging the Gap Between Anchor-based and Anchor-free Detection via Adaptive Training Sample Selection. In Proceedings of the IEEE/CVF Conference on Computer Vision and Pattern Recognition, Virtual, 16–18 June 2020; pp. 9759–9768.
24. Think Autonomous. Finally Understand Anchor Boxes in Object Detection (2D and 3D). Available online: https://www.thinkautonomous.ai/blog/anchor-boxes/ (accessed on 6 October 2024).

25. Jiang, T.; Yu, Q.; Zhong, Y.; Shao, M. PlantSR: Super-Resolution Improves Object Detection in Plant Images. *J. Imaging* **2024**, *10*, 137. [CrossRef]
26. Lin, H.; Chen, Z.; Qiang, Z.; Tang, S.-K.; Liu, L.; Pau, G. Automated Counting of Tobacco Plants Using Multispectral UAV Data. *Agronomy* **2023**, *13*, 2861. [CrossRef]
27. Chandra, N.; Vaidya, H. Automated detection of landslide events from multi-source remote sensing imagery: Performance evaluation and analysis of yolo algorithms. *J. Earth Syst. Sci.* **2024**, *133*, 1–17. [CrossRef]
28. Redmon, J.; Farhadi, A. YOLO9000: Better, Faster, Stronger. In Proceedings of the IEEE Conference on Computer Vision and Pattern Recognition, Honolulu, HI, USA, 21–26 July 2017; pp. 278–282.
29. Wang, N.; Cao, H.; Huang, X.; Ding, M. Rapeseed Flower Counting Method Based on GhP2-YOLO and StrongSORT Algorithm. *Plants* **2024**, *13*, 2388. [CrossRef]
30. Hastie, T.; Tibshirani, R.; Friedman, J.H. *The Elements of Statistical Learning: Data Mining, Inference, and Prediction*, 2nd ed.; Springer: New York, NY, USA, 2009; pp. 347–369.
31. Feng, S.; Qian, H.; Wang, H.; Wang, W. Real-time object detection method based on yolov5 and efficient mobile network. *J. Real-Time Image Process.* **2024**, *21*, 56. [CrossRef]
32. Bai, Y.; Yu, J.; Yang, S.; Ning, J. An improved yolo algorithm for detecting flowers and fruits on strawberry seedlings. *Biosyst. Eng.* **2024**, *237*, 1–12. [CrossRef]
33. Guan, H.; Deng, H.; Ma, X.; Zhang, T.; Zhang, Y.; Zhu, T.; Zhou, H.; Gu, Z.; Lu, Y. A corn canopy organs detection method based on improved DBi-YOLOv8 network. *Eur. J. Agron.* **2024**, *154*, 127076. [CrossRef]
34. Xu, D.; Xiong, H.; Liao, Y.; Wang, H.; Yuan, Z.; Yin, H. EMA-YOLO: A Novel Target-Detection Algorithm for Immature Yellow Peach Based on YOLOv8. *Sensors* **2024**, *24*, 3783. [CrossRef]
35. Wang, C.; Yeh, I.; Liao, H. Yolov9: Learning what you want to learn using programmable gradient information. *arXiv* **2024**, arXiv:2402.13616.
36. Badgujar, C.; Poulose, A.; Gan, H. Agricultural object detection with You Only Look Once (YOLO) Algorithm: A bibliometric and systematic literature review. *Comput. Electron. Agric.* **2024**, *223*, 109090. [CrossRef]
37. Zhan, W.; Sun, C.; Wang, M.; She, J.; Zhang, Y.; Zhang, Z.; Sun, Y. An improved Yolov5 real-time detection method for small objects captured by UAV. *Soft Comput.* **2022**, *26*, 361–373. [CrossRef]
38. Li, S.; Tao, T.; Zhang, Y.; Li, M.; Qu, H. YOLO v7-CS: A YOLO v7-based model for lightweight bayberry target detection count. *Agronomy* **2023**, *13*, 2952. [CrossRef]
39. Terven, J.; Córdova-Esparza, D.-M.; Romero-González, J.-A. A Comprehensive Review of YOLO Architectures in Computer Vision: From YOLOv1 to YOLOv8 and YOLO-NAS. *Mach. Learn. Knowl. Extr.* **2023**, *5*, 1680–1716. [CrossRef]
40. Wu, W.; Liu, H.; Li, L.; Long, Y.; Wang, X.; Wang, Z.; Chang, Y. Application of local fully Convolutional Neural Network combined with YOLOv5 algorithm in small target detection of remote sensing image. *PLoS ONE* **2021**, *16*, e0259283. [CrossRef]
41. Wu, D.; Jiang, S.; Zhao, E.; Liu, Y.; Zhu, H.; Wang, W.; Wang, R. Detection of Camellia oleifera fruit in complex scenes by using YOLOv7 and data augmentation. *Appl. Sci.* **2022**, *12*, 11318. [CrossRef]
42. Wang, G.; Chen, Y.; An, P.; Hong, H.; Hu, J.; Huang, T. UAV-YOLOv8: A small-object-detection model based on improved YOLOv8 for UAV aerial photography scenarios. *Sensors* **2023**, *23*, 7190. [CrossRef]
43. Ashraf, A.H.; Imran, M.; Qahtani, A.M.; Alsufyani, A.; Almutiry, O.; Mahmood, A.; Attique, M.; Habib, M. Weapons detection for security and video surveillance using CNN and YOLO-v5s. *CMC-Comput. Mater. Contin.* **2022**, *70*, 2761–2775.
44. Zhao, L.; Zhu, M. MS-YOLOv7: YOLOv7 based on multi-scale for object detection on UAV aerial photography. *Drones* **2023**, *7*, 188. [CrossRef]
45. Contributors, M. YOLOv8 by MMYOLO. 2023. Available online: https://github.com/open-mmlab/mmyolo/tree/main/configs/yolov8 (accessed on 10 March 2024).
46. Chien, C.T.; Ju, R.Y.; Chou, K.Y.; Chiang, J.S. YOLOv9 for fracture detection in pediatric wrist trauma X-ray images. *Electronics Lett.* **2024**, *60*, e13248. [CrossRef]
47. Neubeck, A.; Van Gool, L. Efficient Non-Maximum Suppression. In Proceedings of the 18th International Conference on Pattern Recognition (ICPR'06), Hong Kong, China, 20–24 August 2006.
48. Zaghari, N.; Fathy, M.; Jameii, S.; Shahverdy, M. The improvement in obstacle detection in autonomous vehicles using YOLO non-maximum suppression fuzzy algorithm. *J. Supercomput.* **2021**, *77*, 13421–13446. [CrossRef]
49. Zhang, Y.F.; Ren, W.; Zhang, Z.; Jia, Z.; Wang, L.; Tan, T. Focal and efficient IOU loss for accurate bounding box regression. *Neurocomputing* **2022**, *506*, 146–157. [CrossRef]
50. Candiago, S.; Remondino, F.; De Giglio, M.; Dubbini, M.; Gattelli, M. Evaluating Multispectral Images and Vegetation Indices for Precision Farming Applications from UAV Images. *Remote Sens.* **2015**, *7*, 4026–4047. [CrossRef]
51. Santana, L.S.; Ferraz, G.A.e.S.; Santos, G.H.R.d.; Bento, N.L.; Faria, R.d.O. Identification and Counting of Coffee Trees Based on Convolutional Neural Network Applied to RGB Images Obtained by RPA. *Sustainability* **2023**, *15*, 820. [CrossRef]

remote sensing

Article

Microwave Coincidence Imaging with Phase-Coded Stochastic Radiation Field

Hang Lin, Hongyan Liu *, Yongqiang Cheng, Ke Xu, Kang Liu and Yang Yang

College of Electronic Science and Technology, National University of Defense Technology, Changsha 410073, China; linhangh@nudt.edu.cn (H.L.); nudtyqcheng@gmail.com (Y.C.); xuek@nudt.edu.cn (K.X.); liukang1117@126.com (K.L.); yangyang16@nudt.edu.cn (Y.Y.)

* Correspondence: lhyan1993@126.com

Abstract: Microwave coincidence imaging (MCI) represents a novel forward-looking radar imaging method with high-resolution capabilities. Most MCI methods rely on random frequency modulation to generate stochastic radiation fields, which introduces the complexity of radar systems and imposes limitations on imaging quality under noisy conditions. In this paper, microwave coincidence imaging with phase-coded stochastic radiation fields is proposed, which generates spatio-temporally uncorrelated stochastic radiation fields with phase coding. Firstly, the radiation field characteristics are analyzed, and the coding sequences are designed. Then, pulse compression is applied to achieve a one-dimensional range image. Furthermore, an azimuthal imaging model is built, and a reference matrix is derived from the frequency domain. Finally, sparse Bayesian learning (SBL) and alternating direction method of multipliers (ADMM)-based total variation are implemented to reconstruct targets. The methods have better imaging performance at low signal-to-noise ratios (SNRs), as shown by the imaging results and mean square error (MSE) curves. In addition, compared with the SBL and ADMM-based total variation methods based on the direct frequency-domain solution, the proposed method's computational complexity is reduced, giving it great potential in forward-looking high-resolution scenarios, such as autonomous obstacle avoidance with vehicle-mounted radar and terminal guidance.

Keywords: microwave coincidence imaging; wavefront modulation; phase modulation; forward-looking imaging; complex target

1. Introduction

Synthetic aperture radar (SAR) and inverse synthetic aperture radar (ISAR) are two main techniques used for high-resolution radar imaging. The two methods rely on the relative motion between the radar and targets to form a virtual aperture [1–3]. However, in the forward-looking region, the viewing angle of the radar remains nearly unchanged, thus making it difficult to form a large virtual aperture, which limits the imaging performance [4,5].

Microwave coincidence imaging (MCI) is a forward-looking imaging method with high azimuthal resolution [6–8]. It generates stochastic radiation fields in the imaging region by modulating transmitted signals [9–11]. The targets can be distinguished by the stochastic radiation fields. Then, they can be reconstructed by correlating echo signals with reference signals. In contrast to other forward-looking radar imaging methods [12–15], microwave coincidence imaging no longer relies on the relative motion between the radar and targets, breaking the azimuthal resolution limitations of a real aperture.

MCI in the literature mainly adopts random frequency modulation. Different array elements transmit random frequency signals to generate spatio-temporally uncorrelated stochastic radiation fields [16–18]. However, this method increases the system complexity. Furthermore, the presence of different frequency signals at the same moment makes it difficult to perform pulse compression on echo signals, which means that the imaging distance is limited and the quality is sensitive to noise.

Sparse recovery [19–21] and compressed sensing (CS) [22–24] can be applied in microwave coincidence forward-looking imaging; these methods have high imaging resolution for simple targets composed of strong scattering points, such as angular antipodal, metal balls, etc. However, for complex-structure targets, such as aircraft and tanks, the imaging reconstruction capability is inadequate, making it challenging to reconstruct the contours and target structural features accurately.

In this paper, a microwave coincidence imaging method based on phase modulation is proposed. This method can utilize a phased array system to generate spatio-temporally uncorrelated stochastic radiation fields. Wavefront entropy and effective rank are selected to analyze the radiation field characteristics, which are used to guide the design of the phase encoding. Then, a one-dimensional range image in the frequency domain is obtained by applying pulse compression to the echo signal. At this point, the range dimension has been segregated from the azimuthal dimension. Consequently, an azimuthal dimensional imaging equation in the frequency domain can be established, and a reference matrix in the azimuthal dimension can be derived. The reconstruction of the target is achieved by stitching together azimuthal scattering coefficients at different distances, which reduces the computational complexity. Finally, the imaging equations are solved by using sparse Bayesian learning (SBL) and alternating direction method of multipliers (ADMM)-based total variation methods. The results of the simulations demonstrate that the method proposed in this paper can achieve better imaging performance at low signal-to-noise ratios (SNRs) [25] with lower complexity.

2. Radiation Field Generation

The characteristics of the stochastic radiation field determine the imaging quality. The traditional radiation field generated by random modulation has better random characteristics, but it poses problems, such as great difficulty in solving imaging equations and poor imaging robustness. Thus, a stochastic radiation field generation method based on phase modulation is proposed, and the phase code of the transmitted signals is designed in this paper.

2.1. Radiation Field Characterization

Radar systems transmit phase-coded linear frequency modulation signals through multiple antennas. Transmitted signals are superimposed on the target area to form a spatio-temporally uncorrelated radiation field, which is the basis of imaging. Assuming that the target is located in the region of the radiation field, an XOY coordinate system is established at the center of the target, as illustrated in Figure 1.

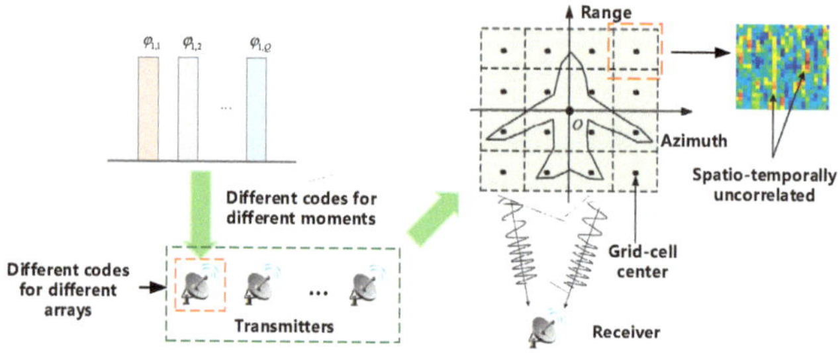

Figure 1. Principle of radar coincidence imaging.

The radar system comprises N transmitting antennas and one receiving antenna. The imaging plane is assumed to be divided into L grids. A total of M modulations are performed. Each antenna transmits a distinct coded signal at different moments. At the mth modulation, the signal transmitted by the nth transmitter can be expressed as

$$St_n(m,\hat{t}) = rect\left[\frac{t}{T_p}\right] A\exp\left[j2\pi(f_c\hat{t} + \frac{1}{2}k\hat{t}^2)\right]\exp(j\varphi_{m,n}) \tag{1}$$

where T_p is the pulse width, A is the amplitude of the signal, f_c is the carrier frequency, k is the frequency modulation coefficient, and $\varphi_{m,n}$ is the phase of the nth array element at the mth modulation $\hat{t}$ is fast time.

A cyclic code means that when any code element in a code group is shifted left or right by any number of bits, the resulting code element still belongs to the code group. The structural characteristics of the cyclic matrix have the advantage of reducing computational complexity [26], as some fast algorithms can be employed to solve the equations. Furthermore, the relatively high rank of the cyclic matrix renders it suitable for generating the radiation field [27,28].

When a cyclic code is used to encode the transmitted signals of N array elements, the coding matrix following M modulations can be expressed as

$$\psi = \begin{bmatrix} \varphi_{1,1} & \varphi_{1,2} & \cdots & \varphi_{1,N} \\ \varphi_{2,1} & \varphi_{2,2} & \cdots & \varphi_{2,N} \\ \vdots & \vdots & \cdots & \vdots \\ \varphi_{M,1} & \varphi_{M,2} & \cdots & \varphi_{M,N} \end{bmatrix} \tag{2}$$

where each row represents the encoding of a different array element under one modulation, each column represents the encoding of an array element under a different number of modulations, and each row of this encoding matrix is shifted in the order of the previous row.

The design of the coding matrix is primarily concerned with the coding sequence of the first row [29]. Note that the sequence code in the first column is $\Phi_1 = [\varphi_{1,1} \quad \varphi_{1,2} \quad \cdots \quad \varphi_{1,N}]$. Φ_1 is designed as a random code in the following analysis of radiation field characteristics.

The radar system transmits the signals defined in (1), and the scattering point is an ideal scattering point with a scattering coefficient of 1. The reference signals can be expressed as a superposition of the received signals in each grid. When the radar system undergoes M modulations, the expression of the reference signals is as follows:

$$\begin{aligned} S(m,l,\hat{t}) &= \sum_{n=1}^{N} rect\left[\frac{t-\tau(n,l)}{T_p}\right] A \\ &\exp\left\{j2\pi\left\{f_c[\hat{t}-\tau(n,l)] + \frac{1}{2}k[\hat{t}-\tau(n,l)]^2\right\}\right\}\exp(j\varphi_{m,n}) \end{aligned} \tag{3}$$

where $\tau(n,l) = \frac{|r_l - R_n|}{c} + \frac{|r_l - R_0|}{c}$, r_l is the position vector of the lth grid, R_n is the position vector of the nth transmitting antenna, R_0 is the position vector of the receiving antenna, and c is the propagation speed of electromagnetic waves.

When $\hat{t}$ satisfies $\hat{t} = [\hat{t}_1 \quad \hat{t}_2 \quad \ldots \quad \hat{t}_s]$, the reference signals' matrix form can be written as

$$\mathbf{S} = \begin{bmatrix} S(1,1,\hat{t}_1) & S(1,2,\hat{t}_1) & \cdots & S(1,L,\hat{t}_1) \\ \vdots & \vdots & & \vdots \\ S(1,1,\hat{t}_s) & S(1,2,\hat{t}_s) & \cdots & S(1,L,\hat{t}_s) \\ \vdots & \vdots & & \vdots \\ S(M,1,\hat{t}_1) & S(M,2,\hat{t}_1) & \cdots & S(M,L,\hat{t}_1) \\ \vdots & \vdots & & \vdots \\ S(M,1,\hat{t}_s) & S(M,2,\hat{t}_s) & \cdots & S(M,L,\hat{t}_s) \end{bmatrix} \tag{4}$$

where $\hat{t}_s$ denotes the number of samples.

Assuming that the target scattering coefficient of each divided grid cell is σ_l, the echo of the superimposed reference signals at the mth modulation is given by

$$S_r(m,\hat{t}) = \sum_{l=1}^{L} \sigma_l \cdot S(m,l,\hat{t}) \tag{5}$$

In order to analyze the characteristics of the radiation field, it is first necessary to define the radiation field evaluation metric. Since the wavefront modulation changes the randomness of the radiation field by changing the undulation of the wavefront, the wavefront entropy is defined. It represents the magnitude of wavefront undulation using the statistical average of the radiation field amplitude over a period of time. Its expression is as follows [30]:

$$\mathbf{H_S} = -\frac{1}{Q}\sum_{q=1}^{Q}\sum_{l=1}^{L} p_{q,l}\log_2 p_{q,l} \tag{6}$$

where $\mathbf{S}$ is the signal reference matrix, L denotes the number of grids, and Q denotes the number of modulations. $p_{q,l} = \frac{\mathbf{S}_{q,l}}{\mathbf{S}_q}$ is the probability of the statistical distribution of elements in the qth row of $\mathbf{S}$, $\mathbf{S}_q = \sum_{l=1}^{L} \mathbf{S}_{q,l}$.

The second evaluation metric is the effective rank [31]. The $\mathbf{S}$ singular value decomposition can be.expressed as follows:

$$\mathbf{S}=\mathbf{U}\boldsymbol{\Sigma}\mathbf{V}^H \tag{7}$$

The effective rank is typically determined through the Frobenius paradigm ratio, where the previous matrix is approximated by a large singular value matrix whose singular value is the original matrix decomposition. This approximated matrix is then compared to the Frobenius paradigm ratio to $\mathbf{S}$.

$$\begin{aligned} v(r) &= \frac{\|\mathbf{S}_r\|_F}{\|\mathbf{S}\|_F} \\ &= \frac{\sqrt{\sigma_1^2+\sigma_2^2+\cdots+\sigma_r^2}}{\sqrt{\sigma_1^2+\sigma_2^2+\cdots+\sigma_L^2}} \end{aligned} \tag{8}$$

The largest value of r satisfying $v(r) \geq \varepsilon$ is taken as the effective rank of the matrix $\mathbf{S}$, where ε is a value close to 1.

The following section will analyze the variation characteristics of evaluation metrics for different bandwidths, array spacing, and number of codes.

The transmitting array is a 16×1 linear array with a spacing of 0.15 m. f_c is 90 GHz; the bandwidths are 200 MHz, 400 MHz, 600 MHz, and 800 MHz; the pulse width T_p is 200 ns; and the imaging plane is divided into a grid of 30×30 with a size of $0.5\ \text{m} \times 0.5\ \text{m}$.

Figure 2 illustrates the undulations of the radiation field at the same moment at varying bandwidths, and Table 1 is the statistical average of wavefront entropy. It can be observed that as the bandwidth increases, the undulations of the radiation field become increasingly sharp at this moment, and the value of wavefront entropy also increases gradually. The value of wavefront entropy demonstrates that the randomness of the radiation field can be increased by expanding the bandwidth.

Table 1. Value of wavefront entropy.

Bandwidths	200 MHz	400 MHz	600 MHz	800 MHz
$\mathbf{H_S}$	2.4052	2.7805	2.9573	3.1301

Figure 3a depicts the trend of the effective rank for varying bandwidths. The number of sampling points is 400. Other parameters are consistent with the preceding analysis. It can be observed that the effective rank increases as the bandwidth increases.

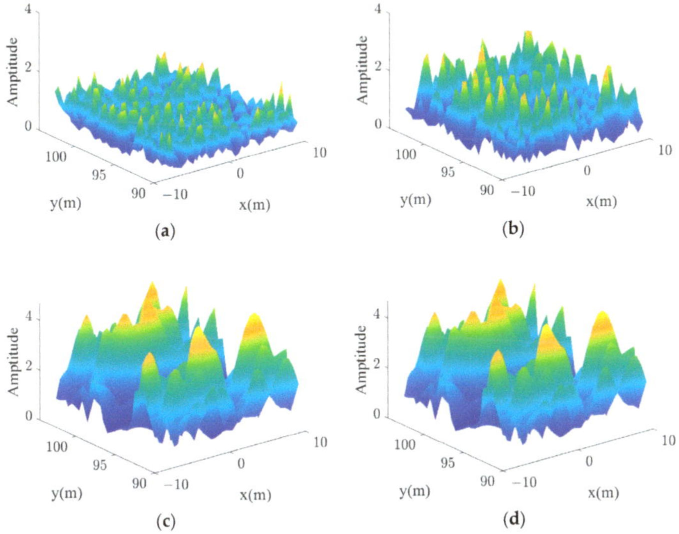

Figure 2. Wavefront undulations of different bandwidths at the same moment. (**a**) B = 200 MHz. (**b**) B = 400 MHz. (**c**) B = 600 MHz. (**d**) B = 800 MHz.

In order to analyze the influence of the array element spacing and the number of codes on the effective rank, a simulation was conducted using different array element spacings and numbers of codes. The results are presented in Figure 3b. It can be seen that increasing either the array element spacing or the number of codes can effectively increase the effective rank and increase the randomness of the radiation field.

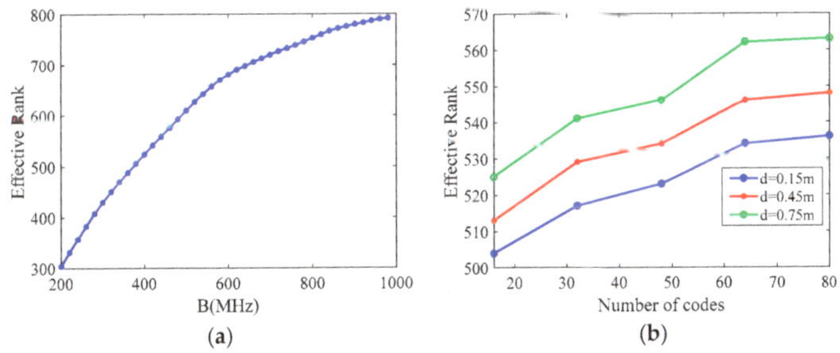

Figure 3. Effective-rank change curves for different influencing factors. (**a**) Effective-rank change curves at different bandwidths. (**b**) Effective-rank change curves with different array spacings and numbers of codes.

2.2. Design of Code Sequence

While Φ_1 produces radiation fields with high randomness when it is randomly coded, this introduces uncertainty. The effective rank of the reference matrix will change with each image acquisition, and its size fluctuates widely. When the effective rank is large, the imaging quality is high, while when the effective rank is small, the imaging quality is poor. This uncertainty introduces variations that are challenging to grasp, resulting in an imaging quality that is not robust.

Chaotic sequences exhibit both intrinsic determinism and extrinsic randomness. For a given mapping, the generated code elements are deterministic, while the generated code elements exhibit random properties among themselves. Therefore, Φ_1 is considered to be encoded as a chaotic sequence, and the logistic mapping is selected to generate the chaotic sequence. The mapping equation is as follows:

$$x(n+1) = \mu x(n)(1 - x(n)) \tag{9}$$

where μ is a system parameter, $\mu \in (0, 4]$ [32], and $x(n) \in (0, 1)$.

When $3.569 < \mu \leq 4$, the logistic mapping enters a chaotic state, producing chaotic sequences with high randomness.

In order to enhance the randomness of the chaotic sequences generated by the logistic mapping, the generated sequence will be converted using the inverse sine function [33].

$$x(n) = \frac{2\arcsin x(n)}{\pi} \tag{10}$$

The system parameter values are designated as $\mu = 4$ and initial value $x(0) = 0.1$. The 20th chaotic sequence is designated as the starting value, and every 10th sequence value is extracted as the desired code, resulting in a total of 16 codes for Φ_1.

Figure 4 illustrates the undulations of the radiation field when Φ_1 is a chaotic sequence generated by logistic mapping. It can be observed that a radiation field with considerable undulations can still be generated, with a wavefront entropy of 3.1214 and an effective rank of 752.

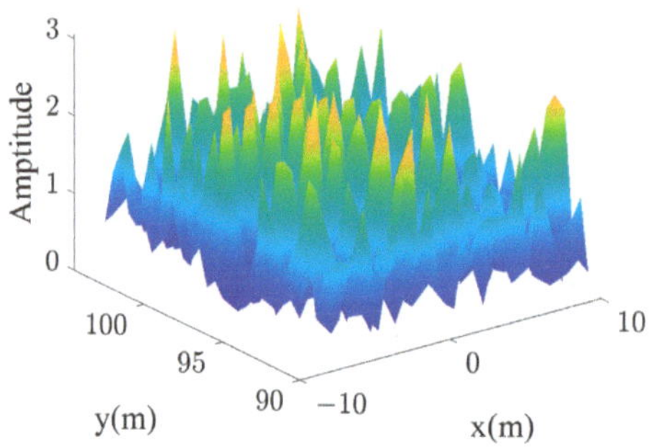

Figure 4. Radiation field distribution of chaotic sequences generating cyclic codes.

Figure 5 illustrates the variation in effective rank when Φ_1 is a chaotic sequence and a random code for different imaging times at different bandwidths. The value of effective rank when Φ_1 is coded as a chaotic sequence is comparable to the high effective rank when Φ_1 is coded as a random code. And, it exhibits extrinsic deterministic properties that contribute to robust imaging when Φ_1 is a chaotic sequence.

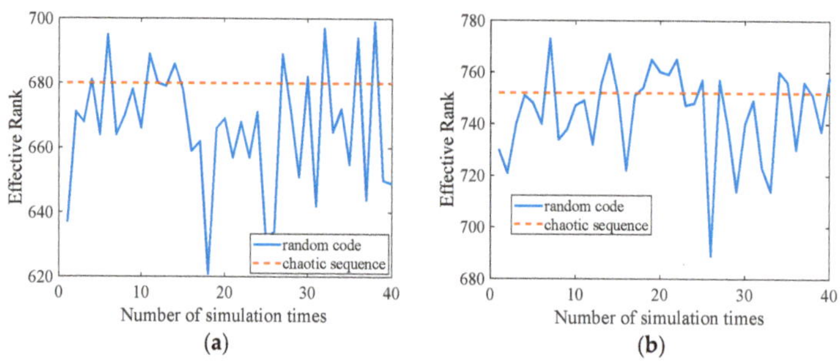

Figure 5. Variation in the effective rank corresponding to random codes and chaotic sequences for different imaging times. (**a**) B = 600 MHz. (**b**) B = 800 MHz.

3. Imaging Model

Wavefront-modulated forward-looking imaging with phase coding offers two significant advantages. Firstly, unlike frequency modulation, the echo signals with phase modulation have the same frequency and differ only in phase. This enables echoes to be processed by pulse compression, which can improve imaging performance at low SNRs. Secondly, phase coding employs a coding sequence that is both deterministic and random, thereby increasing the robustness of the imaging process.

For transmitted signals with phase coding, as illustrated in (1), the echo signals obtained by the superposition of all array elements under M modulations can be expressed as follows:

$$\begin{aligned} S_r(m,\hat{t}) &= \sum_{l=1}^{L}\sum_{n=1}^{N} rect\left[\frac{t-\tau(n,l)}{T_p}\right]\cdot A\cdot\sigma_l \\ &\cdot\exp\left\{j2\pi\left\{f_c[\hat{t}-\tau(n,l)]+\frac{1}{2}k[\hat{t}-\tau(n,l)]^2\right\}\right\}\cdot\exp(j\varphi_{m,n}) \end{aligned} \tag{11}$$

where σ_l is the scattering coefficient in the lth grid.

The echo signals have the same frequency at the same time, and only a phase difference exists. Therefore, the dechirping method can be used for echoes. The reference signal for dechirping is defined as

$$S_{ref}(\hat{t}) = \exp\left[j2\pi\left(f_c\hat{t}+\frac{1}{2}k\hat{t}^2\right)\right] \tag{12}$$

The differential frequency output is obtained by multiplying the echo signals conjugated to the reference signal:

$$\begin{aligned} &S_{if}(m,\hat{t}) \\ &= S_r(m,\hat{t})\cdot S^*{}_{ref}(\hat{t}) \\ &= \sum_{n=1}^{N}\sum_{l=1}^{L}\sigma_l\cdot A\cdot rect\left[\frac{t-\tau(n,l)}{T_p}\right] \\ &\cdot\exp\left\{j2\pi\left[-k\tau(n,l)\hat{t}-f_c\tau(n,l)+\frac{1}{2}k\tau^2(n,l)\right]\right\}\exp(j\varphi_{m,n}) \end{aligned} \tag{13}$$

The one-dimensional range image of the target in the frequency domain is obtained by Fourier transformation. The different frequency points thus correspond to different distances:

$$\begin{aligned} &\mathcal{F}\left\{S_{if}(m,\hat{t})\right\} \\ &= F_R(m,f) \\ &= \sum_{n=1}^{N}\sum_{l=1}^{L}\sigma_l\cdot A\cdot T_p\cdot sinc\{T_p[f+k\tau(n,l)]\} \\ &\cdot\exp\left\{-j2\pi\left[\tau(n,l)f+f_c\tau(n,l)-\frac{1}{2}k\tau^2(n,l)\right]\right\}\exp(j\phi_{m,n}) \end{aligned} \tag{14}$$

At this point, the range dimension of the target has been segregated from the azimuthal dimension. The same operation described above is performed on the reference matrix to obtain its representation in the frequency domain:

$$\begin{aligned} &\mathcal{F}\{S(m,l,\hat{t})\} \\ &= F_S(m,l,f) \\ &= \sum_{n=1}^{N} A\cdot T_p\cdot \mathrm{sinc}\{T_p[f+k\tau(n,l)]\} \\ &\cdot\exp\left\{-j2\pi\left[\tau(n,l)f+f_c\tau(n,l)-\frac{1}{2}k\tau^2(n,l)\right]\right\}\exp(j\phi_{m,n}) \end{aligned} \tag{15}$$

The frequency-domain reference signal matrix form can be expressed as

$$\mathbf{F_s} = \begin{bmatrix} F_S(1,1,f_1) & F_S(1,2,f_1) & \cdots & F_S(1,L,f_1) \\ \vdots & \vdots & & \vdots \\ F_S(1,1,f_N) & F_S(1,2,f_N) & \cdots & F_S(1,L,f_N) \\ \vdots & \vdots & & \vdots \\ F_S(M,1,f_1) & F_S(M,2,f_1) & \cdots & F_S(M,L,f_1) \\ \vdots & \vdots & & \vdots \\ F_S(M,1,f_N) & F_S(M,2,f_N) & \cdots & F_S(M,L,f_N) \end{bmatrix} \tag{16}$$

where f_N denotes the number of samples in a pulse.

In order to reduce the computational complexity, the azimuthal scattering coefficients of targets are reconstructed at different distances separately. Subsequently, the scattering coefficients of all distances are spliced together to obtain the complete scattering coefficients of the target.

Assuming that $f_{eff} = [f_{T_1}, f_{T_2}, \cdots, f_{T_N}]$ is the peak frequency of the one-dimensional range image obtained by pulse compression, the frequency-domain echoes at a distance corresponding to the frequency f_{T_1} can be expressed as follows:

$$\begin{aligned} & F_{Rf}(m, f_{T_1}) \\ & = \sum_{n=1}^{N} \sum_{l=1}^{L} \sigma_l \cdot A \cdot T_p \cdot \mathrm{sinc}\{T_p[f_{T_1} + k\tau(n,l)]\} \\ & \cdot \exp\left\{-j2\pi\left[\tau(n,l)f_{T_1} + f_c\tau(n,l) - \frac{1}{2}k\tau^2(n,l)\right]\right\}\exp(j\varphi_{m,n}) \end{aligned} \tag{17}$$

Then, the reference signals at a distance corresponding to f_{T_1} can be expressed as

$$\begin{aligned} & F_{Sf}(m, l_A, f_{T_1}) \\ & = \sum_{n=1}^{N} A \cdot T_p \cdot \mathrm{sinc}\{T_p[f_{T_1} + k\tau_A(n,l_A)]\} \\ & \cdot \exp\left\{-j2\pi\left[\tau_A(n,l_A)f_{T_1} + f_c\tau_A(n,l_A) - \frac{1}{2}k\tau_A^2(n,l_A)\right]\right\}\exp(j\varphi_{m,n}) \end{aligned} \tag{18}$$

where $\tau_A = \frac{|r_{lA} - R_n| + |r_{lA} - R_0|}{c} + \tau_R$, $\tau_R = \frac{|r_{lR} - R_n| + |r_{lR} - R_0|}{c}$, τ_R denotes the distance delay, τ_A denotes the delay of the azimuthal grid at this distance, r_{lA} is the azimuthal position vectors of the lth grids, and r_{lR} is the range position vectors of the lth grids.

The reference matrix grid is a patchwork of azimuthal grids at different distances when computed directly, but when solving from the azimuthal dimension, the number of grids in the reference matrix becomes the number of azimuthal grids at one distance. A new reference matrix is formed by taking out the reference signals at that distance under each modulation. Its form at a single distance can be expressed as follows:

$$\mathbf{F}_{Sf} = \begin{bmatrix} F_{Sf}(1,1) & F_{Sf}(1,2) & \ldots & F_{Sf}(1,L_A) \\ F_{Sf}(2,1) & F_{Sf}(1,2) & \ldots & F_{Sf}(2,L_A) \\ \vdots & \vdots & \ldots & \vdots \\ F_{Sf}(M,1) & F_{Sf}(M,2) & \ldots & F_{Sf}(M,L_A) \end{bmatrix} \tag{19}$$

where L_A represents the number of azimuthal grids. Each row denotes the reference signals in different azimuthal grids for a single pulse, while each column denotes the reference signals for different pulses in a single azimuthal grid.

Thus, when modulated M times, the imaging equation at a single distance can be expressed as

$$\begin{bmatrix} F_{Rf}(1) \\ F_{Rf}(2) \\ \vdots \\ F_{Rf}(M) \end{bmatrix} = \begin{bmatrix} F_{Sf}(1,1) & F_{Sf}(1,2) & \cdots & F_{Sf}(1,L_A) \\ F_{Sf}(2,1) & F_{Sf}(2,2) & \cdots & F_{Sf}(2,L_A) \\ \vdots & \vdots & \cdots & \vdots \\ F_{Sf}(M,1) & F_{Sf}(M,2) & \cdots & F_{Sf}(M,L_A) \end{bmatrix} \begin{bmatrix} \sigma_1 \\ \sigma_2 \\ \vdots \\ \sigma_A \end{bmatrix} + \mathbf{n} \tag{20}$$

where $\mathbf{n}$ is Gaussian white noise.

The azimuthal target scattering coefficients are obtained by solving (20). Target reconstruction can be achieved by traversing each distance and stitching the azimuthal target scattering coefficients from each solution.

4. Imaging Algorithm and Results

4.1. Sparse Bayesian Learning

SBL [34] first assumes that the target scattering coefficient distribution satisfies the prior. The reconstructed target is then achieved by solving the maximum a posteriori estimate in the Bayesian framework using prior information. At this point, the question form becomes

$$\hat{\boldsymbol{\sigma}}_A = \underset{\boldsymbol{\sigma}_A}{\mathrm{argmax}} p(\boldsymbol{\sigma}_A|\mathbf{F}_{Rf}) \tag{21}$$

The problem can be transformed by the Bayesian formulation:

$$\hat{\boldsymbol{\sigma}}_A = \underset{\boldsymbol{\sigma}_A}{\mathrm{argmax}} p(\mathbf{F}_{Rf}\big|\boldsymbol{\sigma}_A)p(\boldsymbol{\sigma}_A) \tag{22}$$

Assume that the noise follows a Gaussian distribution, where the mean is 0 and the variance is $\xi^2\mathbf{I}_M$.

$$p(\mathbf{F}_{Rf}|\boldsymbol{\sigma}_A) = (2\pi\xi^2) - \frac{M}{2}\exp[-\frac{1}{2\xi^2}(\mathbf{F}_{Rf} - \mathbf{F}_{Sf}\boldsymbol{\sigma}_A)^H(\mathbf{F}_{Rf} - \mathbf{F}_{Sf}\boldsymbol{\sigma}_A)] \tag{23}$$

Assume that $\boldsymbol{\sigma}_A$ is generated by the hyperparameter

$$p(\boldsymbol{\sigma}_A;\boldsymbol{\gamma}) = (2\pi)^{-\frac{M}{2}}|\Lambda|^{-\frac{1}{2}}\exp\left(-\frac{1}{2}\boldsymbol{\sigma}_A^H\Lambda^{-1}\boldsymbol{\sigma}_A\right) \tag{24}$$

where $\boldsymbol{\gamma} = [\gamma_1, \gamma_2, \cdots, \gamma_L]$, and $\Lambda = diag(\boldsymbol{\gamma})$.

$$\begin{aligned} p(\mathbf{F}_{Rf};\boldsymbol{\gamma}) &= \int p(\mathbf{F}_{Rf}|\boldsymbol{\sigma}_A)p(\boldsymbol{\sigma}_A;\boldsymbol{\gamma})d\boldsymbol{\sigma}_A \\ &= \int (2\pi\xi^2)^{-\frac{M}{2}}(2\pi)^{-\frac{L}{2}}|\Lambda|^{-\frac{1}{2}} \\ &\cdot \exp[-\tfrac{1}{2\xi^2}(\mathbf{F}_{Rf} - \mathbf{F}_{Sf}\boldsymbol{\sigma}_A)^H(\mathbf{F}_{Rf} - \mathbf{F}_{Sf}\boldsymbol{\sigma}_A) - \tfrac{1}{2}\boldsymbol{\sigma}_A^H\Lambda^{-1}\boldsymbol{\sigma}_A]d\omega \end{aligned} \tag{25}$$

Considering the exponential part as a whole, the derivation and collation of the calculations leads to

$$p(\mathbf{F}_{Rf};\boldsymbol{\gamma}) = C\exp\left\{-\frac{1}{2\xi^2}\mathbf{F}_{Rf}^H\left[\mathbf{I} - \mathbf{F}_{Sf}\left(\mathbf{F}_{Sf}^H\mathbf{F}_{Sf} + \xi^2\Lambda^{-1}\right)^{-1}\mathbf{F}_{Sf}^H\right]\mathbf{F}_{Rf}\right\} \tag{26}$$

Further calculations based on (22) give the mean and variance of the scattering coefficient:

$$\boldsymbol{\mu} = \left(\mathbf{F}_{Sf}^H\mathbf{F}_{Sf} + \xi^2\Lambda^{-1}\right)^{-1}\mathbf{F}_{Sf}^H\mathbf{F}_{Rf} \tag{27}$$

$$\boldsymbol{\Sigma}^{-1} = \frac{1}{\xi^2}\mathbf{F}_{Sf}^H\mathbf{F}_{Sf} + \Lambda^{-1}\boldsymbol{\sigma}_A \tag{28}$$

The hyperparameters are updated by the EM algorithm:

$$\gamma^{i+1} = \left|\left|\mu\right|\right|_2^2 + \boldsymbol{\Sigma} \tag{29}$$

$$\left(\xi^2\right)^{i+1} = \frac{\| \mathbf{F}_{Rf} - \mathbf{F}_{Sf}\mu \|_2^2 + (\xi^2)^i (L - \sum_{i=1}^{L} \frac{\boldsymbol{\Sigma}^i}{\gamma_i})}{M} \tag{30}$$

In order to verify the effectiveness of the phase modulation azimuthal dimensional imaging model (AziPM) proposed in this paper, a sparse Bayesian learning imaging algorithm was used to image a point aircraft target, as shown in Figure 6. The imaging results of the traditional frequency modulation (FM) method are compared, and the simulation parameters are shown in Table 2.

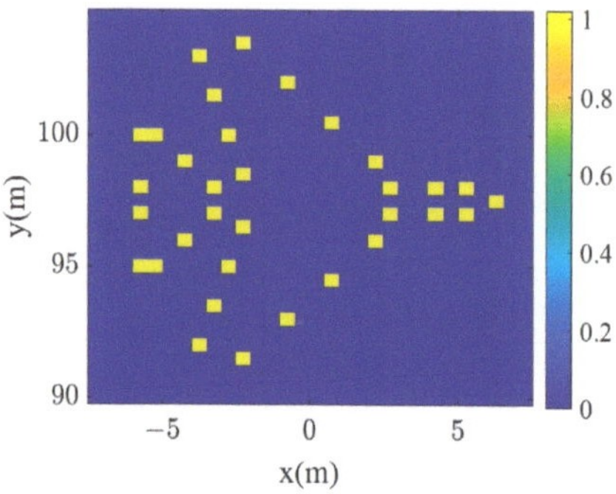

Figure 6. Aircraft point targets.

Figure 7 illustrates that the AziPM method achieves satisfactory reconstruction results at both 0 dB and 5 dB, with superior imaging results compared to the FM method.

Figure 8 plots the azimuthal dimensional slices in the second row of target distances for both methods at 0 dB. It can be seen that the azimuthal slices of the proposed method possess lower sidelobes compared to those of the FM method, which further indicates that the method is able to obtain better imaging performance.

Table 2. Simulation parameters.

Parameters	Values
Carrier frequency	90 GHz
Bandwidth	800 MHz
Number of modulations	16
Array number	16 (16 × 1)
Imaging range	100 m
Array aperture	0.15 m × 0.15 m
Grid size	0.5 m × 0.5 m
Number of grids	30 × 30

In order to quantitatively analyze the quality of imaging, the mean square error (MSE) is defined by the following expression:

$$\rho_{MSE} = \frac{\|\hat{\sigma} - \sigma\|^2}{\|\sigma\|_2^2} \tag{31}$$

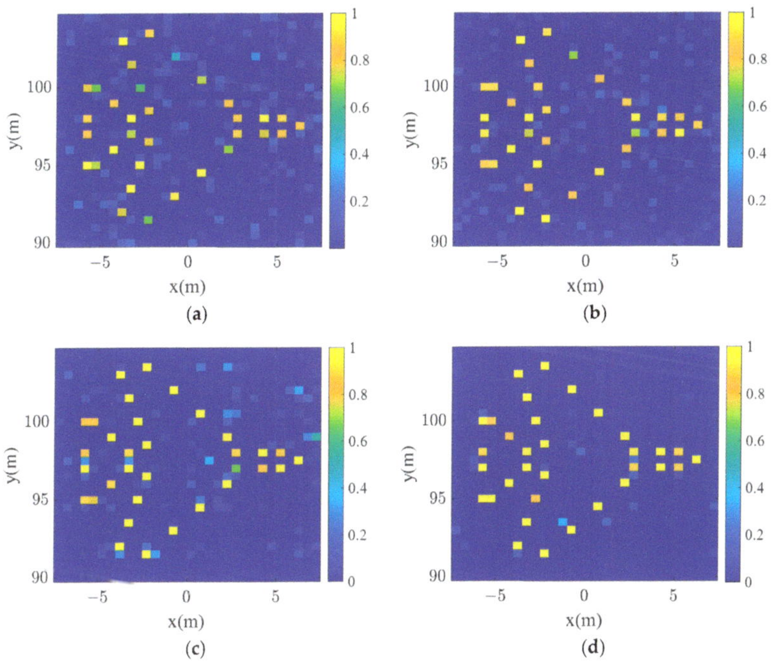

Figure 7. SBL imaging results. (**a**) FM results at 0 dB. (**b**) FM results at 5 dB. (**c**) AziPM results at 0 dB. (**d**) AziPM results at 5 dB.

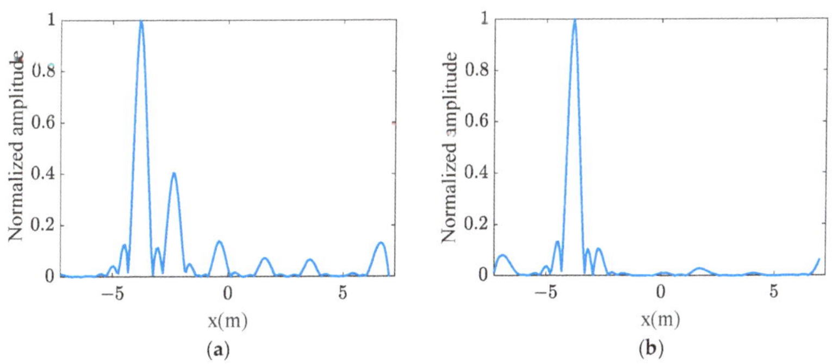

Figure 8. Azimuthal dimensional slices at 0 dB. (**a**) FM method. (**b**) AziPM method.

Figure 9a depicts the MSE-versus-SNR curves of the two methods. It can be observed that the MSE curve of the AziPM method is considerably lower than that of the FM method, which suggests that the imaging performance of the AziPM is superior to that of the FM method.

Figure 9 shows the comparison of the MSE curves of the two methods for different numbers of modulations.

It can be observed that the MSE of the AziPM method is considerably lower than that of the FM method when the number of modulations is limited. As the number of modulations gradually increases, the relative MSE curves of the two methods gradually converge, and the final performance of the conventional FM method approaches that of the AziPM method. The final results demonstrate that while the FM method can generate a radiated field with a higher degree of stochasticity, this must be achieved when the number of modulations is sufficiently large. Conversely, the AziPM method is capable of achieving a lower MSE with a smaller number of modulations.

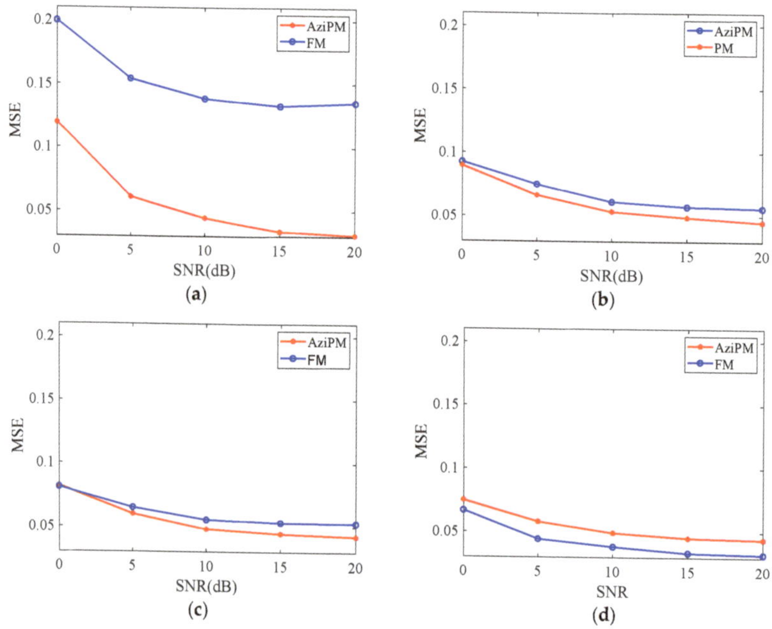

Figure 9. MSE curves of FM and AziPM methods with different modulation counts: (**a**) 16 modulations, (**b**) 48 modulations, (**c**) 64 modulations, (**d**) 96 modulations.

In order to compare the effects of circular coding and random coding on the imaging performance, Figure 10 shows the results of imaging simulations carried out at 0 dB for both codes. As can be seen from the results, there is little difference between the two coded imaging effects. In order to analyze the effects qualitatively, Figure 11 plot the MSE curves at different SNR. It can be seen that, although the reference matrix rank generated by the cyclic code is not as high as that obtained with the random code, the cyclic code MSE is lower than the random code MSE after Monte Carlo due to its robustness.

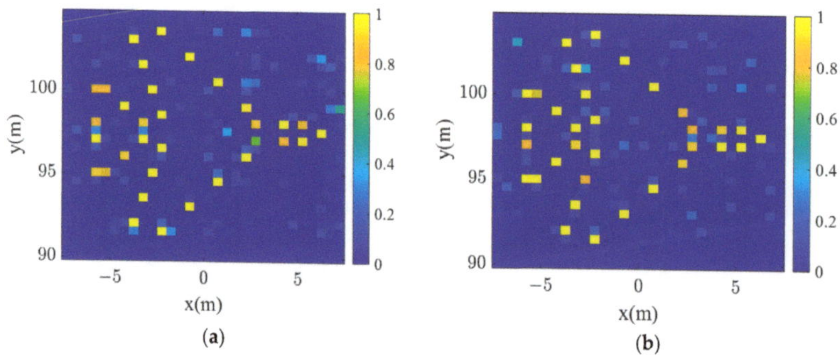

Figure 10. Imaging results with different codes at 0 dB. (**a**) Cyclic code. (**b**) Random code.

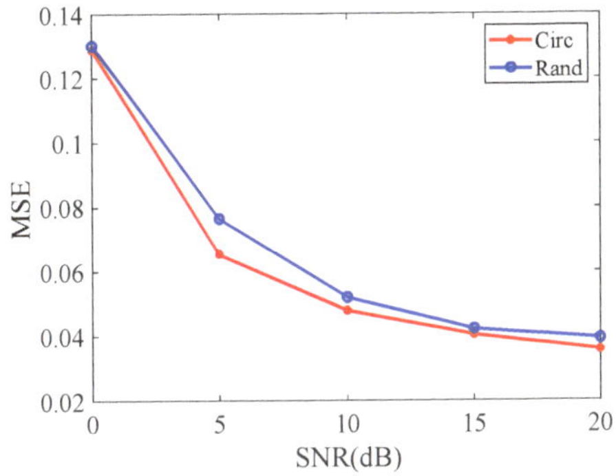

Figure 11. MSE curves of cyclic code and random code methods.

The AziPM method for target reconstruction is effective when the scattering points in the azimuthal direction are sparsely distributed, allowing for effective reconstruction at a low SNR. However, when the number of azimuthal scattering points increases and the complexity of the target increases, the imaging results are gradually drowned in noise. It is already difficult to perform a complete reconstruction of the target using the SBL method. Figure 12 illustrates the imaging results of three targets at 0 dB and 5 dB. It is evident that as the complexity of the target increases, the SBL method is unable to reconstruct the target effectively.

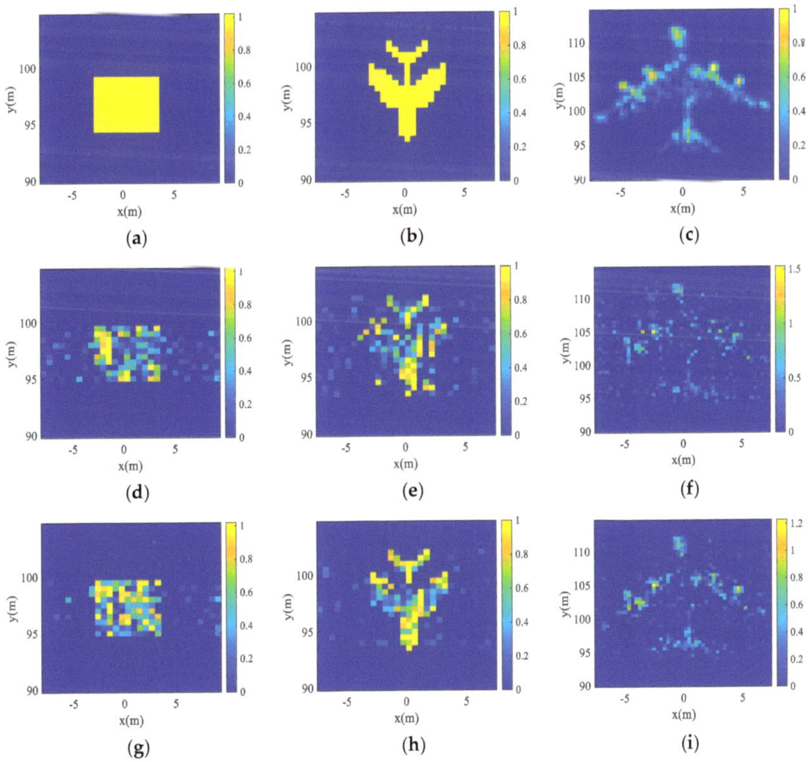

Figure 12. Imaging results for different targets. **(a–c)** Targets. **(d–f)** Imaging results at 0 dB. **(g–i)** Imaging results at 5 dB.

4.2. ADMM-Based Total Variation

The total variation regularization method was initially employed in the field of image processing [35], where the total variation model permitted the solution to be discontinuous [36]. This enabled the retention of edge and contour information in the image while denoising.

The total variable is defined as follows:

$$\min_{\sigma_A} \sum_{l} \|D_i \boldsymbol{\sigma}_A\|_p, \ s.t. \ \mathbf{F}_{Sf}\boldsymbol{\sigma}_A = \mathbf{F}_{Rf} \tag{32}$$

where $D_l \boldsymbol{\sigma}_A$ is the derivative of the reference coefficient $\boldsymbol{\sigma}_A$ in grid l, $\mathbf{F}_{Sf}$ is the reference matrix, and $\mathbf{F}_{Rf}$ is the echo vector. It is anisotropic total variation when $p = 1$ and isotropic total variation when $p = 2$.

By employing the Lagrangian method to introduce convex relaxation, the objective function can be transformed into the following form:

$$\min_{\boldsymbol{\sigma}_A} J(\boldsymbol{\sigma}_A) = \sum_{l} \|D_l \boldsymbol{\sigma}_A\|_p - \lambda_1{}^T(\mathbf{F}_{Sf}\boldsymbol{\sigma}_A - \mathbf{F}_{Rf}) + \frac{\lambda_2}{2}\|\mathbf{F}_{Sf}\boldsymbol{\sigma}_A - \mathbf{F}_{Rf}\|_2^2 \tag{33}$$

Given that $D_l \boldsymbol{\sigma}_A$ is not derivable within the paradigm term $\|\cdot\|$, a slack variable w is introduced to replace $D_l \boldsymbol{\sigma}_A$ [37], resulting in the following model:

$$\min_{w_l, \boldsymbol{\sigma}_A} \sum_{l} \|w_l\|_p, \ s.t. \ \mathbf{F}_{Sf}\boldsymbol{\sigma}_A = \mathbf{F}_{Rf}, D_l \boldsymbol{\sigma}_A = w_l \tag{34}$$

The objective function undergoes a transformation from the quadratic penalty function to

$$\begin{aligned} \min_{\mathbf{w}, \boldsymbol{\sigma}_A} J(\mathbf{w}, \boldsymbol{\sigma}_A) \ &= \sum_{l} \left(\|w_l\|_p - \varepsilon_l{}^T(D_l \boldsymbol{\sigma}_A - w_l) + \frac{\mu}{2}\|D_l \boldsymbol{\sigma}_A - w_l\|_2^2 \right) \\ &-\lambda_1{}^T(\mathbf{F}_{Sf}\boldsymbol{\sigma}_A - \mathbf{F}_{Rf}) + \frac{\lambda_2}{2}\|\mathbf{F}_{Sf}\boldsymbol{\sigma}_A - \mathbf{F}_{Rf}\|_2^2 \end{aligned} \tag{35}$$

Consequently, the alternating iterative multiplier method can be employed to transform the original problem into the resolution of two subproblems, namely, the solution of w and that of $\boldsymbol{\sigma}_A$, in consecutive iterations.

4.2.1. **w** Sub-Problem

When $\boldsymbol{\sigma}_A$ is determined, (35) becomes

$$\begin{aligned} \min_{\mathbf{w}} J(\mathbf{w}) \ &= \min_{\mathbf{w}} \sum_{l} \left(\|w_l\|_p - \varepsilon_l{}^T(D_l \boldsymbol{\sigma}_A - w_l) + \frac{\mu}{2}\|D_l \boldsymbol{\sigma}_A - w_l\|_2^2 \right) \\ &= \min_{\mathbf{w}} \sum_{l} \left(\|w_l\|_p + \frac{\mu}{2}\left\|w_l - \left(D_l \boldsymbol{\sigma}_A + \frac{\varepsilon_l}{\mu}\right)_l\right\|_2^2 \right) \end{aligned} \tag{36}$$

When $p = 1$, it follows that

$$w_l = \max\left\{ \left| D_l \boldsymbol{\sigma}_A - \frac{\varepsilon_l}{\mu} \right| - \frac{1}{\mu}, 0 \right\} \mathrm{sgn}\left(D_l \boldsymbol{\sigma}_A - \frac{\varepsilon_l}{\mu} \right) \tag{37}$$

When $p = 2$, it follows that

$$w_l = \max\left\{ \left\| D_l \boldsymbol{\sigma}_A - \frac{\varepsilon_l}{\mu} \right\|_2 - \frac{1}{\mu}, 0 \right\} \frac{D_l \boldsymbol{\sigma}_A - \frac{\varepsilon_l}{\mu}}{\left\| D_l \boldsymbol{\sigma}_A - \frac{\varepsilon_l}{\mu} \right\|_2} \tag{38}$$

4.2.2. $\boldsymbol{\sigma}_A$ Sub-Problem

When $\mathbf{w}$ is determined, (35) becomes

$$\begin{aligned}\min_{\boldsymbol{\sigma}_A} J(\boldsymbol{\sigma}) &= \min_{\boldsymbol{\sigma}_A} \sum_i \left(-\varepsilon_l{}^T(D_l\boldsymbol{\sigma}_A - w_l) + \frac{\mu}{2}\|D_l\boldsymbol{\sigma}_A - w_l\|_2^2\right) \\ &-\lambda_1{}^T(\mathbf{F}_{Sf}\boldsymbol{\sigma}_A - \mathbf{F}_{Rf}) + \frac{\lambda_2}{2}\|\mathbf{F}_{Sf}\boldsymbol{\sigma}_A - \mathbf{F}_{Rf}\|_2^2\end{aligned} \tag{39}$$

In each iteration, the value of ε_l needs to be updated in accordance with the following equation:

$$\varepsilon_l^{k+1} = \varepsilon_l^k - \mu\left[D_l(\boldsymbol{\sigma}_A)_{k+1} - w_l^{k+1}\right] \tag{40}$$

The solution to (39) can be obtained by means of the fastest descent method. The iterative solution equation is as follows:

$$\boldsymbol{\sigma}_A = \boldsymbol{\sigma}_A - \alpha d(\boldsymbol{\sigma}_A) \tag{41}$$

where $d(\boldsymbol{\sigma}_A)$ is the derivative of (29) and can be expressed as

$$d(\boldsymbol{\sigma}_A) = \sum_l \left[\mu D_l^T(-D_l\boldsymbol{\sigma}_A - w_l) - D_l^T\varepsilon_l\right] + \lambda_2\mathbf{F}_{Sf}{}^T\left(\mathbf{F}_{Sf}\boldsymbol{\sigma}_A - \mathbf{F}_{Rf}\right) \tag{42}$$

The most rapid single-step descent can be applied to $\boldsymbol{\sigma}_A$

$$(\boldsymbol{\sigma}_A)_{k+1} = (\boldsymbol{\sigma}_A)_k - \alpha_k d_k \tag{43}$$

where k denotes the number of iterations, and α_k denotes the step size.

According to the literature [38], α_k can be determined by the following equation:

$$\alpha_k = \frac{(\mathbf{F}_{Sf})_k^T(\mathbf{F}_{Sf})_k}{(\mathbf{F}_{Sf})_k^T(\mathbf{F}_{Rf})_k} \tag{44}$$

where $(\mathbf{F}_{Sf})_k = (\boldsymbol{\sigma}_A)_k - (\boldsymbol{\sigma}_A)_{k-1}$, and $(\mathbf{F}_{Rf})_k = d_k((\boldsymbol{\sigma}_A)_k) - d_k((\boldsymbol{\sigma}_A)_{k-1})$.

The initial value of α_k can be derived from the following equation:

$$\alpha_1 = \frac{d_k^T d_k}{d_k^T\left(\mathbf{F}_{Sf}d_k\right)} \tag{45}$$

The following conditions need to be satisfied between two iterations:

$$J_k((\boldsymbol{\sigma}_A)_k - \alpha_k d_k) \le C_k - \delta\alpha_k d_k^T d_k \tag{46}$$

where C_k is defined as follows:

$$\begin{aligned}P_k &= \zeta P_{k-1} + 1 \\ C_k &= (\zeta P_{k-1}C_{k-1} + J_{k-1}(\boldsymbol{\sigma}_k))/P_k\end{aligned} \tag{47}$$

In the simulation, $P_k = 1$ was selected.

With the above iterations, $\boldsymbol{\sigma}_A$ can be solved.

The same three targets were reconstructed by using the ADMM-based total variation method. The resulting imaging results are presented in Figure 13.

Comparing SBL with the ADMM-based total variation method reveals that the latter has the capacity to reconstruct the structure of the target more completely and to perform better edge reconstruction.

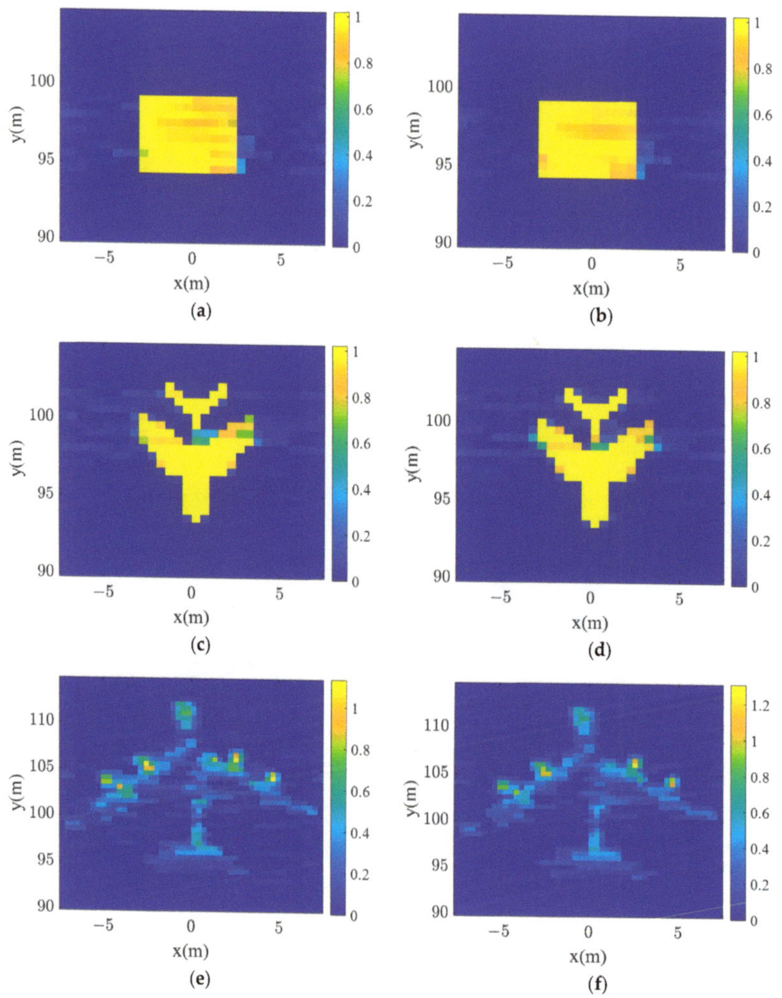

Figure 13. ADMM total variation imaging results. (**a**,**c**,**e**) Imaging results at 0 dB. (**b**,**d**,**f**) Imaging results at 5 dB.

In order to quantitatively analyze the effect of the ADMM-based total variation method on the imaging quality, the MSE curves of this method and the other four methods at different SNRs are plotted in Figure 14. This figure shows that the MSE curves of the five methods all demonstrate a decreasing trend as the SNR increases. The MSEs of the AziPM and PM methods are roughly the same when imaging, whether using the SBL or ADMM-based total variation algorithm. The MSE of FM-SBL is relatively high, while AziPM-TV has the lowest MSE among the different methods, with the curve located at the bottom. This further indicates the effectiveness of the method proposed in this paper.

Figure 15 compares the runtimes of the four methods, demonstrating that the total variation method yields a more accurate solution with a longer runtime. However, by solving the azimuthal dimensions at each distance sequentially, the computational complexity can be greatly reduced. The runtime can be reduced by about 3.8 compared with solving directly with the SBL algorithm and can be reduced by about 5.5 compared with solving directly with ADMM-based total variation.

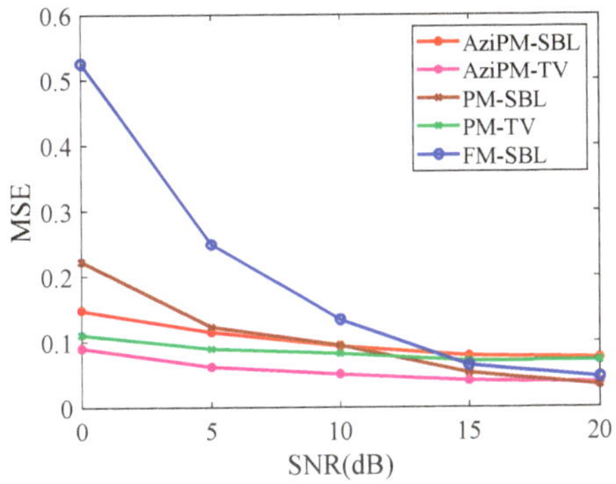

Figure 14. MSE curves for different methods.

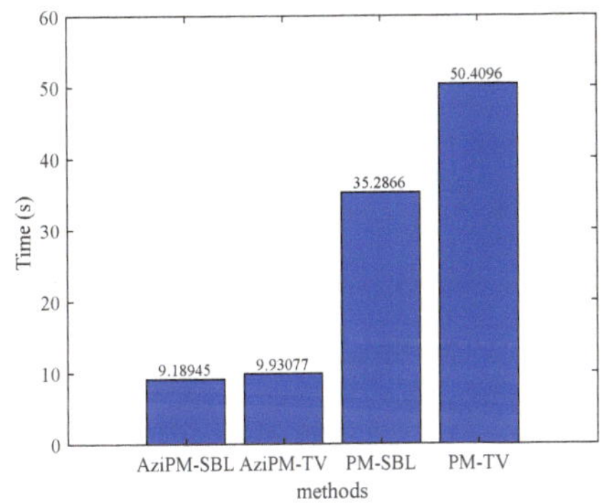

Figure 15. Comparison of runtimes of different methods.

5. Conclusions

In this paper, microwave coincidence imaging with phase-coded stochastic radiation fields is proposed. Firstly, due to the characteristics of phase modulation, the pulse compression technique is utilized in the microwave coincidence imaging process to obtain a one-dimensional range image in the frequency domain. Then, the equation in the azimuthal dimension is established, and the reconstruction of the target can be achieved by correlating echo signals with reference signals at different distances in the frequency domain. SBL and ADMM-based total variation methods were used to reconstruct targets. The two algorithms achieved good imaging results at low SNRs, but ADMM-based total variation could achieve better reconstruction of the edge structure of complex targets. The runtime of the proposed method can be reduced compared with the direct frequency-domain solution. The method can be applied to forward-looking radar imaging scenarios, such as autonomous obstacle avoidance with vehicle-mounted radar and terminal guidance.

Author Contributions: Writing—original draft preparation, H.L. (Hang Lin); writing—review and editing, H.L. (Hang Lin), H.L. (Hongyan Liu), Y.C., K.X., K.L. and Y.Y.; supervision, H.L. (Hongyan Liu); conceptualization and methodology, H.L. (Hang Lin); investigation, H.L. (Hongyan Liu). All authors have read and agreed to the published version of the manuscript.

Funding: This work was supported in part by the National Key Research and Development Program of China under Grant 2022YFB3902400, in part by the Distinguished Youth Science Foundation of Hunan Province under Grant 2022JJ10063, and in part by the National Natural Science Foundation of China under Grant 62322122.

Data Availability Statement: The original contributions presented in the study are included in the article, further inquiries can be directed to the corresponding author.

Conflicts of Interest: The authors declare no conflicts of interest.

References

1. Sun, M.; Qi, F.; Zhao, Y.; Ji, C. Resolution enhancement based on antenna aperture compensation in synthetic aperture radar imaging with hyper-laplacian prior deconvolution. *IEEE Sens. J.* **2019**, *23*, 21679–21686. [CrossRef]
2. Wu, H.; Cheng, Y.; Chen, X. Information geometry-based track-before-detect algorithm for slow-moving fluctuating target detection. *IEEE Trans. Signal Process.* **2023**, *71*, 1833–1848. [CrossRef]
3. Wu, H.; Cheng, Y.; Chen, X. Power spectrum information geometry-based radar target detection in heterogeneous clutter. *IEEE Trans. Geosci. Remote Sens.* **2024**, *62*, 5102016. [CrossRef]
4. Wu, J.; Li, Z.; Huang, Y. Focusing bistatic forward-looking SAR with stationary transmitter based on keystone transform and nonlinear chirp scaling. *IEEE Geosci. Remote Sens. Lett.* **2014**, *11*, 148–152. [CrossRef]
5. Yang, J. Multi-directional evolution trend and law analysis of radar ground imaging technology. *J. Radars.* **2019**, *8*, 669–692.
6. Li, D.; Li, X.; Qin, Y. Radar coincidence imaging: An instantaneous imaging technique with stochastic signals. *IEEE Trans. Geosci. Remote Sens.* **2014**, *52*, 2261–2277.
7. Zhu, S.; Zhang, A.; Xu, Z. Radar coincidence imaging with random microwave source. *IEEE Antennas Wirel. Propag. Lett.* **2015**, *14*, 1239–1242. [CrossRef]
8. Guo, Y.; He, X.; Wang, D. A novel super-resolution imaging method based on stochastic radiation radar array. *Meas. Sci. Technol.* **2013**, *24*, 074013. [CrossRef]
9. He, X.; Liu, C.; Liu, B. Sparse frequency diverse MIMO radar imaging for off-grid target based on adaptive iterative MAP. *Remote Sens.* **2013**, *5*, 631–647. [CrossRef]
10. Zhao, M. Frequency-diverse metacavity cassegrain antenna for differential coincidence imaging. *IEEE Trans. Antennas Propag.* **2023**, *71*, 9054–9059. [CrossRef]
11. Cao, K.; Cheng, Y.; Liu, Q. Microwave coincidence imaging based on attributed scattering model. *IEEE Signal Process. Lett.* **2022**, *29*, 1694–1698. [CrossRef]
12. Wu, J.; Sun, Z.; Yang, J. Star-machine dual-base SAR imaging based on GF-3 irradiation and experimental validation. *Radar Sci. Technol.* **2021**, *19*, 241–247.
13. Zhang, Y. Fast inverse-scattering reconstruction for airborne high squint radar imagery based on doppler centroid compensation. *IEEE Trans. Geosci. Remote Sens.* **2022**, *60*, 5205517. [CrossRef]
14. Zhu, Y.; He, M.; He, Z. Joint minimum variance and maximum likelihood spectral estimation for forward-looking imaging methods. *Syst. Eng. Electron.* **2023**, *45*, 3108–3115.
15. Li, W.; Zhang, B.; Li, K.; Yan, J. LRSD-ADMM-NET: Simultaneous super-resolution imaging and target detection for forward-looking scanning radar. *Remote Sens.* **2024**, *17*, 4052–4061. [CrossRef]
16. Liu, Q.; Cheng, Y.; Cao, K. Radar 3-D forward-looking imaging for extended targets based on attribute scattering model. *IEEE Geosci. Remote Sens. Lett.* **2023**, *20*, 3502305. [CrossRef]
17. Cao, K.; Cheng, Y.; Liu, K. Reweighted-dynamic-grid-based microwave coincidence imaging with grid mismatch. *IEEE Trans. Geosci. Remote Sens.* **2022**, *60*, 5102510. [CrossRef]
18. Zhu, S.; He, Y.; Chen, X. Resolution Threshold Analysis of the Microwave Radar Coincidence Imaging. *IEEE Trans. Geosci. Remote Sens.* **2020**, *58*, 2232–2243. [CrossRef]
19. Zhou, X.; Fan, B.; Wang, H. Sparse Bayesian perspective for radar coincidence imaging with array position error. *IEEE Sens. J.* **2017**, *17*, 5209–5219. [CrossRef]
20. Tian, B.; Zhang, X.; Tang, X. A joint sparse recovery algorithm for coprime adjacent array synthetic aperture radar 3D sparse imaging. *Int. J. Remote Sens.* **2021**, *42*, 6560–6580. [CrossRef]
21. Lu, X.; Yang, J.; Tan, K. Range-Doppler image reconstruction for collocated MIMO noise radar by sparse recovery. *Remote Sens. Lett.* **2022**, *13*, 279–289. [CrossRef]
22. Candès, J.; Wakin, M. An introduction to compressive sampling: A sensing sampling paradigm that goes against the common knowledge in data acquisition. *IEEE Signal Process.* **2008**, *25*, 21–30. [CrossRef]
23. Xu, G.; Zhang, B.; Yu, H.; Chen, J. Sparse Synthetic Aperture Radar Imaging from Compressed Sensing and Machine Learning: Theories, applications, and trends. *IEEE Geosci. Remote Sens.* **2022**, *10*, 32–69. [CrossRef]
24. Seidi, M.; Shah, U.; Oberhammer, J. Analysis of a Minimalistic Imaging Radar Concept Employing Beam Shape Switching and Compressed Sensing. *IEEE Trans. Geosci. Remote Sens.* **2024**, *62*, 1–12. [CrossRef]
25. Bai, X.; Wang, G.; Liu, S. High-Resolution Radar Imaging in Low SNR Environments Based on Expectation Propagation. *IEEE Trans. Geosci. Remote Sens.* **2021**, *59*, 1275–1284. [CrossRef]
26. Hariprasad, M. Circulant decomposition of a matrix and the eigenvalues of Toeplitz type matrices. *Appl. Math. Comput.* **2024**, *468*, 128473. [CrossRef]
27. Heng, Z.; Wang, X.; Li, X. Constructions of cyclic codes and extended primitive cyclic codes with their applications. *Finite Fields Appl.* **2023**, *89*, 102208. [CrossRef]

28. Babur, G.; Aubry, P. Space-time radar waveforms: Circulating codes. *Electr. Comput. Eng.* **2013**, *2013*, 9. [CrossRef]
29. Zhu, J.; Xie, Z.; Jiang, N. Delay-Doppler map shaping through oversampled complementary sets for high speed target detection. *Remote Sens.* **2024**, *16*, 2898. [CrossRef]
30. Zou, L. Study of wavelet entropy for airport pavement inspection using a multistatic ground-penetrating radar system. *Geophysics* **2021**, *86*, 69–78. [CrossRef]
31. Zhang, X. Singular value analysis. In *Matrix Analysis and Application*, 2nd ed.; Tsinghua Univ. Press: Beijing, China, 2013; pp. 301–312.
32. Mankar, V.; Das, T.; Sarkar, S. Discrete chaotic sequence based on logistic map in digital communications. *arXiv* **2012**, arXiv:1207.2694.
33. Li, C.; Li, Y. Research on statistical characteristics of chaotic pseudorandom sequence for one-dimensional Logistic map. *Appl. Res. Comput.* **2014**, *5*, 1403–1406.
34. Fan, B.; Zhou, X.; Chen, S. Sparse bayesian perspective for radar coincidence imaging with model errors. *Math. Probl. Eng.* **2020**, *2020*, 9202654. [CrossRef]
35. Xiao, Y.; Yang, J. A fast algorithm for total variation image reconstruction from random projections. *Mathematics* **2010**, *6*, 547–563.
36. Wang, Y. A new alternating minimization Algorithm for total variation image reconstruction. *Imaging Sci.* **2008**, *1*, 248–272. [CrossRef]
37. Liu, X. Augmented lagrangian method for total generalized variation based poissonian image restoration. *Comput. Math. Appl.* **2013**, *71*, 1649–1705. [CrossRef]
38. Eltiyeb, A.; Salem, M. Adaptive hybrid mixed two-point step size gradient algorithm for solving non-linear systems. *Mathematics* **2023**, *11*, 2102. [CrossRef]

remote sensing

MDPI

Article

High-Precision Sub-Wavelength Motion Compensation Technique for 3D Down-Looking Imaging Sonar Based on an Acoustic Calibration System

Jun Wang [1,2,3,†], Peihui Liang [1,2,†], Junqiang Song [1,2], Pan Xu [1,2], Yongming Hu [1,2], Peng Zhang [1,2], Kang Lou [1,2], Rongyao Ren [3] and Wusheng Tang [1,2,*]

1 College of Meteorology and Oceanography, National University of Defense Technology, Changsha 410073, China; wwjj103@nudt.edu.cn (J.W.); peinudt@nudt.edu.cn (P.L.); junqiang@nudt.edu.cn (J.S.); xupan09@nudt.edu.cn (P.X.); sdss@21cn.com (Y.H.); zhangpeng23a@nudt.edu.cn (P.Z.); loukang17@nudt.edu.cn (K.L.)
2 Hunan Key Laboratory for Marine Detection Technology, Changsha 410073, China
3 Tianjin Institute of Advanced Technology, Tianjin 300450, China; renry9710@163.com
* Correspondence: tangwusheng@nudt.edu.cn; Tel.: +86-152-0087-0306
† These authors contributed equally to this work.

Abstract: Three-dimensional hydro-acoustic imaging is a research hot spot in the underwater acoustic signal processing field, which has a wide range of application prospects in marine environmental resource surveying, seabed topography and geomorphological mapping, and underwater early warning and monitoring. To solve the problem that the resolution of the current imaging sonar reduces rapidly with increase in distance and a scanning gap exists in side-scan sonar, we designed a down-looking 3D-imaging sonar with a linear array structure. The imaging scheme adopts a time-domain spatial beam-forming method with the Back Projection (BP) algorithm as the core, and the formation of a virtual plane array can effectively improve the along-track resolution. To cope with the interference of the carrier motion error on the imaging, we proposed a high-precision sub-wavelength motion compensation method based on a real-time acoustic calibration system. Simulation and real data experiments show that the motion compensation method can effectively eliminate the influence of motion error and make the imaging energy more focused, leading to higher-quality acoustic images. Under equal average energy, the maximum superimposed sound intensity values in the imaging results increased by 20.75 dB and 6.57 dB, respectively, for simulation and real data. After motion compensation, the resolution of this imaging system reached 3 cm × 3 cm × 2.5 cm @ Depth = 17 m, TBP = 30 s · Hz.

Keywords: 3D hydro-acoustic imaging; down-looking linear array; motion compensation; acoustic calibration

1. Introduction

In recent years, with the increasing maturity of underwater acoustic transducer technology and the accelerating development of hydro-acoustic imaging technology, high-resolution imaging and high-confidence recognition of small underwater targets have become a research hot spot [1–3]. Three-dimensional (3D)-imaging sonars employ beam imaging technology popularly to reconstruct underwater 3D acoustic images [4,5]. However, the spatial resolution of conventional beam imaging is limited, which is usually determined by the array aperture [6–8]. Adaptive beam imaging is a technique which is capable of improving the spatial resolution. Stoica et al. [9] applied a natural extension of the Capon beamformer to the case of uncertain steering vectors, addressing the issue of performance degradation in the Capon beamformer when the steering vector knowledge is imprecise. Liao et al. [10] proposed a robust adaptive beam imaging technique that controls the peak sidelobe level in adaptive arrays, which is important for maintaining the integrity of the imaging resolution by managing interference. Along with the adaptive beam imaging technique, Wang et al. [6] used the deconvolution algorithm DAMAS2 to enhance 3D sonar resolution.

High-resolution imaging must provide a sufficient processing gain to combat the background noise. One way to combat noise is through the design of image reconstruction [11]. Yazici et al. [12] designed a Wiener filter in the affine group Fourier transform domain to minimize the wideband noise range-Doppler reflectivity, which can enable an operator to precondition transmitted waveforms to reject noise. Limin et al. [13] explored a family of rational orthogonal wavelets in active sonar detection to achieve Doppler robustness and noise mitigation.

Furthermore, motion errors, waveform design schemes and the sensor array structure can affect the resolution of 3D underwater acoustic imaging, with motion errors particularly reducing the imaging resolution significantly [14–16]. Due to various random factors, such as wind, waves, and sonar carrier manipulation, the carrier will deviate from the ideal trajectory. The image quality will be compromised if motion errors are not accounted for during the imaging progress [17,18]. Motion errors exceeding 1/8 of a wavelength will significantly affect the quality of 3D-imaging. General motion error compensation methods include error coarse compensation based on motion sensors [19,20] and motion error compensation based on sonar echo data [21,22]. The former is limited by the performance of the sensors and usually cannot achieve sub-wavelength accuracy. The latter has high compensation accuracy but has strict requirements for the structure and movement of the array.

In our study, the imaging scheme adopts the time-domain spatial beam-forming method with the Back Projection (BP) algorithm as the core. The formation of a virtual plane array can effectively improve the along-track resolution. To cope with the interference of the carrier motion errors on the imaging, we proposed a high-precision motion compensation method based on a real-time acoustic calibration system. The compensation accuracy can reach the sub-wavelength level, which can meet the needs of high-resolution imaging.

2. Moving Linear Array Down-Looking 3D-Imaging Model

2.1. Down-Looking Imaging Sonar System

Given the limited along-track resolution of the down-looking linear array sonar, we create a large virtual aperture along the trajectory by moving the linear array. The signal transceiver mode of the downward view can avoid the gaps problem of the side-scan imaging sonar and has unique advantages in 3D imaging, which is similar but different to the mode of the multi-beam sounder. In addition, adopting the down-looking imaging system, the spatial coordinate relationship is relatively simple, which is convenient for spatiotemporal registration and conducive to the fusion imaging of sonar data from multiple voyages.

In this paper, we studied a 3D image reconstruction method for the moving down-looking linear array imaging sonar. Figure 1 shows the working diagram of the down-looking linear array 3D-imaging system.

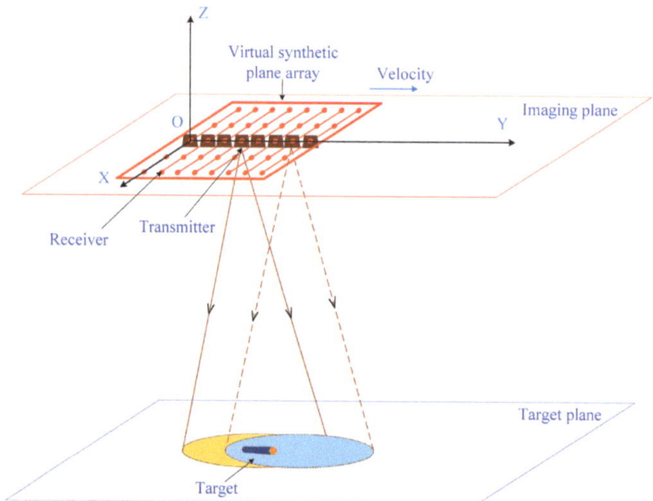

Figure 1. Schematic diagram of 3D-imaging sonar operation.

As shown in Figure 1, the carrier coordinate system is established with the center point of the linear array as the coordinate origin, also called the relative coordinate system. The extension direction of the linear array is the X-direction, the direction perpendicular to the linear array is the Y-direction, and the vertical upward direction is the Z-direction. The establishment of the coordinate system follows the right-hand rule. To better interpret the motion imaging process, the Y-direction in the diagram is redefined as the along-track direction. The X-direction and Z-direction in the diagram are redefined as the across-track direction and the depth direction, respectively.

Different from the general linear array, the specially designed receiving array has two hydrophones with a certain distance on both sides of the along-track direction, which is convenient for receiving three-dimensional underwater acoustic signals for the positioning of the sonar system [23]. The transmitter is a cylindrical acoustic source with a diameter of 3 cm located below the center of the horizontal array. The structure of the down-looking linear array imaging sonar is shown in Figure 2.

In motion imaging of the sonar system, the sound source of the down-looking linear array imaging sonar periodically actively emits a pulsed linear frequency modulation signal, also called a chirp signal. The chirp signal has good anti-interference ability due to its good autocorrelation characteristics, which makes it resistant to noise interference and improves the imaging quality in hydro-acoustic imaging. At the same time, the wide beam of the sound source can continuously irradiate the same target, and the imaging of the target can be completed by forming a virtual synthesis plane array by moving the linear array, which can effectively improve the along-track resolution of the target image [24].

The echo signals received by the linear receiving array, after pulse compression, can be recorded as a two-dimensional matrix $S_{N,M}(i,j;t)$, representing that this set of processed signals comes from M frames of echo signals received by N hydrophone elements. i is the index number of the hydrophone, and j is the frame number of the received signal.

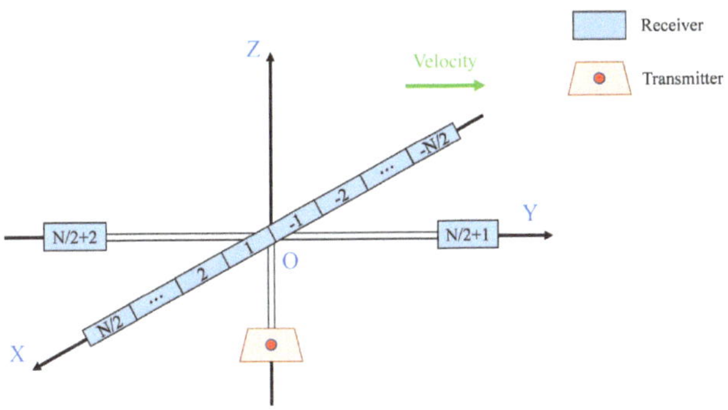

Figure 2. Structure of the transceiver array in the carrier coordinate system.

2.2. 3D BP Algorithm for the Down-Looking Linear Array Sonar

Despite the lower imaging efficiency of time-domain algorithms compared to frequency-domain and beam-domain algorithms, which results in high computational loads when dealing with long echo data and large-scale imaging scenes, both frequency-domain range-Doppler algorithms and beam-domain $w-k$ algorithms require the carrier to move in an almost strictly uniform linear motion. In the context of complex underwater environmental interference and difficult carrier attitude control, the advantages of time-domain imaging algorithms, which do not demand a specific trajectory for the carrier and offer good imaging stability and high precision, become prominent. Moreover, time-domain imaging algorithms are characterized by their simple imaging principles and ease of engineering implementation. Research into time-domain 3D-imaging algorithms has a more direct significance for array imaging on a larger scale [25]. The application of GPU parallel computing technology will greatly enhance the efficiency of time-domain imaging algorithms. And we have applied this technology to data processing.

Therefore, the 3D-imaging scheme adopts the time-domain spatial beam-forming method with the Back Projection (BP) algorithm as the core. The BP algorithm is an exact point-by-point search imaging method within the beam domain [26,27]. The basic principle for achieving 3D-imaging is based on the forward sound propagation model and pulse-compressed sonar echo data. It calculates the round-trip time delay of each grid point in the 3D target space to the sound source and hydrophone. Based on the round-trip time delay and the spatial geometric correspondence, the algorithm coherently superimposes the amplitudes of the processed signals that come from each hydrophone element and back-projects this sum onto the 3D grid point in the target space as the scattering intensity at that point. The described protocol will address the signal of each frame and amalgamate the outcomes of the 3D grid point intensity calculations. Thereby, the 3D image of the target space is reconstructed. It should be noted that when the platform is moving at a high speed, the resolution along the trajectory will be affected by the time delay. If this is not taken into account, it will affect the quality of the imaging.

The fundamental steps for 3D-imaging via the back-projection algorithm are establishing the forward model of sound propagation (i.e., ascertaining the positional information of the sound source, hydrophone elements, and target grid points), calculating the round-trip time delays, and performing back-projection. In the 3D BP imaging algorithm for the down-looking linear array, the imaging space is defined within the absolute coordinate system. And then we need to obtain the absolute position coordinates of the hydrophones and the sound source in each frame. Based on these coordinates, the target's round-trip time delay is computable, and finally, back-projection is conducted to achieve 3D-imaging. The specific process is as follows:

Step I: Set the imaging space

Establish an absolute coordinate system, setting the plane center of the bottom surface of the target imaging space as the origin of coordinate O_{abs}. The imaging space is designated as $A\{x \mid X\min \le x \le X\max, y \mid Y\min \le y \le Y\max, z \mid Z\min \le z \le Z\max\}$, and let the spacing of the imaging grid points along each axis of the coordinate system be Dx, Dy, Dz. Consequently, the coordinates of the imaging grid points in the absolute coordinate system can be determined as $P(Gx, Gy, Gz)$. The number of the imaging grid points can be calculated as the following equation:

$$\mathrm{P_{mum}} = \left[\frac{Xmax - Xmin}{Dx} + 1\right] \times \left[\frac{Ymax - Ymin}{Dy} + 1\right] \times \left[\frac{Zmax - Zmin}{Dz} + 1\right] \quad (1)$$

Step II: Obtain the absolute position coordinates of the hydrophones and the sound source in each frame

Let the absolute coordinates of the i th hydrophone element in the j th frame be denoted as $H(x_H^{(i,j)}, y_H^{(i,j)}, z_H^{(i,j)})$, and the absolute coordinates of the sound source in the j th frame be denoted as $S(x_S^{(j)}, y_S^{(j)}, z_S^{(j)})$. Given the assumptions in the down-looking linear array 3D-imaging scheme, the positional relationship between each hydrophone element and the sound source is fixed. Therefore, by calculating the absolute position coordinates of a specific hydrophone element or the sound source in a frame, one can determine the absolute position coordinates of all hydrophone elements and the sound source in this frame.

To facilitate the calculation of the absolute position coordinates for each hydrophone element and the sound source, the carrier coordinate system (relative coordinate system) has already been established in the text, as shown in Figures 1 and 2. Here, let the relative coordinates of the i th hydrophone element be denoted as $H_0(x_0^{(i)}, y_0^{(i)}, 0)$, and the relative coordinates of the sound source be denoted as $S_0(0, 0, z_0)$. Let Ds represent the spacing between each hydrophone element along the x-axis, and Ls represent the distance from two hydrophone elements along the y-axis to the linear array.

The relative horizontal coordinate of the hydrophone element $x_0^{(i)}$ can be calculated using the following formula:

$$x_0^{(i)} = \begin{cases} (Ds) \times i - \frac{Ds}{2}, 1 \le i \le N/2 \\ (Ds) \times i + \frac{Ds}{2}, -N/2 \le i \le -1 \\ 0, i = N/2+1, N/2+2 \end{cases} \quad (2)$$

The relative vertical coordinate of the hydrophone element can be calculated using the following formula:

$$y_0^{(i)} = \begin{cases} 0, -N/2 \le i \le N/2 \\ Ls, i = N/2+1 \\ -Ls, i = N/2+2 \end{cases} \quad (3)$$

Step III: Calculate the target's round-trip time delay

Given the aforementioned conditions, the round-trip time delays from the target grid coordinate points to each hydrophone element and the sound source can be calculated in the j th frame.

The time delay from the target grid point to the sound source can be calculated using the following formula:

$$\tau_{TS} = \frac{1}{c}\sqrt{\left(Gx - x_S^{(j)}\right)^2 + \left(Gy - y_S^{(j)}\right)^2 + \left(Gz - z_S^{(j)}\right)^2} \quad (4)$$

The time delay from the target grid point to the hydrophone elements can be calculated using the following formula:

$$\tau_{TH} = \frac{1}{c}\sqrt{\left(Gx - x_H^{(i,j)}\right)^2 + \left(Gy - y_H^{(i,j)}\right)^2 + \left(Gz - z_H^{(i,j)}\right)^2} \quad (5)$$

Therefore, the round-trip time delays from the target grid coordinate points to the i th hydrophone element and the sound source in the j th frame can be written as:

$$\tau_{i,j} = \tau_{TS} + \tau_{TH} \quad (6)$$

Step IV: Delay and sum to achieve 3D-imaging

Let the signal sampling rate be denoted as fs. After pulse compression, the time-domain signal $S_{N,M}(i,j;t)$ is sampled to obtain the discrete-time echo signal $S_{N,M}(i,j;n)$, and the round-trip time delay $\tau_{i,j}$ is sampled to obtain the round-trip delay points number $\Gamma_{i,j}$. For a target point P in space, the imaging result at this point is the accumulation of the amplitudes of the echo signals received by each hydrophone element in each frame at the time delay of that point. The imaging result for this point can be represented by the following formula:

$$\text{Im}\,P(Gx, Gy, Gz) = \left| \sum_{i=-N/2}^{N/2} \sum_{j=1}^{M} s_{N,M}(i,j;n - \Gamma_{i,j}) \right| \quad (7)$$

Applying the aforementioned processing to each grid point in the target imaging space, the 3D acoustic map of the target imaging space can be obtained.

2.3. Simulation of the 3D BP Algorithm

In order to validate the effectiveness of the time-domain BP 3D-imaging algorithm used in our paper, the point target simulation experiment under typical parameters is designed and carried out.

In the simulation, the pulse length of the linear frequency modulation signal $T_r = 1$ ms, the carrier center frequency $f_c = 85$ kHz, the signal bandwidth $B_r = 30$ kHz, and the frequency range is from 70 kHz to 100 kHz. The signal amplitude is 0.8, and the initial phase is 0. A total of frame number $M = 50$ chirp signals are emitted, and the sampling rate of the echo signal $f_s = 1$ MHz. The sound speed in water $c = 1449.93$ m/s.

In the simulation, the receiving array consists of a total of $(N+2) = 64$ hydrophone elements. The spacing between the N hydrophone elements along the x-axis $Ds = 9.5$ cm, and the distance from the latest two hydrophone elements along the y-axis to the linear array $Ls = 0.5$ m. The velocity of the carrier $V_0 = 0.05$ m/s.

The relative coordinate of the sound source is $S_0(0,0,0.04)$. The imaging space is designed as $A\{x \mid -0.5\text{ m} \le x \le 0.5\text{ m}; y \mid -0.5\text{ m} \le y \le 0.5\text{ m}; z \mid -0.1\text{ m} \le z \le 0.1\text{ m}\}$, and the spacing of the imaging grid points along each axis of the coordinate system $Dx = Dy = Dz = 0.01$ m. The single target grid point is set at the origin of the absolute coordinate system; its coordinate is $P(0, 0, 0)$. The distance from the target grid point to the imaging plane is $D_T = 10$ m.

The simulated echoes received by the hydrophone array of the down-looking linear array sonar are processed using the time-domain 3D back-projection imaging algorithm. The imaging results are shown in Figure 3.

As the 2D-imaging results show in Figure 3a, apart from the slight side lobes in the across-track and along-track directions where the target point is located, the point target focusing imaging effect is good, truly reflecting the position of the underwater target. By setting a reasonable threshold, the imaging results of the point target can be highlighted from the target imaging space. Figure 3b presents the 3D-imaging result of the point target after adjusting the threshold.

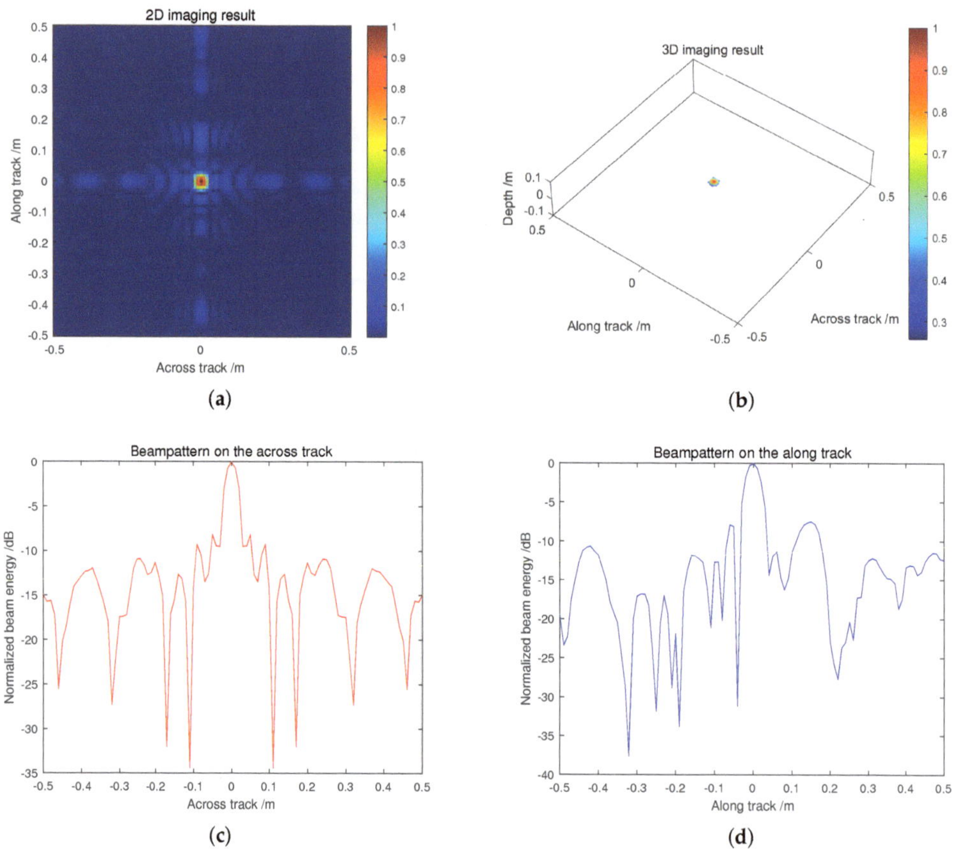

Figure 3. *Cont.*

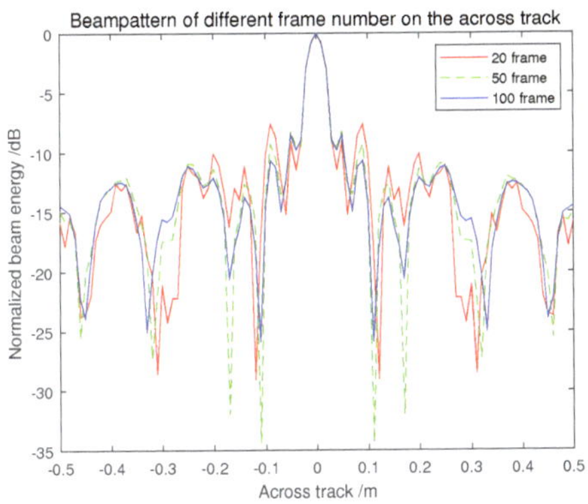

(e)

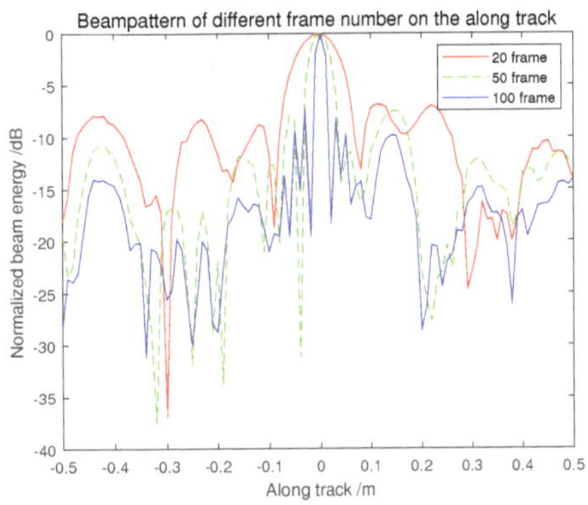

(f)

Figure 3. Simulation of single point target: (**a**) 2D-imaging result of the point target; (**b**) 3D-imaging result of the point target; (**c**) beampattern on the across track; (**d**) beampattern on the along track; (**e**) beampattern of different frame number on the across track; (**f**) beampattern of different frame number on the along track.

In Figure 3c, the beam pattern of the point target in the across-track direction has a 3 dB bandwidth of 0.04 m. The imaging result of the down-looking linear array sonar in the across-track direction is obtained by processing the echo signals received by N hydrophone elements in the across-track direction after pulse compression. The resolution capability cannot surpass the limitation of the physical aperture. The resolution capability in the across-track direction is determined by the across-track receiving beam width, which is related to the real aperture of the N-element linear array, the wavelength of the central wavelength of the sound wave λ_c, and the target depth D_T. Therefore, the resolution capability of the sonar system in the across-track direction is a constant value at the same depth.

In Figure 3d, the beam pattern of the point target in the along-track direction also has a 3 dB bandwidth of 0.04 m. However, the along-track resolution of the imaging system surpasses the limitation imposed by the physical aperture. It is determined by the along-track receiving beam width, which is related to the virtual synthetic aperture $L = V_0 t$ formed by the down-looking linear array during the motion imaging process, the central wavelength of the sound wave λ_c, and the target depth D_T.

For the down-looking linear array imaging system, the across-track imaging resolution is almost fixed, but the along-track imaging resolution can be improved by shortening the along-track receiving beam width, which is caused by increasing the length of the synthetic virtual aperture along the track. Different virtual aperture lengths can be chosen and the echo signals of the varied frames processed to obtain the beampattern in the across-track and along-track directions at the target depth, which is shown in Figure 3e,f, respectively. More imaging results can be found in Figure A1.

From Figure 3e,f, it can be observed that as the length of the synthetic virtual aperture increases, the resolution of the sonar system's acoustic image in the across-track direction (x-axis) remains constant, while the resolution in the along-track direction (y-axis) gradually improves. The results of the simulation experiment are consistent with the theoretical conclusions. However, due to the limitation of the transmitted beam width, the length of the virtual synthetic aperture will not increase indefinitely.

3. Motion Compensation Based on Real-Time Acoustic Calibration System

3.1. Effect of Motion Errors

Down-looking array imaging sonar is typically mounted on underwater Remote Operated Vehicles (ROVs), Autonomous Underwater Vehicles (AUVs), or fixed to the bottom of a ship. Due to the influence of dynamic factors such as wind and waves, as well as the limited motion control capabilities of the carrier, the carrier's movement will inevitably deviate from the set trajectory, making it challenging to obtain the absolute position coordinates of each hydrophone element and the sound source during motion. According to theoretical analysis, motion errors exceeding 1/8 of a wavelength will significantly affect the quality of 3D-imaging. Using only GPS positioning or inertial navigation often cannot meet the accuracy requirement.

The carrier motion error refers to the discrepancy between the attitude and trajectory of the carrier during motion and the established mathematical model. The main sources of motion error for the sonar carrier can be summarized as follows:

1. Carrier trajectory measurement error: When performing sonar imaging, it is necessary to accurately measure parameters such as the position, direction, and velocity of the sonar carrier. However, due to the precision limitations of various sensing devices or environmental interference, there is often some error in the measurement results;
2. Insufficient carrier attitude stability: The sonar carrier is easily disturbed by natural factors such as waves, wind, and ocean currents during travel, causing changes in its attitude, which, in turn, affects the reception direction and propagation path of the sonar signals;
3. Human operation error: Mistakes made by operators when adjusting the attitude of the sonar carrier or failing to adjust the attitude in time at specified locations can also lead to motion errors.

The motion error of the carrier manifests as a time delay difference when the sonar receives the echo signals, which will affect the phase information of the received echoes and thus affect the quality of the acoustic image. In practice, however, hydrophone elements at different locations will experience a time delay error τ_e due to the carrier's motion error. Therefore, the actual expression of the echo signal received by a single hydrophone element is:

$$s_{re}(t) = rect(\frac{t-\tau-\tau_e}{T_r}) \exp\left\{j(2\pi f_c(t-\tau-\tau_e) + \pi K_r(t-\tau-\tau_e)^2)\right\} \quad (8)$$

From the above formula, it can be seen that motion errors will simultaneously affect the envelope and phase shift of the echo signal. Motion compensation actions are required to process the echoes to obtain a relatively ideal acoustic image.

When the sonar carrier is in motion, it typically has six degrees of freedom: three translational degrees of freedom and three rotational degrees of freedom. The translational degrees of freedom include movements along the x-, y-, and z-axes, while the rotational degrees of freedom encompass rotations around the x-, y-, and z-axes. For ease of description, the carrier motion errors introduced by these six degrees of freedom are as follows: swag, surge, heave, pitch, roll, and yaw [28]. A schematic illustration of these six types of motion errors is shown in Figure 4.

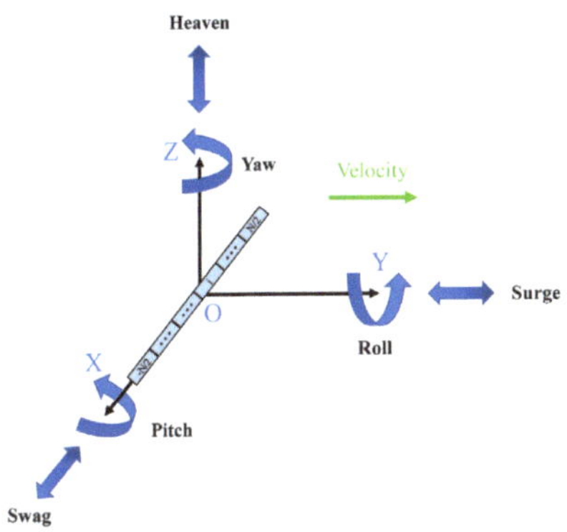

Figure 4. Six types of motion errors.

For the down-looking linear array 3D-imaging sonar, the first three translational motion error components will primarily affect the calculation of the distance from the sound source and the hydrophone elements to the target, thereby affecting the round-trip time delay and phase shift of the echo signals received by the hydrophone, and consequently having a significant impact on the imaging results. The last three rotational motion error components will mainly affect the range of the observation area for 3D sonar, and on the acoustic signal propagation, they will affect the Doppler center frequency of the signal and the phase between the various elements of the receiving array, which, in turn, affects the beamforming results. However, for our downward-looking linear array 3D-imaging sonar system mounted on an Autonomous Underwater Vehicle (AUV), it is required that there should be no significant attitude changes while it is operating. The presence of the three rotational motion error components may result in geometric distortion in the sonar imagery, but the impact on image quality is rather smaller. Therefore, our research focuses on compensating for the first three translational motion error components.

Given that the acoustic array is stationary, the three translational motion errors uniformly influence each hydrophone element. Therefore, in the analysis of translational motion errors, the motion error of the array center-point can be used to represent the entire hydrophone array. Let the position of the array center point at any moment be $(x(t), y(t), z(t))$. When the linear array moves from the coordinate origin along the y-axis at a speed of V_0, the ideal position of the array center-point can be written as $(0, V_0 t, 0)$. For the imaging target point $P(Gx, Gy, Gz)$ in the target space, the ideal distance from the array center to this point is:

$$R_{ideal}(t) = \sqrt{Gx^2 + (V_0 t - Gy)^2 + Gz^2} \tag{9}$$

However, in reality, there is translational motion error of the carrier, denoted as $\triangle x = x(t) - 0$, $\triangle y = y(t) - V_0 t$, $\triangle z = z(t) - 0$. For the target point P, the actual distance from the array center to this point is:

$$R_{actual}(t) = \sqrt{[x(t) - Gx]^2 + [y(t) - Gy]^2 + [z(t) - Gz]^2} \tag{10}$$

Let $R_0 = \sqrt{(Gx^2 + Gz^2)}$ represent the ideal distance from the target point to the moving trajectory, and let the ideal angle between the line connecting the target point and the array center and the YOZ plane be fixed as θ. Then, we have $Gx = R_0 \sin\theta$ and $Gz = -R_0 \cos\theta$, and (9) can be written as:

$$R_{ideal}(t) = \sqrt{R_0^2 + (V_0 t - Gy)^2} \approx R_0 + \frac{(V_0 t - Gy)^2}{2R_0} \tag{11}$$

Equation (10) can be written as:

$$R_{actual}(t) = \sqrt{[y(t) - Gy]^2 + R_0^2 - 2x(t)R_0 \sin\theta + 2z(t)R_0 \cos\theta} \tag{12}$$

After derivation, the difference between the actual distance and the ideal distance from the array center to the target point can be expressed as:

$$\triangle R(t) = R_{actual}(t) - R_{ideal}(t) \approx \frac{\triangle y(V_0 t - Gy - \triangle y/2)}{R_0} - \triangle x \sin\theta + \triangle z \cos\theta \tag{13}$$

Analyzing the impact of the three error components of swag, surge, and heave separately: the swag and heave, after projection, will directly affect the distance measurement accuracy and have a significant impact on the imaging results, whereas the surge, compared to the distance from the array center to the target point, is a small account, and its impact on the imaging results is relatively minor. Therefore, the compensation accuracy for the carrier motion errors of swag and heave needs to reach the millimeter level to ensure that the carrier's motion error is controlled at the sub-wavelength level, thereby obtaining high-quality acoustic images.

3.2. *Scheme of Motion Compensation*

3.2.1. Motion Compensation System

In this section, the acoustic calibration system is first introduced. Figure 5 shows a schematic diagram including the acoustic calibration system. While the hydrophone elements of the down-looking linear array imaging sonar receive the target scattering echo signals, they also receive the calibration signals emitted by the underwater preset calibration sound source.

By reasonably setting the sound emission period and delay of the calibration sound source to match the emission period of the sound source in the sonar array and the target depth, the calibration sound source signal and the target echo signal can fall within the same sound signal acquisition window. This enables the calibration sound source signal to be used for positioning each hydrophone element. It should be noted that the calibration sound source has a limited range of effectiveness. When the calibration sound source is far from the sonar array, the calibration signal received by the hydrophone elements will be submerged in the background noise and cannot be effectively extracted.

There are two different processing methods for the received calibration sound source signals. One is the sonar positioning measurement algorithm based on three-point positioning, which has high computational efficiency but is relatively weak in adaptability to the entire hydrophone system. It can only reflect the position coordinates of the selected three hydrophone elements well, and the method will fail when the data of the selected hydrophone elements for positioning are abnormal. The other is the sonar positioning measurement algorithm based on multi-point positioning, which fully utilizes the calibration signals received by each hydrophone element and positions the sonar system based on the idea of least squares. It has good adaptability to the system but has lower computational efficiency. In engineering applications, it is essential to choose the appropriate method based on the specific circumstances. Both kinds of sonar positioning measurement algorithms will be introduced and compared with each other.

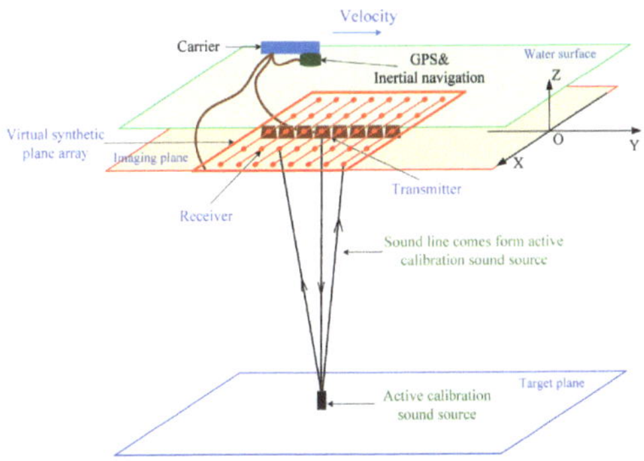

Figure 5. Schematic diagram of motion compensation system.

3.2.2. Motion Compensation Algorithm Based on Three-Point Positioning

The principle of the motion compensation algorithm based on three-point positioning involves selecting three hydrophone elements that are not collinear and using the calibration signals they receive to calculate the distances from these hydrophone elements to the calibration sound source. Through geometric relationships, the relative position coordinates of the calibration sound source can be calculated. Since the absolute position coordinates of the calibration sound source are known, the absolute position coordinates of each hydrophone element and the carrier sound source can be obtained through coordinate transformation, thereby compensating for the sonar echo signals.

Step I: Select the hydrophone elements for positioning

To attain optimal outcomes in acoustic positioning predicated on geometric relationships, it is crucial to select hydrophone elements with favorable spatial characteristics. Generally, the three hydrophone elements used for positioning will be chosen from the hydrophone elements at the ends of the linear array and the two hydrophone elements perpendicular to the linear array.

Step II: Calculate the distance from the selected hydrophone elements to the calibration sound source

During the down-looking linear array sonar imaging process that includes a calibration sound source, the echo signals received by the hydrophone elements consist of two parts: one is the imaging signal containing target information, and the other is the calibration signal containing the position information of the calibration sound source. The echo signals received by the sonar can be expressed in the following form:

$$s_r(n) = s_{imag}(n) + s_{carli}(n) \tag{14}$$

where $s_r(n)$ represents the sampled echo signal, while $s_{imag}(n)$ and $s_{carli}(n)$ denote the sampled imaging signal and calibration signal, respectively.

Perform pulse compression on the calibration signal and select the sampling point corresponding to the maximum amplitude of the calibration signal N_{max}; the following condition should be met:

$$s_{carli}(N_{max}) = \max\{|s_{carli}(n)|\} \tag{15}$$

The distance from the hydrophone element receiving the calibration signal to the calibration sound source can be calculated using the following formula:

$$R_{HC} = (N_{max} + N_{start} - N_{delay})\frac{c}{fs} \tag{16}$$

where N_{start} is the delay point of the channel transmission, and N_{delay} is the delay point of the calibrated sound source.

For the three hydrophone elements used for positioning, the distances to the calibration sound source are denoted as $R_{HC}^{(j,n)}, n = 1,2,3$.

Step III: Calculate the relative position coordinates of the calibration sound source

Let the relative coordinates of the calibration sound source in the j th frame be denoted as $C(x_C^{(j)}, y_C^{(j)}, z_C^{(j)})$. Based on the distances from the three selected hydrophone elements to the calibration sound source R_{HC} and the relative position coordinates of the selected hydrophones $H_0(x_0^{(i)}, y_0^{(i)}, z_0^{(i)}), i = 1,2,3$, the following system of equations can be established:

$$\begin{cases} \sqrt{\left(x_C^{(j)} - x_0^{(1)}\right)^2 + \left(y_C^{(j)} - y_0^{(1)}\right)^2 + \left(z_C^{(j)} - z_0^{(1)}\right)^2} = R_{HC}^{(j,1)} \\ \sqrt{\left(x_C^{(j)} - x_0^{(2)}\right)^2 + \left(y_C^{(j)} - y_0^{(2)}\right)^2 + \left(z_C^{(j)} - z_0^{(2)}\right)^2} = R_{HC}^{(j,2)} \\ \sqrt{\left(x_C^{(j)} - x_0^{(3)}\right)^2 + \left(y_C^{(j)} - y_0^{(3)}\right)^2 + \left(z_C^{(j)} - z_0^{(3)}\right)^2} = R_{HC}^{(j,3)} \end{cases} \tag{17}$$

Using mathematical tools, the numerical solution of the aforementioned equations can be calculated. This numerical solution includes two solutions that are symmetrical with respect to the plane. By analyzing the geometric relationship, the solution that corresponds to the desired criteria is selected as the relative position coordinates of the calibration sound source.

Step IV: Solve the absolute coordinates of each hydrophone element and the carrier sound source by coordinate transformation, and calculate the motion error compensation values

Since the operational steps after the calculation of the relative position coordinates of the calibration sound source are the same in both the three-point positioning and multi-point positioning algorithms, these two calculation methods will be described with a specific frame calculation in the motion compensation algorithm based on multi-point positioning.

3.2.3. Motion Compensation Algorithm Based on Multi-Point Positioning

The basic principle of the motion compensation algorithm based on multi-point positioning is similar to that of the ultra-short baseline (USBL) acoustic positioning method. In the USBL acoustic positioning method, the shipborne transceiver emits a pulse interrogation signal to the underwater transponder. After receiving the interrogation signal, the transponder emits a response pulse signal that is different from the interrogation signal. By receiving the response pulse signal through several hydrophone elements spaced a few centimeters apart, the distance and bearing of the transponder are calculated based on the round-trip time and phase difference of the sound wave propagation in water, thereby calculating the coordinates of the transponder. With the absolute coordinates of the transponder known, the precise coordinates of the carrier can be determined through coordinate transformation.

The down-looking linear array sonar's $(N+2)$ hydrophone elements are utilized to simultaneously receive imaging signals and calibration signals. Since the calibration sound source emits signals slightly later than the carrier sound source, the calibration signals can be separated from the signals used for imaging during the data processing. At the same time, the echo ranging is performed using the calibration signals received by all of the hydrophone elements, and the relative position coordinates of the underwater calibration sound source are calculated using the least squares method. Similarly to the three-point positioning method, since the absolute coordinates of the calibration sound source have been determined during deployment, the absolute coordinates of each hydrophone element and the carrier sound source can be obtained through coordinate transformation. Then, the motion errors embodied in the echo signal can be calculated.

Step I: Calculate the distance from each hydrophone element to the calibration sound source

Using the same method as in the three-point positioning to calculate the distance from the hydrophone element to the calibration sound source, extract the calibration signal received by each hydrophone element, and select the maximum amplitude value among them. The distance from each hydrophone element to the calibration sound source can be calculated using Equation (16)

Step II: Calculate the relative coordinates of the calibration sound source using the Newton–Raphson iterative method

During the motion imaging process of the down-looking linear array sonar, it is necessary to calculate the relative position coordinates of the calibration sound source for each frame. Let the distance from the $(N+2)$ hydrophones to the calibration sound source be $R_{HC}^{(j,n)}, n = 1,2,\ldots,N+2$. The relative coordinates of the calibration sound source $C(x_C^{(j)}, y_C^{(j)}, z_C^{(j)})$ in the j th frame can be obtained by solving a nonlinear system of $(N+2)$ equations. Based on the principle of least squares, the Newton–Raphson iterative method is used to calculate the relative coordinates of the calibration sound source. The algorithm flow is as follows:

(1) Set the initial parameters

Let the initial relative coordinates of the calibration sound source be denoted as $C^{(k)}(x_C^{(j,k)}, y_C^{(j,k)}, z_C^{(j,k)}), k = 0$, the maximum number of iterations be N_{break}, and the maximum allowable error be $\sigma_{\max}$. In order to minimize the number of iterations, the relative coordinates of the calibration sound source should be initialized to approximate the true values as closely as feasible.

(2) Calculate the Jacobian matrix of the system of equations

The i th equation in the system can be represented by the following formula:

$$\sqrt{\left(x_C^{(j)} - x_H^{(i,j)}\right)^2 + \left(y_C^{(j)} - y_H^{(i,j)}\right)^2 + \left(z_C^{(j)} - z_H^{(i,j)}\right)^2} = R_{HC}^{(j,n=i)} \quad (18)$$

This system of nonlinear equations can be written in the following simplified form:

$$\begin{cases} f_1(x_C^{(j)}, y_C^{(j)}, z_C^{(j)}) = 0 \\ f_2(x_C^{(j)}, y_C^{(j)}, z_C^{(j)}) = 0 \\ \vdots \\ f_{N+2}(x_C^{(j)}, y_C^{(j)}, z_C^{(j)}) = 0 \end{cases} \tag{19}$$

Then, the Jacobian matrix of the system of nonlinear equations can be denoted as:

$$F'(C) = \begin{bmatrix} \frac{\partial f_1(X)}{x_c^{(j)}} & \frac{\partial f_1(Y)}{y_c^{(j)}} & \frac{\partial f_1(Z)}{z_c^{(j)}} \\ \frac{\partial f_2(X)}{x_c^{(j)}} & \frac{\partial f_2(Y)}{y_c^{(j)}} & \frac{\partial f_2(Z)}{z_c^{(j)}} \\ \vdots & \vdots & \vdots \\ \frac{\partial f_{N+2}(X)}{x_c^{(j)}} & \frac{\partial f_{N+2}(Y)}{y_c^{(j)}} & \frac{\partial f_{N+2}(Z)}{z_c^{(j)}} \end{bmatrix} \tag{20}$$

(3) Iteratively solve the relative coordinates of the calibrated sound source

The iterative format of the Newton–Raphson iterative method is:

$$C^{(k+1)} = C^{(k)} - F'(C^{(k)})^{-1} F(C^{(k)}) \tag{21}$$

The termination condition for the iteration is $\left\| C^{(k+1)} - C^{(k)} \right\|_\infty < \sigma_{\max}$, or the iteration reaches the maximum number N_break. At the end of the iteration, the relative coordinates of the calibrated sound source in the j th frame $C(x_C^{(j)}, y_C^{(j)}, z_C^{(j)})$ can be obtained.

Step III: Solve for the absolute coordinates of each hydrophone element and the carrier sound source by coordinate transformation

Once the system deployment is completed, the absolute position coordinates of the calibration sound source are thereby determined. The coordinates in the relative coordinate system can be transformed into the coordinates in the absolute coordinate system through a linear transformation. Let the absolute position coordinates of the calibration sound source be C_{abs}. In the j th frame, the absolute coordinates of each hydrophone element and the carrier sound source can be calculated using the following formula:

$$\begin{cases} H = (C_{abs} - C) + H_0 \\ S = (C_{abs} - C) + S_0 \end{cases} \tag{22}$$

Step IV: Calculate the compensation values for the echo motion error

The absolute coordinates of the aforementioned hydrophone elements and the carrier sound source are obtained through the joint calculation of signals from $(N+2)$ hydrophone elements, which can be considered as the most ideal results obtained under the condition of limited data. The difference between the calculated distance R^* from the hydrophone element to the calibration sound source and the distance R_{HC} from the hydrophones to the calibration sound source measured by the echo signal is the error value $\varepsilon_{i,j}$ generated in the sonar positioning process. The compensation value for the sonar motion error is the opposite number of the error value.

In the j th frame, the ideal distance R^* from the i th hydrophone element to the calibration sound source can be calculated using the following formula:

$$R^* = \sqrt{\left(x_H^{(i,j)} - x_C^{(j)}\right)^2 + \left(y_H^{(i,j)} - y_C^{(j)}\right)^2 + \left(z_H^{(i,j)} - z_C^{(j)}\right)^2} \tag{23}$$

The error value can be calculated by the following formula:

$$\varepsilon_{i,j} = R^* - R_{HC} \tag{24}$$

The compensation value can be calculated by the following formula:

$$\rho_{i,j} = -\varepsilon_{i,j} \tag{25}$$

In addition, depending on the sound speed c and the sampling rate fs, the number of echo delay points N_{comp} used to compensate can be obtained. After compensation, the echo signals used for imaging can be expressed as:

$$s_{used}(t) = rect(\frac{t-\tau-\tau_e+\rho/c}{T_r})\exp\left\{j(2\pi f_c(t-\tau-\tau_e+\rho/c)+\pi K_r(t-\tau-\tau_e+\rho/c)^2)\right\} \tag{26}$$

In mathematical terms, the three types of translational motion errors can be separated through approximate assumptions, looking at Equation (13). However, in the actual signal processing process, these three types of motion errors are deeply coupled together, such as Equation (8). Our motion error compensation method does not focus on measuring individual motion error components but compensates for motion errors in the time delay of the echos, ingeniously resolving the issue of the spatial coupling of motion errors, which has been shown in Equation (26). Additionally, the compensation accuracy and the different effects between the motion compensation algorithm based on three-point positioning and multi-point positioning can be found in Figure A2.

4. Experiments and Results

4.1. Validation Test of Simulation Data

This section tests the feasibility of the motion error compensation method based on the acoustic calibration system that we propose, using simulated data. The setting of simulation parameters is all based on the actual experimental system and experimental environment. Therefore, the simulated data can more closely approximate the real imaging environment.

The parameters of the acoustic signal in the simulation are as follows: the a pulse length of the linear frequency modulation signal $T_r = 1$ ms, the carrier center frequency $f_c = 85$ kHz, the signal bandwidth $B_r = 30$ kHz, and the frequency range from 70 kHz to 100 kHz. The signal amplitude is 0.8, and the initial phase is 0. A total of frame number $M = 100$ chirp signals are emitted, and the sampling rate of the echo signal $f_s = 1$ MHz. The sound speed in water $c = 1449.93$ m/s.

The parameters related to the transducer array in the simulation are as follows: the receiving array consists of a total of $(N+2) = 64$ hydrophone elements. The spacing between the N hydrophone elements along the x-axis $Ds = 9.5$ cm, and the distance from the latest two hydrophone elements along the y-axis to the linear array $Ls = 0.5$ m. The velocity of the carrier $V_0 = 0.05$ m/s. The relative coordinate of the sound source is $S_0(0, 0, 0.04)$.

The parameters related to the imaging space are as follows: the imaging space is designed as $A\{x \mid -0.5\text{ m} \le x \le 0.5\text{ m}; y \mid -0.5\text{ m} \le y \le 0.5\text{ m}; z \mid -0.1\ m \le z \le 0.1\text{ m}\}$, and the spacing of the imaging grid points along each axis of the coordinate system $Dx = Dy = Dz = 0.01$ m.

The parameters related to the target are as follows: the interval between the target points is 0.2 m, and the distance from the target plane to the imaging plane is 10 m. The distribution of the simulation targets can be found in Figure 6a. Under ideal conditions, the imaging results of these point targets are shown in Figure 6b.

To verify the correction effect of the motion compensation algorithm, we added the sinusoidal errors with an amplitude of 0.2 m to the ideal trajectory in the along-track, across-track, and depth directions. At the same time, we used the motion error compensation method proposed in our paper to estimate the trajectory and compensate for the motion error. Under real experimental conditions, such a large translational motion error would not occur, but to vividly demonstrate the effect of our motion compensation method, we deliberately magnified the motion error in the simulation. In Figure 7a, the blue dashed line represents the ideal straight trajectory, the red solid line represents the actual trajectory with errors, and a series of green dots represent the predictions of our proposed motion error compensation method, which is highly consistent with the actual trajectory. Figure 7b shows the difference between the actual motion error and the estimated value. We can observe that our method achieves a compensation accuracy of $-4{\sim}4 \times 10^{-3}$ m for the sway motion error on the x-axis and the heave motion error on the z-axis. Based on the analysis in Section 3.1, the compensation accuracy for the sway and heave components of the motion error is relatively high, requiring precision to the millimeter level. As Figure 7b shows, the motion error compensation method we proposed meets the corresponding accuracy requirements for the compensation of these two motion error components, resulting in high-quality imaging results.

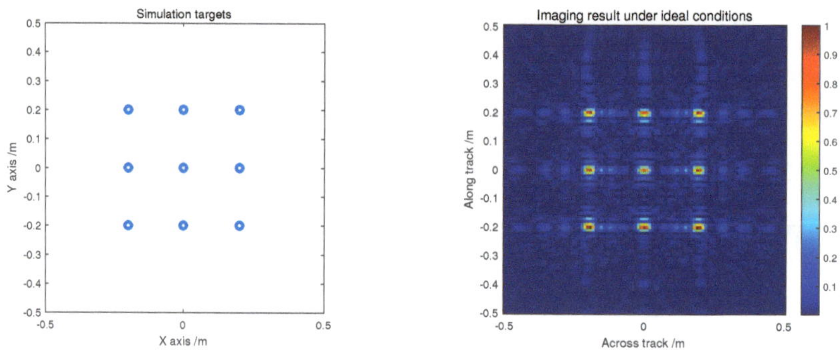

(a) (b)

Figure 6. Target and ideal imaging result of simulation data: (**a**) distribution of the simulation targets; (**b**) ideal imaging result.

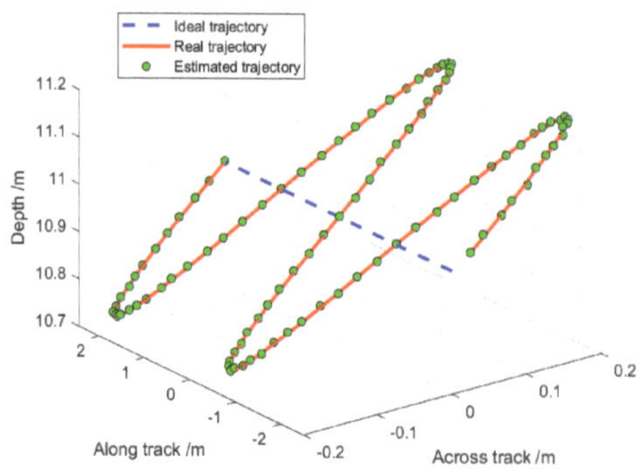

(a)

Figure 7. *Cont.*

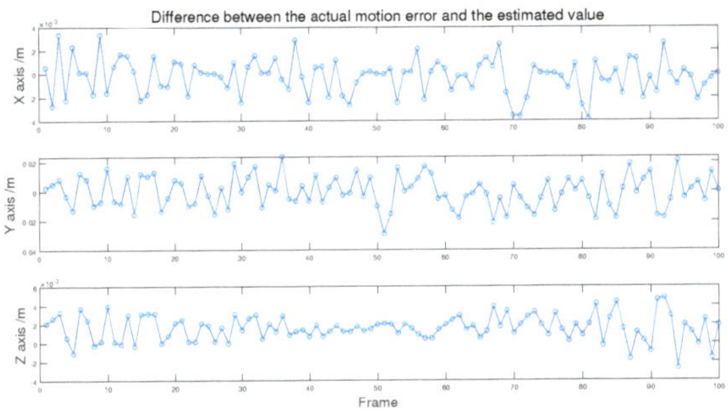

(**b**)

Figure 7. Effect of the motion compensation: (**a**) trajectory estimation result, (**b**) the difference between the actual motion error and the estimated value.

Ultimately, we can use the motion-compensated trajectory of the carrier for imaging processing. We compared the visual effects of the targets in the normalized imaging results and the changes in the maximum superimposed sound intensity value before and after motion compensation. Figure 8a shows the normalized imaging result of the point group by down-looking linear array sonar without motion error compensation. The imaging results exhibit severe defocus, making it impossible to discern the nine target points in the sonar image. And the maximum superimposed sound intensity value is 42.68 dB. Figure 8b displays the normalized imaging result after processing with the motion error compensation algorithm. The imaging results indicate that the motion error compensation algorithm effectively addressed the defocus issue, leading to a significant enhancement in the imaging quality of the sonar image from the downward-looking 3D-imaging sonar. At this time, the maximum superimposed sound intensity value in the sound field increases to 63.42 dB.

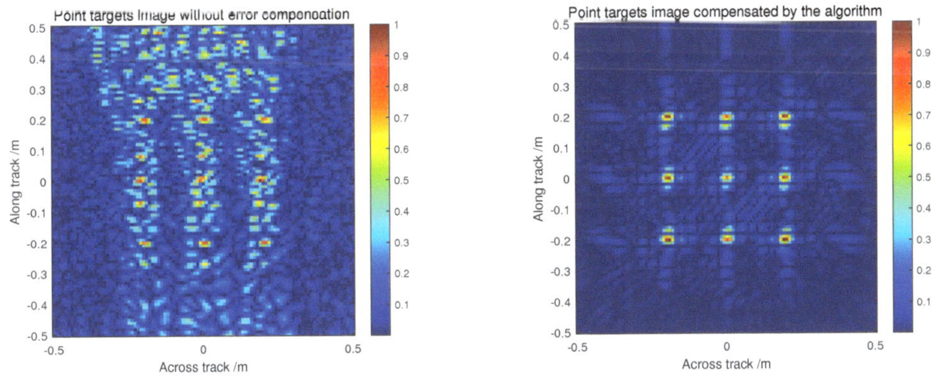

(**a**) (**b**)

Figure 8. Impact of motion compensation on imaging results: (**a**) point targets image without error compensation; (**b**) point targets image compensated by the algorithm.

4.2. Validation Test of Real Data

In this section, we use the real data collected by the downward-looking linear array 3D imaging sonar during the lake trial to verify the motion error compensation method based on the real-time acoustic calibration system. The parameters related to the transducer array and the acoustic signal are the same as those in simulation. However, due to the variability of the underwater environment, in our experiments with real data, we have measured the sound velocity profile of the experimental water area before imaging to correct the sound velocity values used in the imaging process. And the total of frame number $M = 701$. The parameters related to the imaging space are changed: the imaging space is expanded to $A\{x \mid -1\ \text{m} \leq x \leq 1\ \text{m}; y \mid -1\ \text{m} \leq y \leq 1\ \text{m}; z \mid -1\ \text{m} \leq z \leq 1\ \text{m}\}$, and the spacing of the imaging grid points along each axis of the coordinate system $Dx = Dy = Dz = 0.01$ m.

The target depth is 17 m, corresponding to the distance from the target plane to the imaging plane. The diameter of the circular point target on the target is 3 cm, and the interval between the point targets is 5 cm, and our emitted sound signal has a central wavelength of 2 cm. This target design is already very close to the limit of the resolution capability of sound waves to showcase the performance of imaging sonar with our motion compensation method. The shape of the target is shown in Figure 9.

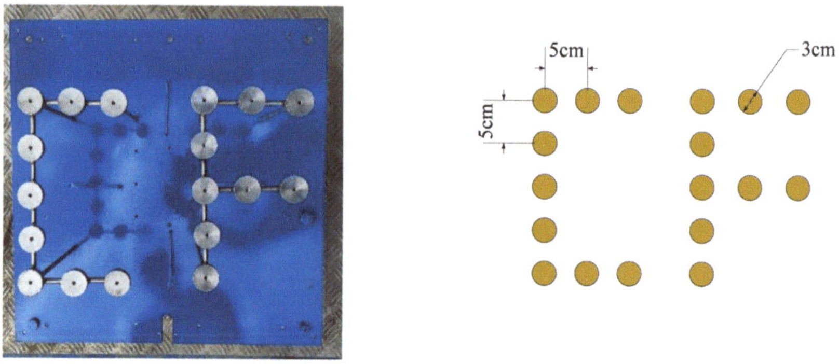

(a) (b)

Figure 9. The shape of the target: (**a**) optical photograph of the target; (**b**) the dimensions of the target.

The imaging system during the lake trial consists of a transducer array, a carrier, and auxiliary equipment. The transducer array is rigidly connected to the carrier, which is also equipped with other auxiliary measurement devices, such as a GPS positioning system and an inertial navigation system. A motor can drive the carrier to move along with the transducer array. The central control platform can control the auxiliary measurement system to work in conjunction with the transducer array, completing the collection of the sonar echo signal data, GPS positioning data, and inertial navigation data during the movement. For the calibration sound source, as the imaging distance increases, the calibration accuracy tends to decrease. Raising the calibration sound source frequency can improve the positioning accuracy, but the effective range of the signal will also decrease. By increasing the beam opening angle of the calibration sound source, the effective range of positioning can be expanded, but the positioning distance will be reduced. Therefore, key factors such as the calibration sound source frequency, signal strength, and beam opening angle, which constrain the range and distance, need to be comprehensively designed according to specific application requirements. Within a bottom imaging height of 30 meters, the calibration sound source we use fully meets the precision requirements for sub-wavelength motion error compensation.

In the experiments with real data, we compared the visual effects of the targets in the normalized imaging results and the changes in the maximum superimposed sound intensity value between the error coarse compensation scheme based on motion sensors and our proposed motion error compensation method. The comparison of the imaging results is shown in Figure 10. Figure 10a presents the imaging result with error coarse compensation based on motion sensors. Figure 10b presents the imaging result with motion compensation we proposed. We also perform an internal comparative analysis in this imaging progress to demonstrate the BP algorithm's quality and clarity, which can be found in Figure A3. Figure 10c presents the 3D-imaging result after motion compensation.

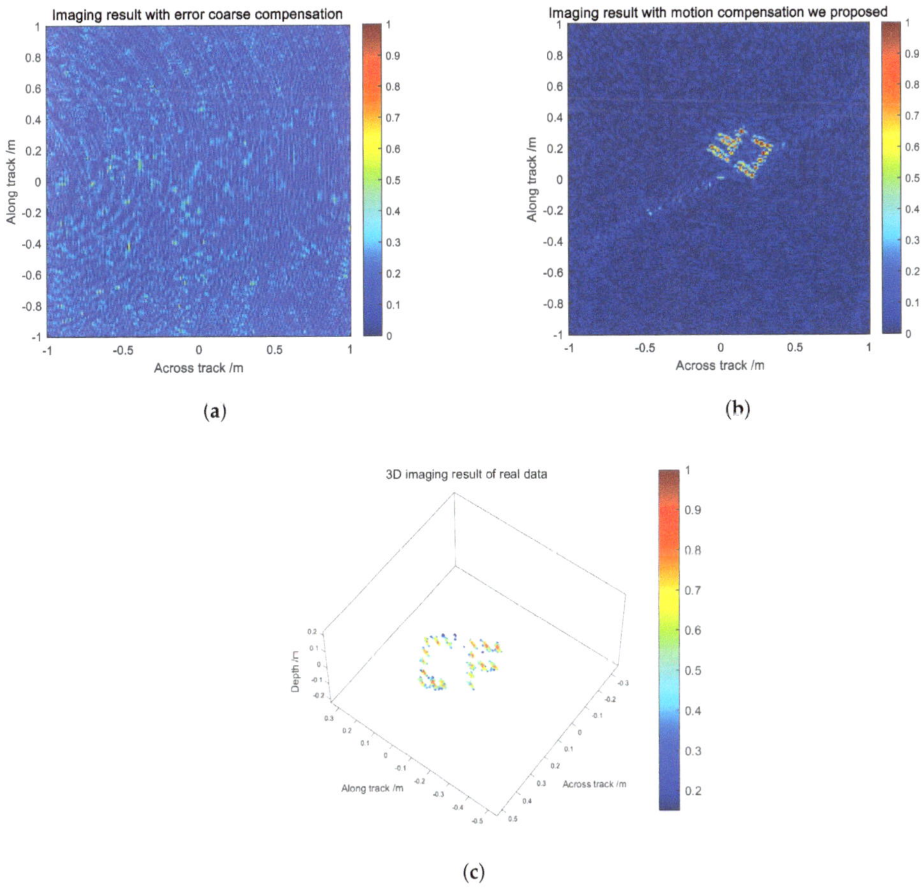

Figure 10. Imaging results of the lake trial: (**a**) imaging result with error coarse compensation based on motion sensors; (**b**) imaging result with motion compensation we proposed; (**c**) 3D-imaging result after motion compensation we proposed.

From Figure 10a, it can be seen that relying solely on the error coarse compensation scheme based on GPS positioning and inertial navigation systems, although the positioning error can be controlled within 2 cm, such positioning accuracy is far from sufficient for acoustic imaging. The acquired normalized imaging result is completely out of focus, and the target objects cannot be observed. Under this condition, the maximum superimposed sound intensity value is 31.43 dB. In contrast, Figure 10b presents the normalized imaging result with motion compensation we proposed. At the target depth of 17 m, the synthetic aperture length along-track is 5.18 m, and the physical aperture length across-track is

5.78 m, theoretically achieving a resolution of 3 cm. Not only the shape of the target can be clearly seen in the figure, but also the points on the target can be distinguished. And the maximum superimposed sound intensity value increases to 38.02 dB. In the comparison, discernible targets could no longer be identified in Figure 10a, but clear resolution results are presented in Figure 10b. This demonstrates the effectiveness of our proposed motion compensation method compared to the error coarse compensation. For the motion error compensation scheme based on sonar echo data, it is not applicable to the sonar array structure used in this study according to references [21,22].

5. Discussion

According to our knowledge of public publications, there are no scholars or research institutions that have proposed motion error compensation methods based on acoustic calibration systems in underwater 3D acoustic imaging. In experiments, compared to the poor imaging result with error coarse compensation, the resolution of the imaging system after motion compensation we proposed has already reached 3 cm $\times$ 3 cm $\times$ 2.5 cm @ Depth = 17 m, TBP = 30 s $\cdot$ Hz. By employing the motion error compensation method proposed in this paper, the error compensation values for the signals received by most hydrophone elements are less than 1.5 mm, equivalent to $1/10\lambda$, demonstrating that the motion compensation accuracy has reached the sub-wavelength level. Furthermore, under the condition of equal average energy, the maximum superimposed sound intensity values in the imaging results increased by 20.75 dB and 6.57 dB, respectively, for simulated and actual data, resulting in better visual imaging results. These experimental data and results demonstrated the effectiveness of our proposed plan. However, it should be noted that although the acoustic calibration system can bring high motion error compensation accuracy, the useful range of a single calibration sound source is limited and cannot meet the needs of large-area imaging. In the future, distributed acoustic calibration systems will be one of the directions for research. Additionally, we must consider the implementation complexity and cost of the scheme based on the requirements of the task. If sub-wavelength level motion error compensation is required within a small range, our proposed motion compensation method not only costs less but also performs better compared to the error coarse compensation scheme based on motion sensors. However, for large-range but low-precision motion error compensation requirements, the error coarse compensation scheme based on motion sensors is sufficient, and at this time, the cost of large-range high-precision acoustic calibration is enormous.

6. Conclusions

In this paper, a moving linear array 3D down-looking imaging model is established, and the back-projection algorithm for point-by-point imaging in the time domain is adopted to achieve clear 3D-imaging of point targets. To address the challenge of high-precision motion error compensation, we proposed a sub-wavelength motion compensation method based on a real-time acoustic calibration system, establishing two motion compensation models of three-point positioning and multi-point positioning. To verify the key role of the motion compensation method in practical hydro-acoustic imaging, we added sinusoidal motion errors to the simulation data, proving that the proposed motion compensation method meets the theoretical accuracy requirements and has achieved good results in simulation experiments. In addition, we compared the imaging results before and after motion compensation using real data. The experimental results show that compared with the measurement method of GPS combined with inertial navigation, after using the motion error compensation method based on the real-time acoustic calibration system, the focusing quality of the down-looking 3D-imaging was significantly improved. Ultimately, our system is capable of presenting high-resolution 3D target images with fine details.

Author Contributions: Conceptualization, J.W. and P.L.; methodology, J.W. and J.S.; software, W.T.; validation, J.W., P.L. and W.T.; formal analysis, P.L.; investigation, P.X.; resources, Y.H.; data curation, W.T.; writing—original draft preparation, P.L.; writing—review and editing, J.W.; visualization, K.L. and R.R.; supervision, J.S.; project administration, J.W. and W.T.; funding acquisition, W.T. and P.Z. All authors have read and agreed to the published version of the manuscript.

Funding: This research was funded by the National Natural Science Foundation of China under Grant No. 62401601; the Science Foundation of National University of Defense Technology under Grant No. ZK21-30.

Data Availability Statement: The original contributions presented in the study are included in the article; further inquiries can be directed to the corresponding author.

Conflicts of Interest: The authors declare no conflicts of interest.

Appendix A

As a supplement to Section 2.3, these are the imaging results for a single point target in the simulation with 20 frames and 100 frames.

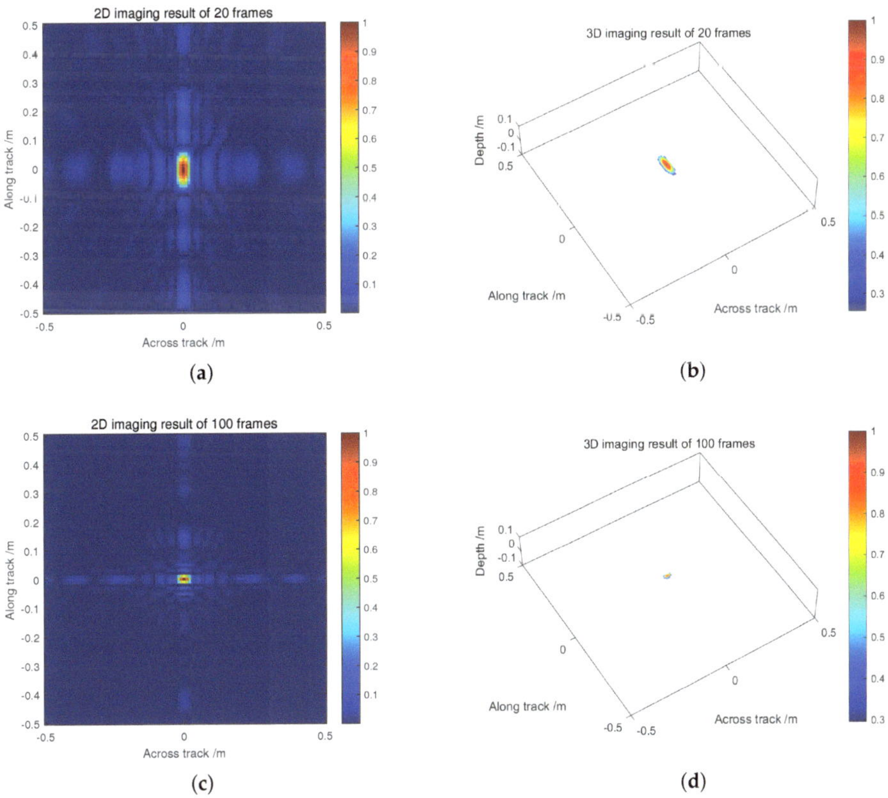

Figure A1. Imaging results of different frame number in simulation: (**a**) 2D-imaging result of 20 frames; (**b**) 3D-imaging result of 20 frames; (**c**) 2D-imaging result of 100 frames; (**d**) 3D-imaging result of 100 frames.

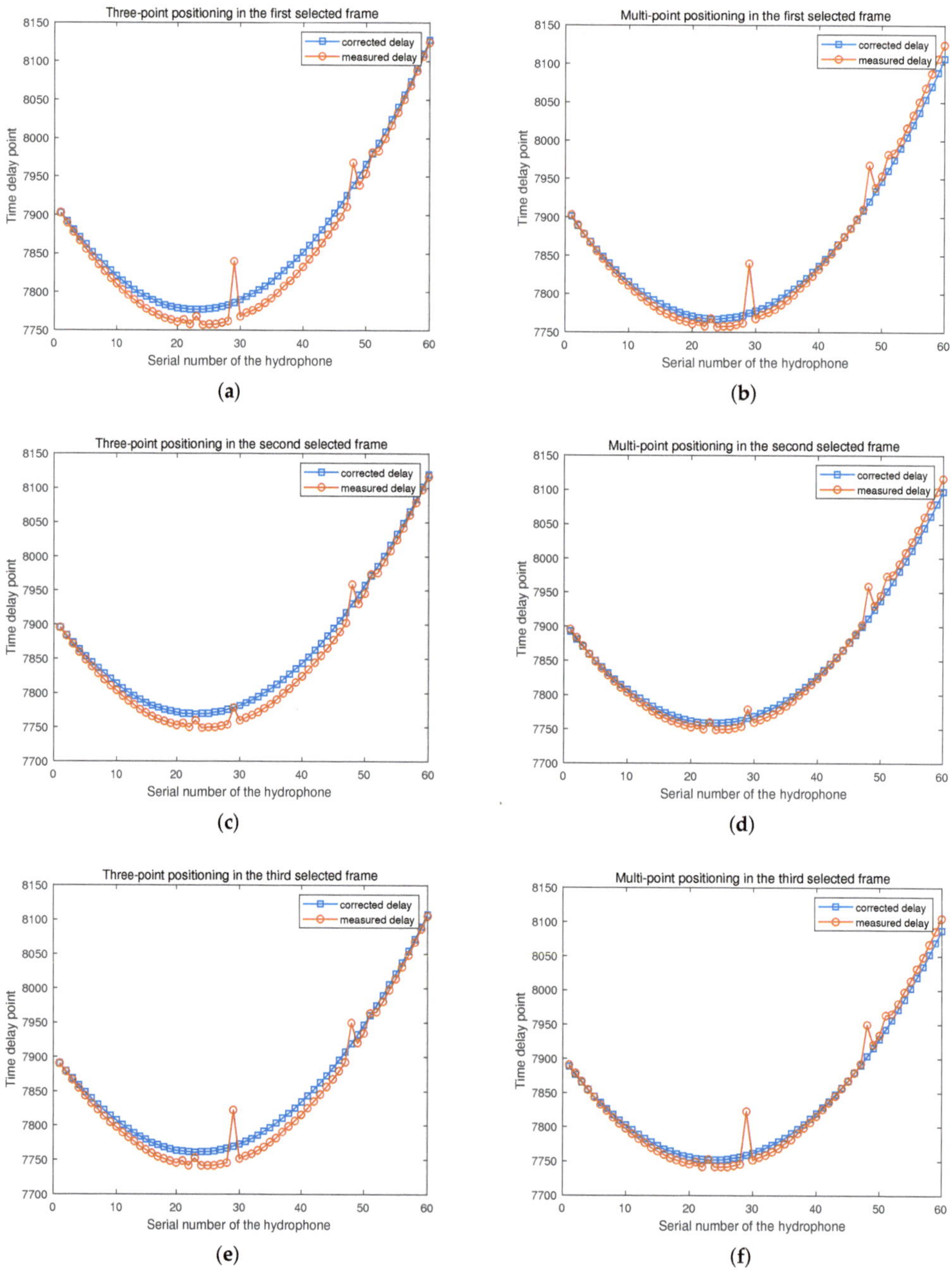

Figure A2. Comparison of the effects of three-point positioning and multi-point positioning: (**a**) three-point positioning in the first selected frame; (**b**) multi-point positioning in the first selected frame; (**c**) three-point positioning in the second selected frame; (**d**) multi-point positioning in the second selected frame; (**e**) three-point positioning in the third selected frame; (**f**) multi-point positioning in the third selected frame.

To illustrate the differences between three-point positioning and multi-point positioning, we randomly selected three consecutive frames from a segment of data and plotted their time delay point of each hydrophone in Figure A2. It can be seen from the figure that, since three-point positioning uses the hydrophones at the ends of the linear array for positioning, the time delay correction results fit well at both ends, with increased error at the center of the linear array. In contrast, multi-point positioning, which fully utilizes the echo signals received by all hydrophone elements, results in a corrected time delay that matches the measured time delay more closely. Additionally, the motion compensation accuracy of both positioning algorithms has reached the sub-wavelength level in experiments, which can be seen from the correction of the time delay points.

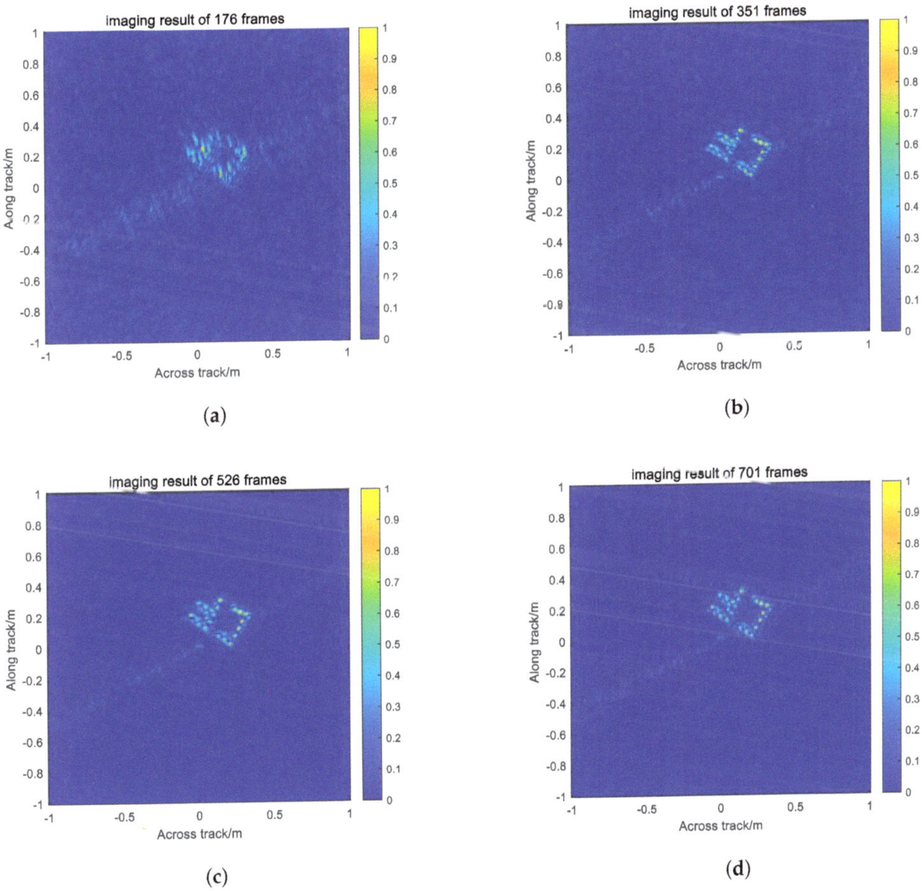

Figure A3. Reconstruction of different frame number: (**a**) imaging result of 176 frames; (**b**) imaging result of 351 frames; (**c**) imaging result of 526 frames; (**d**) imaging result of 701 frames.

In Figure A3, we performed an internal comparative analysis in the imaging progress of Figure 10b to demonstrate the BP algorithm's quality and clarity. We obtained a series of imaging results through iterative reconstruction of data from different frames, corresponding to different virtual synthetic apertures. In Figure A3a, the shape of the target cannot be discerned, and the maximum sound intensity value is 27.62 dB. As shown in Figure A3b, when the number of frames is increased to 351, the reconstructed result allows the shape of the target to be identified, and the noise points around the target are significantly reduced, with the maximum sound intensity value being 32.63 dB. In Figure A3c, the target becomes even clearer, and the surrounding noise points are further reduced, with a maximum sound intensity value of 37.31 dB. Compared to Figure A3c, Figure A3d shows almost no visual difference, but the maximum sound intensity value has increased to 38.02 dB. With the increase in the number of frames used, the visual effects of the iterative reconstruction results are better, and the maximum sound intensity value is also increasing, highlighting the quality and clarity of the BP algorithm. However, as can be seen, continuing to increase the number of data frames has a limited effect on the enhancement of the imaging results. Therefore, we chose the imaging result of 701 frames as the final result presented in Figure 10b.

References

1. Li, H.; Wei, B.; Du, W. Technical Progress in Research of Multibeam Synthetic Aperture Sonar. *Acta Geod. Cartogr. Sin.* **2017**, *46*, 1760–1769. [CrossRef]
2. Chu, X.; Fu, Z.; Yu, S.; Tu, X.; Huang, Y.; Ding, X. Underwater Image Enhancement and Super-Resolution Using Implicit Neural Networks. In Proceedings of the 2023 IEEE International Conference on Image Processing (ICIP), Kuala Lumpur, Malaysia, 8–11 October 2023; pp. 1295–1299. [CrossRef]
3. Zhu, J.; Yin, T.; Guo, W.; Zhang, B.; Zhou, Z. An Underwater Target Azimuth Trajectory Enhancement Approach in BTR. *Appl. Acoust.* **2025**, *230*, 110373. [CrossRef]
4. Menakath, M.M.; Raveendranatha Panicker, M.; Hareesh, G. Exploring time domain beamforming algorithm for underwater 3D acoustic imaging. In Proceedings of the OCEANS 2022—Chennai, Chennai, India, 21–24 February 2022; pp. 1–6. [CrossRef]
5. Qin, L.; Huang, H.; Wang, P.; Pengfei, Z.; Liu, J. Broadband downward-looking multi-beam 3D imaging algorithm in frequency domain. *Acta Acust.* **2019**, *44*, 604–612. [CrossRef]
6. Wang, P.; Chi, C.; Liu, J.; Huang, H. Improving performance of three-dimensional imaging sonars through deconvolution. *Appl. Acoust.* **2021**, *175*, 107812. [CrossRef]
7. Chen, J.; Xiong, R.; Yu, H.; Xu, G.; Xing, M. Nonparametric Full-Aperture Autofocus Imaging for Microwave Photonic SAR. *IEEE Trans. Geosci. Remote. Sens.* **2024**, *62*, 5214815. [CrossRef]
8. Chen, J.; Li, M.; Yu, H.; Xing, M. Full-Aperture Processing of Airborne Microwave Photonic SAR Raw Data. *IEEE Trans. Geosci. Remote. Sens.* **2023**, *61*, 5218812. [CrossRef]
9. Stoica, P.; Wang, Z.; Li, J. Robust Capon beamforming. *IEEE Signal Process. Lett.* **2003**, *10*, 172–175. [CrossRef]
10. Liao, Z.; Duan, K.; He, J.; Qiu, Z.; Li, B. Robust Adaptive Beamforming Based on a Convolutional Neural Network. *Electronics* **2023**, *12*, 2751. [CrossRef]
11. Li, C.X.; Guo, M.F.; Zhao, H.F. An Iterative Deconvolution-Time Reversal Method with Noise Reduction, a High Resolution and Sidelobe Suppression for Active Sonar in Shallow Water Environments. *Sensors* **2020**, *20*, 2844. [CrossRef]
12. Yazici, B.; Xie, G. Wideband Extended Range-Doppler Imaging and Waveform Design in the Presence of Clutter and Noise. *IEEE Trans. Inf. Theory* **2006**, *52*, 4563–4580. [CrossRef]
13. Yu, L.; Ma, F.; Lim, E.; Cheng, E.; White, L.B. Rational-Orthogonal-Wavelet-Based Active Sonar Pulse and Detector Design. *IEEE J. Ocean. Eng.* **2019**, *44*, 167–178. [CrossRef]
14. Myers, V.; Quidu, I.; Zerr, B.; Sæbø, T.O.; Hansen, R.E. Synthetic Aperture Sonar Track Registration With Motion Compensation for Coherent Change Detection. *IEEE J. Ocean. Eng.* **2020**, *45*, 1045–1062. [CrossRef]
15. Zhu, J.; Xie, Z.; Jiang, N.; Song, Y.; Han, S.; Liu, W.; Huang, X. Delay-Doppler Map Shaping through Oversampled Complementary Sets for High-Speed Target Detection. *Remote Sens.* **2024**, *16*, 2898. [CrossRef]
16. Chi, C. Basic Theory for Underwater Real-Time 3-D Acoustical Imaging. In *Underwater Real-Time 3D Acoustical Imaging: Theory, Algorithm and System Design*; Springer: Singapore, 2019; pp. 11–19. [CrossRef]
17. Wu, M.; Yan, S. Motion Compensation for OFDM Inverse Synthetic Aperture Sonar Imaging Based on Compressed Sensing. *IEEE Trans. Instrum. Meas.* **2021**, *70*, 1–13. [CrossRef]
18. Wu, H.; Zhou, F.; Xie, Z.; Tang, J.; Zhong, H.; Zhang, J. Two-Dimensional Space-Variant Motion Compensation Algorithm for Multi-Hydrophone Synthetic Aperture Sonar Based on Sub-Beam Compensation. *Remote Sens.* **2024**, *16*, 2144. [CrossRef]
19. Zhang, Y.; Li, G.; Zhang, P.; Wei, L.; Liu, J. Motion Compensation Algorithm of Synthetic Aperture Sonar Based on Multisensor Data Fusion. *J. Beijing Univ. Posts Telecommun.* **2017**, *40*, 82–86. [CrossRef]

20. Yang, F.; Lu, X.; Li, J.; Guo, J. Correction of Imperfect Alignment of MRU for Multibeam Bathymetry Data. *Geomat. Inf. Sci. Wuhan Univ.* **2010**, *35*, 816–820.
21. Bellettini, A.; Pinto, M. Theoretical accuracy of synthetic aperture sonar micronavigation using a displaced phase-center antenna. *IEEE J. Ocean. Eng.* **2002**, *27*, 780–789. [CrossRef]
22. Pinto, M.A. Long Term Accuracy of Synthetic Aperture Sonar Micronavigation using a Displaced Phase Centre Antenna. In Proceedings of the OCEANS 2018 MTS/IEEE Charleston, Charleston, SC, USA, 22–25 October 2018; pp. 1–6. [CrossRef]
23. Liu, X.; Zhou, F.; Zhou, H.; Tian, X.; Jiang, R.; Chen, Y. A Low-Complexity Real-Time 3-D Sonar Imaging System with a Cross Array. *IEEE J. Ocean. Eng.* **2016**, *41*, 262–273. [CrossRef]
24. Zhang, X.; Yang, P.; Cao, D. Synthetic aperture image enhancement with near-coinciding Nonuniform sampling case. *Comput. Electr. Eng.* **2024**, *120*, 109818. [CrossRef]
25. Chen, P.; Tian, X.; Chen, Y. Optimization of the Digital Near-Field Beamforming for Underwater 3-D Sonar Imaging System. *IEEE Trans. Instrum. Meas.* **2010**, *59*, 415–424. [CrossRef]
26. Yegulalp, A. Fast backprojection algorithm for synthetic aperture radar. In Proceedings of the Proceedings of the 1999 IEEE Radar Conference. Radar into the Next Millennium (Cat. No. 99CH36249), Waltham, MA, USA, 22 April 1999; pp. 60–65. [CrossRef]
27. Guo, Y.; Suo, Z.; Jiang, P.; Li, H. A Fast Back-Projection SAR Imaging Algorithm Based on Wavenumber Spectrum Fusion for High Maneuvering Platforms. *Remote Sens.* **2021**, *13*, 1649. [CrossRef]
28. Zhang, J.; Cheng, G.; Tang, J.; Wu, H.; Tian, Z. A Subaperture Motion Compensation Algorithm for Wide-Beam, Multiple-Receiver SAS Systems. *J. Mar. Sci. Eng.* **2023**, *11*, 1627. [CrossRef]

remote sensing

MDPI

Article

Situation Awareness and Tracking Algorithm for Countering Low-Altitude Swarm Target Threats

Nannan Zhu [1], Fuli Zhong [2,3,4,*], Xueyue Lei [5], Guo Niu [6], Hongtu Xie [5] and Yue Zhang [5]

1 School of Systems Science and Engineering, Sun Yat-sen University, Guangzhou 510275, China; zhunn25@mail.sysu.edu.cn
2 School of Mathematics and Computing Science, Guilin University of Electronic Technology, Guilin 541004, China
3 Center for Applied Mathematics of Guangxi (GUET), Guilin 541004, China
4 Guangxi Colleges and Universities Key Laboratory of Data Analysis and Computation, Guilin 541004, China
5 School of Electronics and Communication Engineering, Sun Yat-sen University, Shenzhen 518107, China; leixy9@mail2.sysu.edu.cn (X.L.); xiehongtu@mail.sysu.edu.cn (H.X.); zhangyue8@mail.sysu.edu.cn (Y.Z.)
6 School of Electronics and Information Engineering, Foshan University, No. 18 Jiangwan 1st Road, Foshan 528225, China; niuguo@mail2.sysu.edu.cn
* Correspondence: zhongfli@mail.sysu.edu.cn

Abstract: The escalating threat posed by low-altitude swarm targets underscores the critical need for precise tracking and situation awareness to secure key areas. While existing tracking methods based on random matrix theory offer promising opportunities, they face significant challenges. The high similarity among swarm targets, combined with radar resolution limitations, often leads to instabilities in target counts and measurements due to occlusion, environmental factors, and other disturbances, significantly increasing tracking complexity. To address these challenges, we design a digital staring radar system integrated with an adaptive random matrix method for efficient tracking of low-altitude swarm targets. The system achieves full spatiotemporal coverage without beam scanning or complex resource scheduling, enabling simultaneous detection and tracking of multiple targets. Algorithmically, the random matrix model is enhanced by introducing extension parameters to accurately capture the dynamic changes in swarm shape. Leveraging an adaptive Rao-Blackwellized Particle Filter (RBPF), the presented method jointly estimates the motion and extension states of swarm targets. Extensive simulation experiments and real-data validation demonstrate that the proposed method significantly improves the estimation accuracy for swarm extension states under complex shape variations while maintaining high precision in motion state estimation. This work provides a practical and effective solution for countering low-altitude swarm threats, with strong potential for real-world security applications.

Keywords: low-altitude threat tracking; UAV swarm target; adaptive random matrix method; Rao-Blackwellized Particle Filter (RBPF); cluster target tracking

1. Introduction

Low-altitude threat swarm targets, mainly represented by unmanned aerial vehicle (UAV) swarms, have become crucial security concerns. Their low cost, high flexibility, and large-scale deployability endow them with remarkable potential in reconnaissance, intelligence gathering, and low cost strike operations. Moreover, they can carry out complex cooperative tasks [1,2], presenting unprecedented challenge to traditional defense systems.

During some regional conflicts, UAVs have been extensively deployed for reconnaissance, target identification, and electronic warfare, which highlights the urgent need for advanced UAV swarm detection systems and underscore the critical demand for upgrading existing air defense infrastructure for security purposes.

Common detection methods for low-altitude threat targets include radar surveillance, multispectral recognition, and spectrum reconnaissance. Multispectral recognition technology captures high-resolution images across multiple spectral bands simultaneously, offering rich and detailed target information [3–12]. However, its detection range is limited, and it cannot provide precise distance information. Spectrum reconnaissance, a passive detection technique, monitors the direction and intensity of radiation emissions from sources. It provides advantages such as low detectability and the ability to detect radiation sources, including enemy radar and electronic interference, at long ranges. This method can also distinguish different types of UAVs. However, large-scale UAV swarms operating on similar communication frequencies may cause signal overlap, complicating signal separation and recognition. Additionally, multipath effects and signal attenuation due to distance further degrade detection performance. Radar detection, on the other hand, offers precise three-dimensional target coordinates (range, azimuth, and elevation), making it highly effective for all-weather monitoring, long-range detection, and multi-target tracking. As a result, radar has become the primary tool for the detection of low-altitude threat targets [13–24]. However, radar detection of low-altitude targets is often compromised by environmental factors, such as ground clutter. The diversity of low-altitude threat targets, along with their low flight altitudes, high maneuverability, small radar cross-sections, and the complex detection environments, presents significant challenges. These issues have become global challenges in defense technology.

Recent advancements in radar technology have prompted the development of several new radar systems specifically tailored for low-altitude threat detection. Notable systems include the U.S. Merlin radar, the U.S. Navy Research Laboratory's holographic radar, the Canadian Accipiter radar, and the Dutch Robin radar. The Merlin radar employs an S-band horizontal scanning system to monitor low-altitude airspace around airports, complemented by an X-band vertical scanning radar for tracking aircraft during takeoff and landing. The Accipiter radar, equipped with an X-band parabolic antenna, features a narrow 4° beam that allows for precise altitude measurements, although its detection range is limited. In addition to the S-band horizontal and X-band vertical scanning radars, the Robin radar integrates a frequency-modulated continuous wave (FMCW) radar, optimized for classifying low-altitude threat targets. The Fraunhofer Institute for Applied Research (FHR) in Germany uses a Ku-band MIMO radar system to tackle the challenge of UAV target signals blending with clutter. By training a detector on environmental data, the system dynamically adjusts detection thresholds in both cluttered and non-cluttered environments, enhancing radar performance for detecting low-speed targets. Saab, a Swedish company, has enhanced its "Giraffe" radar to provide conventional air surveillance while also detecting, classifying, and tracking small, low-speed UAVs. This radar has been validated for tracking up to six UAV targets simultaneously in complex environments [25].

In contrast to traditional multi-target tracking algorithms, cluster target tracking algorithms treat UAV clusters as a single entity, extracting features such as shape and size from radar observation data for classification, recognition, and state estimation of the cluster target [9,26]. Drummond et al. [27] modeled the cluster's shape as an ellipse and estimated parameters such as the center of the ellipse. Koch et al. presented the cluster target tracking algorithm based on random matrices [28]. This algorithm collectively refers to features like shape and size as extended information. They modeled the cluster shape as an extended ellipse and represented it using a positive definite random matrix, known as the extension matrix. This approach integrated the extended ellipse into the motion and measurement equations of the targets, creating a state-space model that simultaneously described both the motion and extension states using random matrices. This method simplified the model representation by avoiding the need to separately model the cluster shape. Additionally, the target's motion state was modeled as a Gaussian random vector, and the extension matrix was modeled as an inverse Wishart random matrix [29], enabling the derivation of posterior estimation recursive formulas similar to Kalman filtering within a Bayesian framework.

Feldmann et al. pointed out that the introduction of measurement noise can cause the measurement update formulas to fail [30]. To address this problem, Feldmann proposed simulating the distribution of targets within the cluster using a uniform distribution and incorporating measurement noise into the measurement model. Additionally, Feldmann derived approximate measurement update formulas by assuming the extension ellipse as a constant matrix. In contrast, Orguner [31] employed variational Bayesian inference [32] for measurement updates. Lan [33] proposed an enhanced random matrix model and employed matrix matching to derive novel recursive estimation formulas for both motion and extension states. This approach utilized rotation matrices to represent the rotation of extended ellipses. By using multiple rotation matrices with different angles, a multi-model algorithm was implemented to switch between random matrix models. In a separate study [34], Lan introduced matching linearization and variational Bayesian methods to address the challenges of cluster target tracking under nonlinear measurements. These methods demonstrated improved state estimation performance compared to traditional transformation-based measurement techniques [35–37] when handling the nonlinear measurements. While the work in [33] accounted for complex variations in the extended ellipses, other methods assumed these ellipses to be approximately time-invariant. This assumption becomes invalid when the cluster formation changes or the cluster executes turning maneuvers, causing the extended ellipses to rotate. However, the method in [33] only considered specific rotation angles and did not cover all possible rotation scenarios for extended ellipses. Furthermore, determining rotation angles requires knowledge of the motion characteristics of tracked targets, which is impractical for non-cooperative targets [38,39].

In recent years, deep learning techniques have been applied in the field of target tracking research by researchers and led to some valuable work [40–42]. Cheng et al. [40] proposed a deep learning-based approach that merges radar data and camera data to improve the accuracy and robustness of multi-object tracking works of autonomous driving systems. They use a Bi-directional Long Short-Term Memory network to fuse the long-term temporal information and enhance the motion prediction, apply a tri-output mechanism which consists of individual outputs for radar and camera sensors and a fusion output, to improve the system's robustness. Huang et al. [41] proposed a deep-reinforcement-learning-based radar parameter adaptation method for multiple-target tracking, to ensure that the targets are not lost and improve the tracking accuracy in multiple-target tracking cases with variable target numbers and multiple target classes. They designed an improved deep reinforcement learning agent that uses a long- and short-term memory network, and a self-attention mechanism to deal with the problem of target number variation and multiple target types. Zhang et al. [42] presented an algorithm for trajectory optimization in airborne radar for extended target tracking. The deep reinforcement learning agent is used to autonomously learns the trajectory planning strategies to maximize extended target tracking performance. Deep learning based radar tracking techniques provide a novel way for target tracking purpose, however, they commonly require proper training with a large amount of data. While the traditional filtering methods do not need to refined training with too much sample data.

To model the complex extension variation process of cluster targets, this paper introduces a digital staring radar system integrated with an adaptive random matrix approach for tracking UAV swarm. By accumulating data over an extended period, the digital staring radar achieves higher gain and Doppler resolution, effectively distinguishing moving targets from clutter. The enhanced Doppler resolution also facilitates the extraction of micro-Doppler features, providing new insights for target classification and recognition. Additionally, the improved model incorporates extension parameters to capture variations in target extension, and utilizes the adaptive Rao-Blackwellized particle filter (RBPF) to estimate both motion and extension states. The proposed algorithm offers the following advantages:

- The designed digital staring radar achieves higher gain and Doppler resolution through long-term data accumulation, enabling effective separation of moving targets from clutter.
- Additional extension parameters to capture the complex extension state variations of cluster targets, building upon the original random matrix model, is introduced.
- By employing the adaptive RBPF algorithm to approximate the posterior estimation of both motion and extension states, the dimensionality of particle sampling is reduced, enhancing sampling efficiency. Compared to conventional particle filtering, RBPF results in lower estimation variance.

Simulation results validated that the proposed method enhances the estimation performance of extension states under complex variations, while maintaining the precision of motion state estimation.

The structure of this paper is as follows: Section 2 reviews the existing works closely related to the proposed algorithm. Section 3 provides a detailed description of the proposed approach. Section 4 presents the comprehensive experimental results based on simulated data. Finally, Section 5 concludes the paper.

2. Related Works

This section reviews three closely related works to provide a better understanding of the innovation behind the proposed algorithm in this study.

2.1. Koch Method

The shape of a UAV cluster contains valuable feature information. Therefore, when tracking a UAV cluster, it is crucial to estimate not only the motion state of the UAVs but also the shape of the cluster. Koch was the first to apply the random matrix model to a cluster target tracking algorithm. This approach models the shape of the cluster as an ellipse, referred to as the extended ellipse, and uses a symmetric positive definite random matrix to represent this ellipse:

$$(\boldsymbol{y} - \boldsymbol{H}_k\boldsymbol{x}_k)^{\mathrm{T}}\boldsymbol{E}_k^{-1}(\boldsymbol{y} - \boldsymbol{H}_k\boldsymbol{x}_k) = 1, \tag{1}$$

where the random matrix $\boldsymbol{E}_k$ represents the extended matrix, x_k denotes the state to be estimated, y is the measurement, and the joint state of the target at time k is given by

$$\xi_k = (x_k, \boldsymbol{E}_k). \tag{2}$$

In the Bayesian framework, to obtain the posterior estimation of the joint state, it is essential to define the motion model and measurement model of the cluster target. The motion model of the cluster target can be assumed to take the following form:

$$\boldsymbol{x}_k = \boldsymbol{F}_{k-1}\boldsymbol{x}_{k-1} + \boldsymbol{w}_k, \quad \boldsymbol{w}_k \sim \mathcal{N}(0, \boldsymbol{E}_k \otimes \boldsymbol{D}_k), \tag{3}$$

where $F_{k-1} = I_{d_z} \otimes \tilde{F}_{k-1}$ is the full-dimensional state transition matrix, I_{d_z} is the unit matrix of size $d_z \times d_z$, representing the state transition matrix on a single dimension. $\tilde{F}_{k-1}$ represents the covariance matrix of the process noise w_k. $E_k \otimes D_k$ is the covariance matrix of the process noise on a single dimension. D_k denotes the Kronecker product, and by utilizing the properties of the Kronecker product, the state transition matrix and the covariance matrix on a single dimension can be extended to multiple dimensions.

Assuming that the shape of the UAV cluster target remains unchanged over a short period, the variation process of the extended ellipse can be described by the following extended state transition density:

$$p(E_k|E_{k-1}) = \mathcal{W}_{d_z}\left(E_k; \varepsilon_{k|k-1}, \frac{E_{k-1}}{\varepsilon_{k|k-1}}\right), \tag{4}$$

where $\mathcal{W}_{d_z}\left(E_k; \varepsilon_{k|k-1}, E_{k-1}/\varepsilon_{k|k-1}\right)$ represents that the extended matrix E_k follows a Wishart distribution with mean E_{k-1} and degrees of freedom d_z dimensional space. The degrees of freedom $\varepsilon_{k|k-1}$ can describe the uncertainty of the extension variation process. As the degrees of freedom decrease, the uncertainty of the extension increases, acting similarly to the introduction of Gaussian white noise in the motion model. Usually, the uncertainty of the cluster target's extended ellipse increases with the increase of time intervals, and it can be assumed that

$$\varepsilon_{k|k-1} = \varepsilon e^{-T/\tau}, \tag{5}$$

where τ is the attenuation factor, and ε is a constant. The values of these two parameters need to be chosen appropriately based on the specific type of cluster target.

Unlike traditional point targets, cluster targets can generate a large number of indefinite target measurements at the radar receiver. Therefore, the sensor measurement model can be assumed to be:

$$z_k^l = H_k x_k + v_k^l, \quad l = 1, \cdots, n_k \tag{6}$$

where $H_k = I_{d_z} \otimes \tilde{H}_k$ is the full-dimensional measurement matrix, $\tilde{H}_k$ is the measurement matrix on a single dimension, and n_k is the number of measurements at time k. $v_k^1, \cdots, v_k^{n_k}$ represents mutually independent Gaussian white noise following the distribution

$$v_k^l \sim \mathcal{N}(0, E_k). \tag{7}$$

Equation (7) indicates that the targets within the extended ellipse of the cluster follow a Gaussian distribution. This assumption simplifies the target's measurement model, making it convenient to derive the recursive joint state estimation formulas.

In the Bayesian theory framework, given the set of measurements $Z_{1:k} = \{Z_i, n_i\}_{i=1}^k$, the posterior probability density $p(\xi_k|Z_{1:k})$ contains all the statistical information about the joint state ξ_k. Here, $Z_k = \{z_k^1, \cdots, z_k^{n_k}\}$ represents the set of n_k measurements obtained by the radar at time k from the same cluster. Koch's objective is to find a set of recursive formulas that obtain the posterior probability density of the joint state through recursion of a few parameters in the probability density function, similar to the Kalman filter, which obtains the posterior probability density of the motion state through recursion of the mean and covariance.

To achieve this, the prior probability density of the joint state must satisfy the property of a conjugate prior. For Gaussian measurements, the conjugate priors with unknown mean and covariance are the Gaussian distribution and the inverse Wishart distribution, respectively. Under this prior probability assumption, and in combination with the target's motion model and measurement model, the posterior probability density of the joint state can be derived as:

$$\begin{aligned} p(\xi_k|\mathbf{Z}_{1:k}) &= p(\mathbf{x}_k|\mathbf{E}_k,\mathbf{Z}_{1:k})p(\mathbf{E}_k|\mathbf{Z}_{1:k}) \\ &= \mathbb{N}\left(\mathbf{x}_k;\hat{\mathbf{x}}_{k|k},\mathbf{E}_k\otimes\tilde{\mathbf{P}}_{k|k}\right) \\ &\times \mathcal{IW}_{d_z}\left(\mathbf{E}_k;\nu_{k|k},\mathbf{V}_{k|k}\right), \end{aligned} \tag{8}$$

where the posterior probability density of the extended state $p(\mathbf{E}_k|\mathbf{Z}_{1:k})$ can be directly obtained, and it follows an inverse Wishart distribution with $\nu_{k|k}$ degrees of freedom and a scale parameter of $\mathbf{V}_{k|k}$. The posterior estimation of the extended state is given by

$$\hat{\mathbf{E}}_{k|k} = E(\mathbf{E}_k|\mathbf{Z}_{1:k}) = \frac{\mathbf{V}_{k|k}}{\nu_{k|k}-2d_z-2}. \tag{9}$$

The posterior probability density of the motion state, on the other hand, needs to be obtained by marginalizing the joint posterior distribution of the states.

$$p(\mathbf{x}_k|\mathbf{Z}_{1:k}) = \int p(\mathbf{x}_k,\mathbf{E}_k|\mathbf{Z}_{1:k})d\mathbf{E}_k. \tag{10}$$

The posterior probability density of the motion state, after integrating out other variables, follows a multivariate Student's *t*-distribution [29].

$$p(\mathbf{x}_k|\mathbf{Z}_{1:k}) = \mathcal{T}\left(\mathbf{x}_k;\nu_{k|k}-2d_z,\hat{\mathbf{x}}_{k|k},\frac{\mathbf{V}_{k|k}\otimes\tilde{\mathbf{P}}_{k|k}}{\nu_{k|k}-2d_z}\right). \tag{11}$$

The posterior estimation of the motion state and its covariance can be obtained as

$$E(\mathbf{x}|\mathbf{Z}_{1:k}) = \hat{\mathbf{x}}_{k|k}, \tag{12}$$

$$E\left[\left(\mathbf{x}_k-\mathbf{x}_{k|k}\right)\left(\mathbf{x}_k-\mathbf{x}_{k|k}\right)^{\mathrm{T}}|\mathbf{Z}_{1:k}\right] = \hat{\mathbf{E}}_{k|k}\otimes\tilde{\mathbf{P}}_{k|k}. \tag{13}$$

The posterior probability density of the joint state can be obtained through recursive parameters $\mathbf{x}_{k|k}$, $\tilde{\mathbf{P}}_{k|k}$, $\nu_{k|k}$, and $\mathbf{V}_{k|k}$. The detailed process can be referred to [28].

2.2. Feldmann Method

The random matrix model in Koch's algorithm assumes that the target measurements follow a Gaussian distribution with mean and covariance. The advantage of this assumption is that the likelihood function of the measurements has a conjugate prior. However, this measurement model does not always align with real-world scenarios, and presents the following two issues: (1) It does not account for the measurement errors of the sensors; (2) Modeling the shape of the cluster as a Gaussian distribution assumes that the targets are densely distributed around the center of the cluster and sparsely distributed at the edges. However, in practical scenarios, targets are more likely to be uniformly distributed within the cluster, as seen in UAV swarms. Modeling the distribution of targets within the cluster as a uniform distribution is therefore more representative of real-world conditions.

To address these two issues, Feldmann proposed a more comprehensive sensor measurement model in [29].

$$\begin{aligned} \mathbf{y}_k^l &= \mathbf{H}_k\mathbf{x}_k+\mathbf{u}_k^l, \quad \mathbf{u}_k^l \sim \mathcal{U}(\mathbf{0},\mathbf{E}_k), \\ \mathbf{z}_k^l &= \mathbf{y}_k^l+\mathbf{v}_k^l, \quad \mathbf{v}_k^l \sim \mathcal{N}(\mathbf{0},\mathbf{R}_k), \end{aligned} \tag{14}$$

where y_k^l represents the precise location of the target within the cluster under the assumption of no measurement noise. $\mathcal{U}(r, G)$ characterizes the uniform distribution within the ellipse defined by $(y - r)G^{-1}(y - r)^{\mathrm{T}} = 1$. r and G are two variables for describing the ellipse, G is a matrix. As deduced from the aforementioned equations, the true target position satisfies $\mathbf{y}_k^l \sim \mathcal{U}(\mathbf{H}_k\mathbf{x}_k, \mathbf{E}_k)$. Notably, this sensor measurement model posits that the target measurements arise from two uncertain sources: The configuration of the clustered targets and the sensor's measurement errors, with R_k representing the covariance matrix of the sensor's measurement noise.

However, modeling the uniform distribution within an ellipse poses mathematical challenges. To overcome this, Feldmann proposed an approximation technique that uses a Gaussian distribution to model the uniform distribution. As a result, the measurement model is expressed as:

$$z_k^l = H_k x_k + v_k^l, \quad v_k^l \sim \mathcal{N}(0, \alpha E_k + R_k), \tag{15}$$

where α serves as the scale parameter, k is the sampling time, and l denotes the l-th measurement. In the case that measurements are taken in a two-dimensional ($d_z = 2$) plane, the value $\alpha = 1/4$ offers a reliable approximation for the uniform distribution within the mentioned ellipse. The term v_k^l denotes the equivalent measurement noise, encompassing the uncertainties arising from both the cluster distribution and the measurement procedure. For more generalized scenarios, the specific value of α is determined as [43]

$$\alpha = \frac{1}{d_z + 2}. \tag{16}$$

For instance, when $d_z = 3$, a suitable choice for α would be $\alpha = 1/5$. By adjusting the scale parameter α, it becomes possible to flexibly characterize different types of clustered targets. In specific bird flocks, the distribution of individual members within the cluster indeed follows a Gaussian distribution. In such cases, setting the scale parameter to $\alpha = 1$, provides a more accurate representation of the cluster's shape, aligning with the assumptions made in the Koch method concerning the distribution of clustered targets. The precise value of the scale parameter α depends on the type of tracked targets.

The sensor measurement model represented by Equation (15) effectively captures real-world scenarios, and its measurement likelihood function is

$$p(Z_k | n_k, x_k, E_k) = \prod_{i=1}^{n_k} \mathcal{N}\left(z_k^i; H_k x_k, \alpha E_k + R_k\right). \tag{17}$$

The measurement likelihood function given by Equation (17) lacks a conjugate prior with respect to the joint states. To tackle this issue, Feldmann assumes that the target's motion state is independent of the extended state, thus approximates the target's motion state as

$$x_k = F_k x_{k-1} + w_k, \quad w_k \sim \mathcal{N}(0, Q_k), \tag{18}$$

where Q_k is the full-dimensional covariance matrix of the process noise.

2.3. Lan Method

Based on the measurement model described in Equation (15), Lan proposed a more general measurement model [33] with the following form

$$z_k^l = H_k x_k + v_k^l, \quad v_k^l \sim \mathcal{N}\left(0, B_k E_k B_k^{\mathrm{T}}\right), \tag{19}$$

where B_k is an arbitrary invertible matrix, which can be used to describe complex measurement noise covariance. For instance, if we define

$$B_k = (\alpha E_k + R_k)^{\frac{1}{2}} E_k^{-\frac{1}{2}}, \tag{20}$$

the covariance matrix of the measurement noise can be obtained as

$$\begin{aligned} B_k E_k B_k^{\mathrm{T}} &= (\alpha E_k + R_k)^{\frac{1}{2}} E_k^{-\frac{1}{2}} E_k E_k^{-\frac{\mathrm{T}}{2}} (\alpha E_k + R_k)^{\frac{\mathrm{T}}{2}} \\ &= \alpha E_k + R_k. \end{aligned} \tag{21}$$

In this case, Equation (19) is exactly the same as Equation (15). However, the extended matrix E_k in Equation (20) is an unknown estimate that can be replaced with the extended predicted value $\hat{E}_{k|k-1}$, resulting in

$$B_k \approx \left(\alpha \hat{E}_{k|k-1} + R_k\right)^{\frac{1}{2}} \hat{E}_{k|k-1}^{-\frac{1}{2}}. \tag{22}$$

The advantage of the measurement model proposed by Lan is that its measurement likelihood function has a conjugate prior, which simplifies computations and improves the accuracy of joint state estimation. Additionally, Lan takes into account the case of the extended ellipse rotating with time. A new dynamic model for the extended state is redefined, and it is represented by the following transition density of the extended state

$$p(E_k|E_{k-1}) = \mathcal{W}_{d_z}\left(E_k; \varepsilon_{k|k-1}, A_{k-1} E_{k-1} A_{k-1}^{\mathrm{T}}\right), \tag{23}$$

where the degree of freedom $\varepsilon_{k|k-1}$ can describe the variation of the extended ellipse's size over time, and the matrix A_{k-1} can account for the change in the orientation of the extended ellipse with time. The matrix A_{k-1} is a rotation matrix constructed from a known rotation angle θ, defined as

$$A_{k-1} = \begin{bmatrix} \cos(\theta) & -\sin(\theta) \\ \sin(\theta) & \cos(\theta) \end{bmatrix} / \sqrt{\varepsilon_{k|k-1}}. \tag{24}$$

3. Proposed Method

Cluster targets that maintain a specific formation of movement over a certain period. Thus, it is assumed that the extended ellipse undergoes minimal changes within short intervals. Koch proposed using Equation (4) to describe the dynamic process of the extended state, assuming a temporal variation of the extended ellipse. However, when cluster targets execute turning maneuvers or experience changes in their formation, the extended ellipse often undergoes rotation. The original dynamic model for the extended state fails to accurately capture this rotational variation.

To overcome this limitation, Lan introduced a more sophisticated extended state transition density, as represented by Equation (23). This density utilizes rotation matrices with known rotation angles to describe the rotation of the extended ellipse for cluster targets. Multiple extended state dynamic models with different rotation matrices are defined, and the Interacting Multiple Model (IMM) algorithm is employed to switch between these models. However, it is worth noting that the rotation angle of the ellipse is discrete, and pre-defining a finite number of rotation angles may not encompass all possible rotations of the extended ellipse. Moreover, setting the rotation angles requires knowledge of the motion characteristics of the tracked targets, which is impractical for non-cooperative targets.

3.1. Improved Extended State Model of Cluster Target Tracking

In light of the aforementioned issues, this study proposes a solution by introducing supplementary extended parameters to account for the rotation of the extended ellipse. This introduces a new parameter known as the extended rotation rate φ_k into the extended state representation. As a result, the extended state for cluster targets is derived $X_k = (E_k, \varphi_k)$, and the joint state is defined as

$$\xi_k = (x_k, X_k). \tag{25}$$

The dynamic model for the extended state can be represented by the following extended state transition density

$$\begin{aligned} p(X_k|X_{k-1}) &= p(E_k, \varphi_k|E_{k-1}, \varphi_{k-1}) \\ &= p(E_k|\varphi_k, E_{k-1}, \varphi_{k-1})p(\varphi_k|E_{k-1}, \varphi_{k-1}), \end{aligned} \tag{26}$$

Assuming that the transition process of the extended rotation rate satisfies the Markov property and that the extended matrix is only dependent on the previous moment's extended matrix and the extended rotation rate, Equation (26) can be simplified to

$$p(X_k|X_{k-1}) = p(E_k|E_{k-1}, \varphi_{k-1})p(\varphi_k|\varphi_{k-1}). \tag{27}$$

The two terms on the right-hand side of Equation (27) are assumed to be

$$p(E_k|E_{k-1}, \varphi_{k-1}) = \mathcal{W}_{d_z}\left(E_k; \varepsilon_{k|k-1}, A_{k-1}E_{k-1}A_{k-1}^{\mathrm{T}}/\varepsilon_{k|k-1}\right), \tag{28}$$

and

$$p(\varphi_k|\varphi_{k-1}) = \mathcal{N}\left(\varphi_k; \varphi_{k-1}, \sigma_\varphi^2\right). \tag{29}$$

where $A_k = A(\varphi_k)$ represents the rotation matrix, and when $d_z = 2$, it is defined as

$$A_k = \begin{bmatrix} \cos(\varphi_k T) & -\sin(\varphi_k T) \\ \sin(\varphi_k T) & \cos(\varphi_k T). \end{bmatrix} \tag{30}$$

The variance σ_φ^2 can be used to quantify the uncertainty of the extended rotation rate.

More complex extended parameters, such as the extended scaling rate [33], can be considered to be included in the extended state. However, for the sake of generality, this study focuses solely on the extended rotation rate, as it is the most common scenario observed in practical applications. Thus, we summarize the random matrix model that includes the additional extended parameter as follows:

$$x_k = F_{k-1}x_{k-1} + w_k, \quad w_k \sim \mathcal{N}(0, E_k \otimes D_k) \tag{31}$$

$$z_k^l = H_k x_k + v_k^l, \quad v_k^l \sim \mathcal{N}(0, \alpha E_k + R_k) \tag{32}$$

$$p(X_k|X_{k-1}) = p(E_k|E_{k-1}, \varphi_{k-1})p(\varphi_k|\varphi_{k-1}) \tag{33}$$

$$p(E_k|E_{k-1}, \varepsilon_{k-1}) = \mathcal{W}_{d_z}\left(E_k; \varepsilon_{k|k-1}, A_{k-1}E_{k-1}A_{k-1}^{\mathrm{T}}/\varepsilon_{k|k-1}\right) \tag{34}$$

$$p(\varepsilon_k|\varepsilon_{k-1}) = \mathcal{N}\left(\varepsilon_k; \varepsilon_{k-1}, \sigma_\varphi^2\right) \tag{35}$$

Due to the introduction of additional extended parameters, it becomes challenging to obtain the posterior probability density $p(\xi_k \mid \mathbf{Z}_{1:k})$ of the joint state in a recursive manner. To address this issue, the sequential Monte Carlo method can be employed to approximate the posterior probability density of the joint state. However, traditional particle filters may suffer from low particle sampling efficiency when dealing with high-dimensional state estimation, resulting in a larger variance in state estimation.

To overcome this limitation, this study proposes the use of RBPF as a replacement for the basic particle filter. The RBPF offers improved performance, particularly in high-dimensional state estimation scenarios, by efficiently incorporating the extended parameters.

3.2. Joint State Estimation Based on RBPF

To introduce the RBPF algorithm, Bayesian theorem is employed to decompose the posterior density of the joint state into a product of two conditional densities:

$$p(\mathbf{x}_{0:k}, \mathbf{X}_{0:k}|\mathbf{Z}_{1:k}) = p(\mathbf{x}_{0:k}|\mathbf{X}_{0:k}, \mathbf{Z}_{1:k})p(\mathbf{X}_{0:k}|\mathbf{Z}_{1:k}) \tag{36}$$

The first term on the right-hand side of Equation (36) represents the posterior probability density of the motion state given the known extended state $\mathbf{X}_{0:k}$, where the motion state follows a linear Gaussian distribution [44]. This posterior probability density can be directly solved by using a Kalman filter.

Given a set of weighted samples $\left\{\mathbf{x}_{k-1}^{(i)}, \mathbf{X}_{k-1}^{(i)}, w_{k-1}^{(i)}\right\}_{i=1}^{N_s}$, after obtaining n_k measurements $\mathbf{Z}_k$ at time step k, the RBPF allows obtaining an equal number of weighted particles $\left\{\mathbf{x}_k^{(i)}, \mathbf{X}_k^{(i)}, w_k^{(i)}\right\}_{i=1}^{N_s}$ in a recursive manner. These particles are used to compute the posterior estimate of the joint state through weighted summation method. The recursive process of RBPF mainly consists of two stages: prediction and update. The specific descriptions of each stage are as follows:

(1) Sampling new extended state particles

$$\mathbf{X}_k^{(i)} = \left(\mathbf{E}_k^{(i)}, \varphi_k^{(i)}\right) \sim q\left(\mathbf{X}_k|\mathbf{X}_{0:k-1}^{(i)}, \mathbf{Z}_{1:k}\right). \tag{37}$$

Although the optimal importance density is theoretically the only one that exists, it is often challenging to handle [45]. We choose the extended state transition density as the importance density

$$\begin{aligned} p\left(\mathbf{X}_k|\mathbf{X}_{0:k-1}^{(i)}, \mathbf{Z}_{1:k}\right) &= p(\mathbf{X}_k|\mathbf{X}_{k-1}) \\ &= p\left(\mathbf{E}_k|\mathbf{E}_{k-1}^{(i)}, \varphi_{k-1}^{(i)}\right)p\left(\varphi_k|\varphi_{k-1}^{(i)}\right) \end{aligned} \tag{38}$$

(2) Perform a one-step prediction of the N_s Kalman filters

$$\hat{\mathbf{x}}_{k|k-1}^{(i)} = \mathbf{F}_k \hat{\mathbf{x}}_{k-1|k-1}^{(i)}, \tag{39}$$

$$\mathbf{P}_{k|k-1}^{(i)} = \mathbf{F}_{k-1}\mathbf{P}_{k-1|k-1}^{(i)}\mathbf{F}_{k-1}^{\mathrm{T}} + \mathbf{E}_k^{(i)} \otimes \mathbf{D}_k, \tag{40}$$

where $\mathbf{x}_{k|k-1}^{(i)}$ and $\mathbf{P}_{k|k-1}^{(i)}$ are the prior estimates and their corresponding covariance matrix of the known extended state sample set $\mathbf{X}_{0:k}^{(i)}$.

(3) The update formula for the particle weights of the extended state can be obtained through Equation (38)

$$w_k^{(i)} \propto p\left(\mathbf{Z}_k|\mathbf{X}_{0:k}^{(i)}, \mathbf{Z}_{1:k-1}\right) w_{k-1}^{(i)} \tag{41}$$

$$w_k^{(i)} = w_k^{(i)} / \sum_{i=1}^{N_s} w_k^{(i)} \tag{42}$$

To obtain the probability density function $p\left(\mathbf{Z}_k|\mathbf{X}_{0:k}^{(i)}, \mathbf{Z}_{1:k-1}\right)$ in Equation (41), the following property of the Gaussian distribution integral is utilized

$$\int \mathcal{N}(z; \boldsymbol{F}\boldsymbol{y}, \boldsymbol{Q})\mathcal{N}(\boldsymbol{y}; \boldsymbol{x}, \boldsymbol{R})d\boldsymbol{y} = \mathcal{N}\left(z; \boldsymbol{F}\boldsymbol{x}, \boldsymbol{F}\boldsymbol{R}\boldsymbol{F}^{\mathrm{T}} + \boldsymbol{R}\right) \tag{43}$$

Based on the random matrix model as shown in Equation (31), and assuming that the number of measurements n_k is independent of the motion state and extended state, we can use the Chapman-Kolmogorov equation to derive the following:

$$p(\mathbf{Z}_k|\mathbf{X}_{0:k}^{(i)}, \mathbf{Z}_{1:k-1}) \propto \int p(\mathbf{Z}_k|n_k, \boldsymbol{x}_k, \mathbf{X}_k^{(i)}) p(\boldsymbol{x}_k|\mathbf{X}_{0:k}^{(i)}, \mathbf{Z}_{1:k-1}) d\boldsymbol{x}_k \tag{44}$$

In Equation (44), the first term of the integral is the measurement likelihood function and can be expressed as [28]:

$$\begin{aligned} p(\mathbf{Z}_k|n_k, \boldsymbol{x}_k, \mathbf{X}_k^{(i)}) &= \prod_{l=1}^{n_k} \mathcal{N}\left(z_k^l; \boldsymbol{H}_k\boldsymbol{x}_k, \alpha\boldsymbol{E}_k^{(i)} + \boldsymbol{R}_k\right) \\ &\propto \mathcal{N}\left(\bar{z}; \boldsymbol{H}_k\boldsymbol{x}_k, \frac{\alpha\boldsymbol{E}_k^{(i)} + \boldsymbol{R}_k}{n_k}\right) \mathcal{W}_{d_z}\left(\bar{\mathbf{Z}}_k; n_k - 1, \alpha\boldsymbol{E}_k^{(i)} + \boldsymbol{R}_k\right) \end{aligned} \tag{45}$$

In Equation (44), the second term of the integral is the predicted probability density of the known extended state set $\mathbf{X}_{0:k}^{(i)}$, which is given by

$$p(\boldsymbol{x}_k|\mathbf{X}_{0:k}^{(i)}, \mathbf{Z}_{1:k-1}) = \mathbf{N}(\boldsymbol{x}_k; \hat{\boldsymbol{x}}_{k|k-1}^{(i)}, \boldsymbol{P}_{k|k-1}^{(i)}) \tag{46}$$

By substituting Equations (45) and (46) into Equation (44), and using the Gaussian density integral property from Equation (43), we obtain

$$\begin{aligned} p(\mathbf{Z}_k|\mathbf{X}_{0:k}^{(i)}, \mathbf{Z}_{1:k-1}) &\propto \mathbf{W}_{d_z}\left(\bar{\mathbf{Z}}_k; n_k - 1, \alpha\boldsymbol{E}_k^{(i)} + \boldsymbol{R}_k\right) \\ &\times \mathcal{N}\left(\bar{z}; \boldsymbol{H}_k\hat{\boldsymbol{x}}_{k|k-1}^{(i)}, \boldsymbol{H}_k\boldsymbol{P}_{k|k-1}^{(i)}\boldsymbol{H}_k^{\mathrm{T}} + \frac{\alpha\boldsymbol{E}_k^{(i)} + \boldsymbol{R}_k}{n_k}\right) \end{aligned} \tag{47}$$

(4) Perform the measurement update for the N_s Kalman filters

$$\boldsymbol{S}_k^{(i)} = \boldsymbol{H}_k\boldsymbol{P}_{k|k-1}^{(i)}\boldsymbol{H}_k^{\mathrm{T}} + (\alpha\boldsymbol{E}_k^{(i)} + \boldsymbol{R}_k)/n_k \tag{48}$$

$$\boldsymbol{K}_k^{(i)} = \boldsymbol{P}_{k|k-1}^{(i)}\boldsymbol{H}_k^{\mathrm{T}}\boldsymbol{S}_k^{(i)^{-1}} \tag{49}$$

$$\hat{\boldsymbol{x}}_{k|k}^{(i)} = \hat{\boldsymbol{x}}_{k|k-1}^{(i)} + \boldsymbol{K}_k^{(i)}(\bar{z}_k - \boldsymbol{H}_k\hat{\boldsymbol{x}}_{k|k-1}^{(i)}) \tag{50}$$

$$\boldsymbol{P}_{k|k}^{(i)} = \boldsymbol{P}_{k|k-1}^{(i)} - \boldsymbol{K}_k\boldsymbol{S}_k^{(i)}\boldsymbol{K}_k^{(i)^{\mathrm{T}}} \tag{51}$$

(5) Resampling

The fundamental idea of resampling is to discard particles with small weights while retaining particles with larger weights. The degree of particle degeneracy can be measured using the effective particle count N_s which is defined as follows [45]:

$$N_{\text{eff}} = 1/\sum_{i=1}^{N_s}\left(w_k^{(i)}\right)^2 \tag{52}$$

Set an effective particle count threshold N_{th}. When $N_{\text{eff}} < N_{\text{th}}$, perform the resampling operation.

(6) Computing the posterior estimates of the motion state and the extended state

$$\begin{aligned}\hat{\boldsymbol{x}}_{k|k} &= E(\boldsymbol{x}_k|\boldsymbol{Z}_{1:k}) \approx \sum_{i=1}^{N_s} w_k^{(i)}\hat{\boldsymbol{x}}_{k|k}^{(i)},\\ \hat{\boldsymbol{E}}_{k|k} &= E(\boldsymbol{E}_k|\boldsymbol{Z}_{1:k}) \approx \sum_{i=1}^{N_s} w_k^{(i)}\boldsymbol{E}_k^{(i)}.\end{aligned} \tag{53}$$

The process of the algorithm proposed in this paper is shown in Table 1.

Table 1. The process of the proposed algorithm.

(1) Initialization
Initialize Kalman filter
$\hat{\boldsymbol{x}}_{0\|0}^{(i)} = E(\boldsymbol{x}_0), \boldsymbol{P}_{0\|0}^{(i)} = E\left[(\boldsymbol{x}_0 - E(\boldsymbol{x}_0))(\boldsymbol{x}_0 - E(\boldsymbol{x}_0))^{\mathrm{T}}\right]$
Initialize particles filter
$\boldsymbol{X}_0^{(i)} = \left(\boldsymbol{E}_0^{(i)}, w_0^{(i)}\right) \sim p(\boldsymbol{X}_0) = p(\boldsymbol{E}_0)p(w_0)$
(2) Recursion
Prediction
Sequential sampling of particles
$\boldsymbol{X}_k^{(i)} = \left(\boldsymbol{E}_k^{(i)}, \varphi_k^{(i)}\right) \sim p\left(\boldsymbol{E}_k\|\boldsymbol{E}_{k-1}^{(i)}, \varphi_{k-1}^{(i)}\right)p\left(\varphi_k\|\varphi_{k-1}^{(i)}\right)$
One step prediction of motion state and covariance
$\hat{\boldsymbol{x}}_{k\|k-1}^{(i)} = \boldsymbol{F}_k\hat{\boldsymbol{x}}_{k-1\|k-1}^{(i)}$
$\boldsymbol{P}_{k\|k-1}^{(i)} = \boldsymbol{F}_{k-1}\boldsymbol{P}_{k-1\|k-1}^{(i)}\boldsymbol{F}_{k-1}^{\mathrm{T}} + \boldsymbol{E}_k^{(i)} \otimes \boldsymbol{D}_k$
Update
Update particle weights
$w_k^{(i)} \quad p\left(\boldsymbol{Z}_k\|\boldsymbol{X}_{0:k}^{(i)}, \boldsymbol{Z}_{1:k-1}\right)w_{k-1}^{(i)}$
Update state of motion
$\boldsymbol{S}_k^{(i)} = \boldsymbol{H}_k\boldsymbol{P}_{k\|k-1}^{(i)}\boldsymbol{H}_k^{\mathrm{T}} + (\alpha\boldsymbol{E}_k^{(i)} + \boldsymbol{R}_k)/n_k$
$\boldsymbol{K}_k^{(i)} = \boldsymbol{P}_{k\|k-1}^{(i)}\boldsymbol{H}_k^{\mathrm{T}}\boldsymbol{S}_k^{(i)^{-1}}$
$\hat{\boldsymbol{x}}_{k\|k}^{(i)} = \hat{\boldsymbol{x}}_{k\|k-1}^{(i)} + \boldsymbol{K}_k^{(i)}(\bar{z}_k - \boldsymbol{H}_k\hat{\boldsymbol{x}}_{k\|k-1}^{(i)})$
$\boldsymbol{P}_{k\|k}^{(i)} = \boldsymbol{P}_{k\|k-1}^{(i)} - \boldsymbol{K}_k\boldsymbol{S}_k^{(i)}\boldsymbol{K}_k^{(i)^{\mathrm{T}}}$
Normalized weights
$w_k^{(i)} = w_k^{(i)} / \sum_{i=1}^{N_s} w_k^{(i)}$
Resampling
Calculate the effective number of particles, and if $N_{\text{eff}} < N_{\text{th}}$, perform the resampling step.
Compute the posterior estimate of the joint state.
$\hat{\boldsymbol{x}}_{k\|k} = E(\boldsymbol{x}_k\|\boldsymbol{Z}_{1:k}) \approx \sum_{i=1}^{N_s} w_k^{(i)}\hat{\boldsymbol{x}}_{k\|k}^{(i)},$
$\hat{\boldsymbol{E}}_{k\|k} = E(\boldsymbol{E}_k\|\boldsymbol{Z}_{1:k}) \approx \sum_{i=1}^{N_s} w_k^{(i)}\boldsymbol{E}_k^{(i)}$

3.3. Improved Resampling with Particles Segmentation Cross Summation of RBPF

3.3.1. Resampling with Particles Grid Segmentation

Particles grid segmentation sampling is a targeted data collection strategy for RBPF algorithm. It sorts the particles according to the values of selected parameters of particles at first. The parameter weight of particle is chosen in this paper. Then it divides particles into N_{seg} segments based on particles' weight values, in which the weights of particles are in ascending order. The higher weight value means the higher the quality of the particles in the area. During the resampling process, particle samples are randomly picked up from N_{seg} segments of particles, and act as the new particles for the next iteration calculation. All of the segments are used, but the particle sampling ratio is adjusted based on preset weight values of the segments dynamically.

A higher proportion of random sampling is used in high particle weight segments, to ensure that the particles in the region are sufficiently resampled, and the obtained particles are of higher quality. On the contrary, in particle grid segments areas with lower particle weights and relatively less importance, a lower proportion of sampling is adopted to balance the particle degradation problem and the resource consumption.

In addition, due to the sampling of low weight segments during the particle grid segmentation sampling process, it can help improve the process of polynomial resampling that only replicates particles with high weights, resulting in particle impoverishment and a single offspring of samples problem.

3.3.2. Resampling with Particles Segmentation Cross Summation of RBPF

On the basis of particles grid segmentation, we apply the cross summation of selected samples to form the new samples, and act as parts of next generation particles, and generate a satisfactory quantity. Let $\bar{P}$ include the system states $\boldsymbol{x}_k$ and extended parameters (extended rotation rate φ_k and random matrix $\boldsymbol{E}_k$). The specific cross calculation method for the new sample is

$$\bar{P}_i^{new} = \eta_i \bar{P}_i^j + \beta_i \bar{P}_i^k \tag{54}$$

where $\bar{P}_i^j$ and $\bar{P}_i^k$ are two different particles randomly selected from the lower weight part and the higher weight (i.e., the weight value related to each particle) part of the i-th particles grid segments, and $\bar{P}_i^{new}$ is a newly generated particle, η_i and β_i are random values within the interval $[0,1]$, $i \in \{1, 2, ..., N_{seg}\}$.

In this sampling method, only a subset of particles is generated using cross-summation, while the remaining particles are sampled with replacement based on their weight values. This implies that particles with higher weights are more likely to be selected multiple times. Compared to particle grid segmentation sampling, the proposed particle segmentation cross summation sampling ensures the replication of high-weight particles while simultaneously enhancing the diversity of offspring particles through cross summation. This approach achieves a better balance between the preservation of significant particles and the maintenance of population diversity.

4. Experimental Results and Analysis

The digital staring radar system designed in this study operates in the S-band, as shown in Figure 1. The digital staring radar featuring a wide transmission beam and lower gain, enhances detection performance by coherently accumulating echo energy from scattering points, thereby improving the signal-to-noise ratio.

We conduct cluster target tracking simulation experiments under different scenarios to evaluate the proposed tracking algorithms. The tracking accuracy performance is compared against the Koch method, the Feldmann method, and the Lan method, demonstrating the effectiveness and robustness of our approach. The root mean square error (RMSE) is adopted as the evaluation criterion to quantitatively assess the performance of the algorithms tested. It is calculated using the follow formula:

$$\mathrm{RMSE}_k = \sqrt{\frac{1}{N_r}\sum_{i=1}^{N_r}(\boldsymbol{x}_k - \hat{\boldsymbol{x}}_k)^{\mathrm{T}}(\boldsymbol{x}_k - \hat{\boldsymbol{x}}_k)}, \tag{55}$$

where $\boldsymbol{x}_k$ and $\hat{\boldsymbol{x}}_k$ are the real value and the estimated value of the cluster targets, N_r is the number of totaling runs. The RMSEs of both cluster position and velocity are respectively considered in the following simulation experiments of two scenarios.

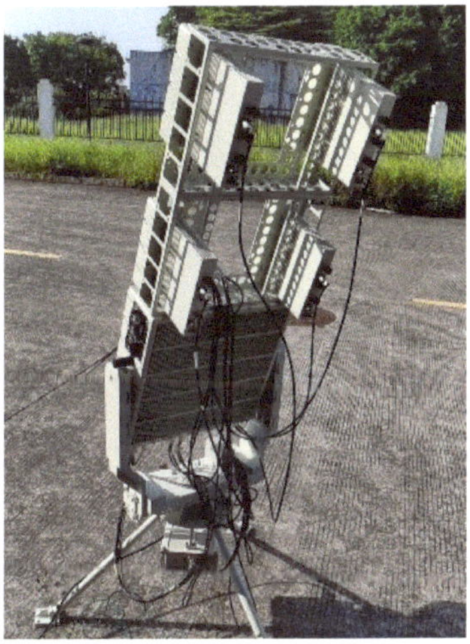

Figure 1. Holographic staring radar sensor application scenarios.

4.1. Cluster Target Tracking Test Experiment of Scenario One

Simulation scenario one: We simulate a formation of 5 individual targets flying with constant speed, 35 m/s, in the (x, y)-plane, the interval between two adjacent targets is set to 5 m; After flying for 19 s, the cluster targets change their formation and move clockwise for 18 s; Finally, they keep the formation and fly in a straight line under constant speed 35 m/s for 15 s. The flying trajectory of the cluster is shown in Figure 2, in which the black solid circle means the real target.

We tested the RBPF with basic resampling method (marked as RBPF-1 in subsequent sections for short), RBPF using Resampling with Particles Grid Segmentation method (marked as RBPF-2 for short), RBPF using Resampling with Particles Segmentation Cross Summation method (marked as RBPF-3), and the compared Koch method, Feldmann method and Lan method. Each cluster target tracking simulation run of the tested methods is repeated 5 times, the target tracking results including targets' position and velocity, and extended parameter values are estimated; The tracking accuracy results RMSE is calculated and listed in Table 2. When calculating the RMSE, we use the position and velocity of the central target of the cluster to form the real state. The cluster target tracking trajectories of Koch method, Feldmann method, Lan method and RBPF-3 are shown in Figure 2. We can see that the achieved green elliptic curves of RBPF-3 cover the cluster targets with proper scale size, and rotated with the adjustable cluster target formation.

The elliptic curves cover all of the targets in a cluster, but with smaller size, reveals the good performance of presented RBPF-3 among the compared methods. In Table 2, the mean values of position/velocity RMSE of tracking targets of presented RBPF-2 and RBPF-3 are smaller than the other compared methods. Figures 3 and 4 also show that the position/velocity RMSE results of the tested RBPF-2 and RBPF-3 methods are better in most situations.

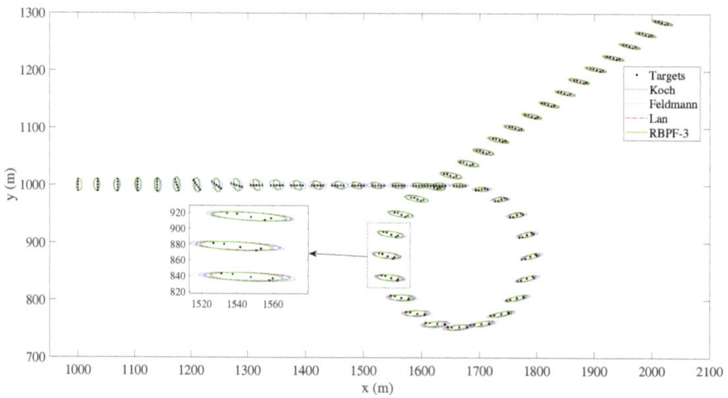

Figure 2. Cluster target tracking trajectory of compared methods in scenario one.

Table 2. Cluster target tracking results (RMSE) in scenario one. (Pos means position (units: m), Vel means velocity (units: m/s), STD means the standard deviation).

Methods	Mean (Pos (m))	STD (Pos (m))	Mean (Vel (m/s))	STD (Vel (m/s))
Koch	3.2602	1.6363	5.6234	2.3776
Feldmann	3.3579	1.7017	6.4933	2.4439
Lan	3.3394	1.7461	6.2769	2.7961
RBPF-1	2.1864	1.2869	4.1325	2.8612
RBPF-2	2.1849	1.2693	4.1088	2.8326
RBPF-3	2.1846	1.2813	4.1207	2.8421

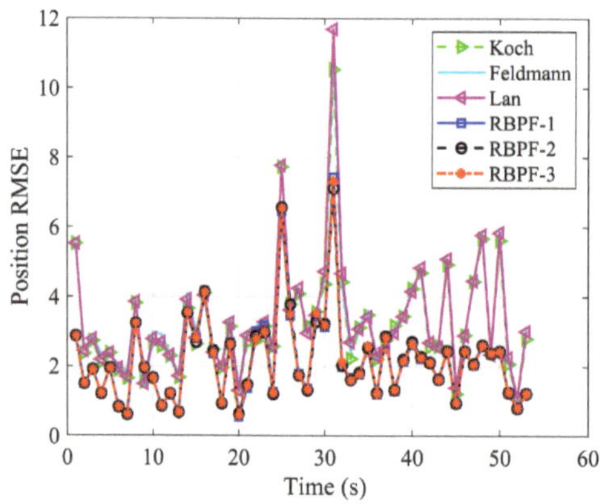

Figure 3. Cluster target tracking RMSE of position of compared methods in scenario one.

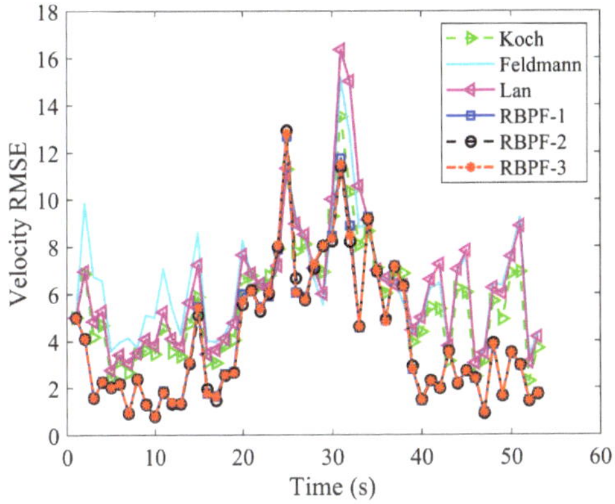

Figure 4. Cluster target tracking RMSE of velocity of compared methods in scenario one.

4.2. Cluster Target Tracking Test Experiment of Scenario Two

Simulation scenario two: We simulate a formation of 5 individual targets flying with constant speed, 30 m/s, in the (x, y)-plane, the interval between two adjacent targets is set to 5 m; After flying for 19 s, the UAV cluster targets change their formation and move counterclockwise for 18 s; Then, they keep the formation and fly in a straight line under constant speed 35 m/s for 10 s; Finally, targets move in a straight line direction with various speeds for 5 s. The flying trajectory of the cluster is shown in Figure 5, in which the black solid circle is the real target.

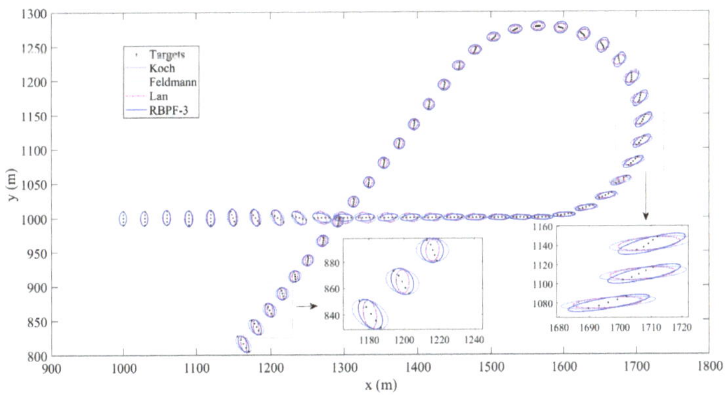

Figure 5. Cluster target tracking trajectory of compared methods in scenario two.

We tested the RBPF-1, RBPF-2, RBPF-3 (using the turn rate parameter), Koch method, Feldmann method and Lan method using the dataset of achieved in scenario two. Each cluster target tracking simulation run of the tested methods is repeated 20 times, the target tracking results including targets' position and velocity, and extended parameter values are estimated; The tracking accuracy results RMSE is calculated and listed in Table 3. In Table 3, the mean values of position/velocity RMSE of tracking targets of presented RBPF-2 and RBPF-3 are smaller than the Koch, Feldmann and Lan methods.

Table 3. Cluster target tracking results (RMSE) in scenario two. (Pos means position (units: m), Vel means velocity (units: m/s), STD means the standard deviation)

Methods	Mean (Pos (m))	STD (Pos (m))	Mean (Vel (m/s))	STD (Vel (m/s))
Koch	2.8768	0.4876	4.3638	1.0702
Feldmann	3.0129	0.5122	5.7570	1.1808
Lan	2.9282	0.4992	4.7849	1.0896
RBPF-1	1.2030	0.5186	2.2970	1.6809
RBPF-2	1.1992	0.5090	2.2874	1.6700
RBPF-3	1.1956	0.4970	2.3914	1.7595

We use the position and velocity of the central target of the cluster to form the real state to calculate the RMSE results. The cluster target tracking trajectories of Koch method, Feldmann method, Lan method and RBPF-3 are shown in Figure 5. We can see that the achieved green elliptic curves of RBPF-3 cover the cluster targets with small scale size, and rotated with the adjustable cluster target formation quickly. The elliptic curves cover the targets well, and reveals the good performance of presented RBPF-3. Figures 6 and 7 show that the position/velocity RMSE results of the tested RBPF-2 and RBPF-3 methods are better.

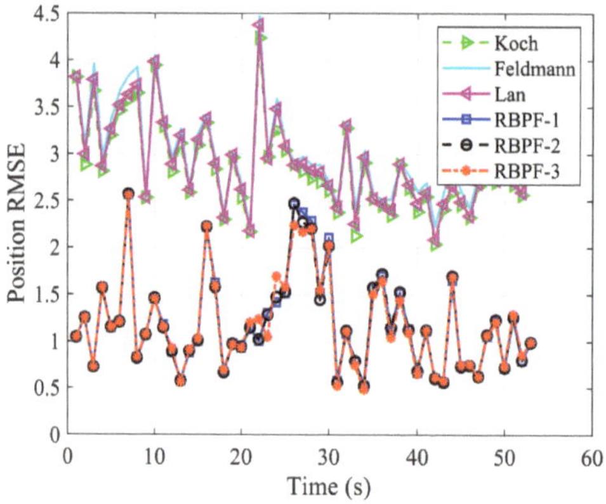

Figure 6. Cluster target tracking RMSE of position of compared methods in scenario two.

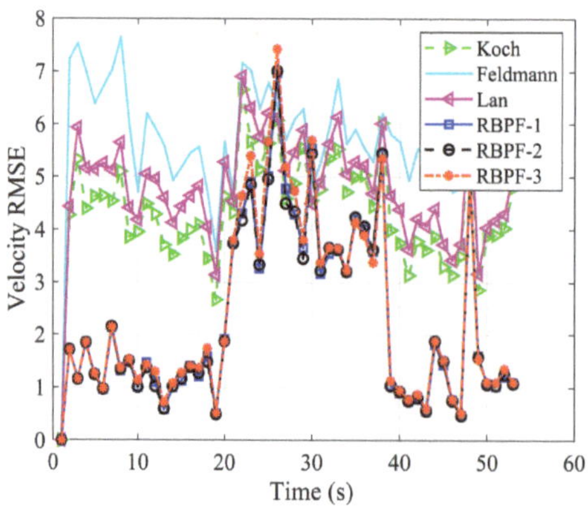

Figure 7. Cluster target tracking RMSE of velocity of compared methods in scenario two.

4.3. Cluster Target Tracking Test Experiment of Scenario Three

Simulation scenario three: We simulate a formation of 5 individual targets flying with constant speed, 30 m/s, in the (x, y)-plane, the interval between two adjacent targets is set to 5 m; After flying for 19 s, the UAV cluster targets change their formation and move counter-clockwise for 18 s; Then, they change their direction and velocity and keep on flying for 8 s. After that, they keep the formation and fly in a straight line under constant speed 35 m/s for 10 s; Finally, targets move in a straight line direction with various speeds for 5 s. The flying trajectory of the cluster is shown in Figures 8 and 9, in which the black solid circle is the real target.

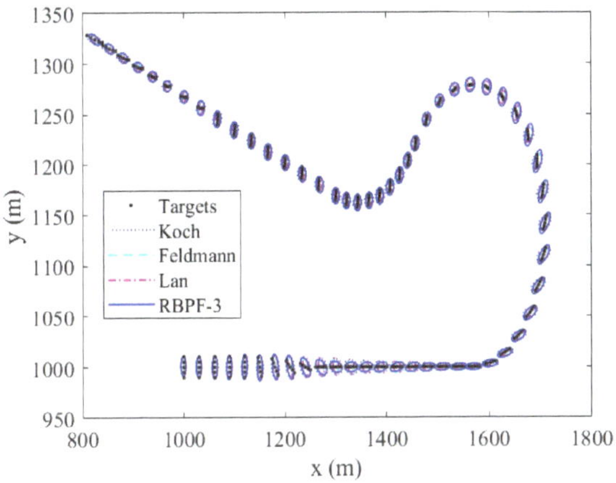

Figure 8. Cluster target tracking trajectory of compared methods in scenario three (full).

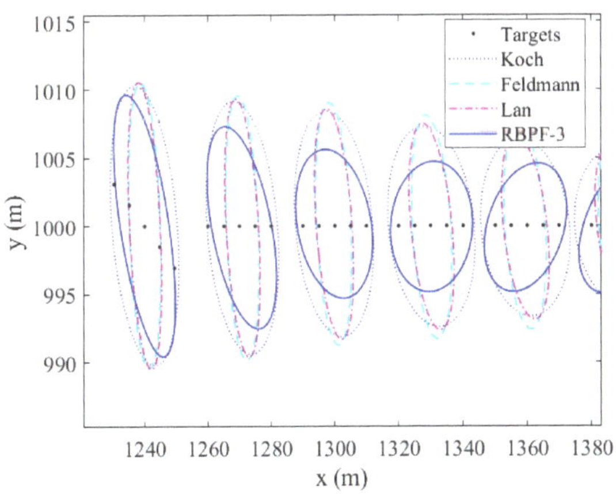

Figure 9. Cluster target tracking trajectory of compared methods in scenario three (partially magnified).

We tested the RBPF-1, RBPF-2, RBPF-3 (using the turn rate parameter), Koch method, Feldmann method and Lan method using the dataset of achieved in scenario three. Each cluster target tracking simulation run of the tested methods is repeated 20 times, the target tracking results including targets' position and velocity, and extended parameter values are estimated. We use the position and velocity of the central target of the cluster to form the real state to calculate the RMSE results. The cluster target tracking trajectories of Koch method, Feldmann method, Lan method and RBPF-3 are shown in Figures 8 and 9. We can see that the achieved green elliptic curves of RBPF-3 cover the cluster targets with small scale size, and rotated with the adjustable cluster target formation quickly. The elliptic curves cover the targets well, and shows the good performance of presented RBPF-3. Figures 10 and 11 show that the position/velocity RMSE results of the tested RBPF-1, RBPF-2 and RBPF-3 methods are better.

The simulation results demonstrate the effectiveness of the proposed improved random matrix and Rao-Blackwellized particle filter algorithm to track the target of the UAV cluster.

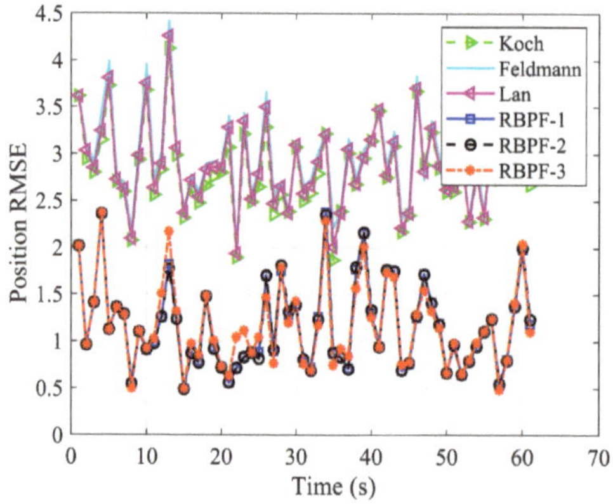

Figure 10. Cluster target tracking RMSE of position of compared methods in scenario three.

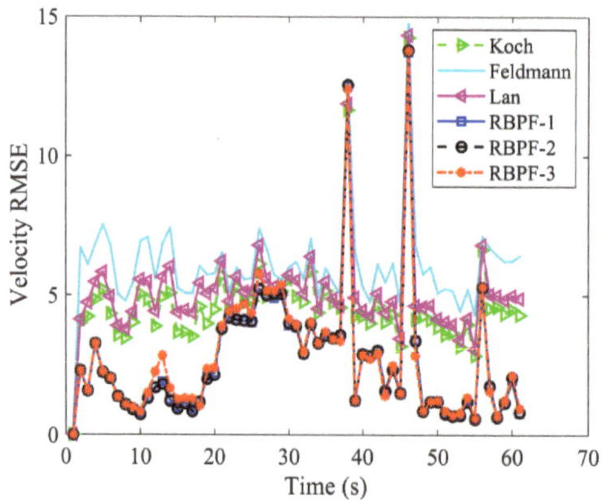

Figure 11. Cluster target tracking RMSE of velocity of compared methods in scenario three.

4.4. Computational Complexity Test

To analyze the computational complexity of the researched methods, simulation experiments were implemented using a laptop of Intel(R) Core(TM) i7 2.30 GHz and RAM 32 GB. Each simulation experiment using the compared methods is repeat 20 times, and each method implemented the target estimation about 60 s. The elapsed time results of each run for cluster target tracking using the compared methods in scenario three are recorded and shown in Figure 12. The average elapsed time of Koch, Feldmann, Lan, RBPF-1, RBPF-2, and RBPF-3 in each run is 0.0886 s, 0.0824 s, 0.0882 s, 4.7556 s, 4.7428 s, and 5.4421 s, respectively, which indicates that the computational complexity of the presented methods is higher than the Koch, Feldmann, and Lan methods. Since the average elapsed time for the target estimation in each second is smaller than 1 when using the presented methods, the RBPF-1, RBPF-2, and RBPF-3 satisfy the time performance required. Compared with the Koch, Feldmann, and Lan methods, the presented method will require more computational resources during the real-world deployment.

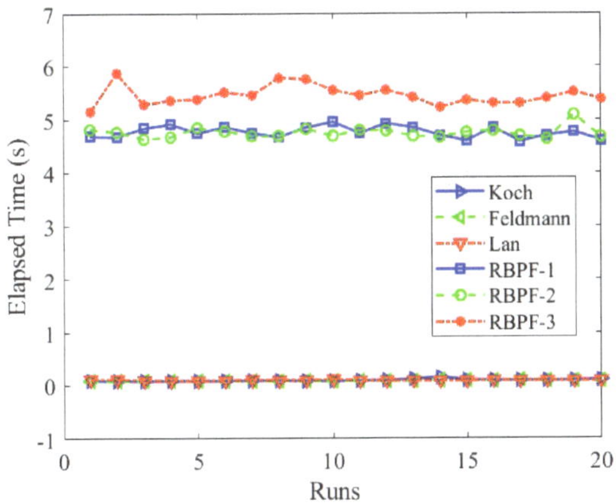

Figure 12. Cluster target tracking Elapsed time of compared methods in scenario three.

5. Conclusions

An adaptive random matrix approach for tracking UAV swarms is integrated into a designed digital staring radar system in this study. In the tracking approaches for UAVs cluster targets, most existing models and methods assume that the extended ellipses of cluster targets remain approximately invariant and do not consider the complex variation processes, such as the rotation of the extended ellipses. This paper proposes the introduction of additional extension parameters to represent the intricate variation processes of the extended state to address this limitation. It employs a sequential Monte Carlo method to approximate the posterior estimation of both the motion state and the extension state. An adaptive Rao-Blackwellized Particle Filter with improved resampling with particles segmentation cross summation, is then employed to estimate the motion and extension states of the UAV swarm jointly. Multiple test experiments were conducted. Based on the simulation results, it is evident that the proposed method can swiftly respond to changes in target extension, particularly when rotation occurs. Furthermore, it ensures an estimation accuracy of the motion state that is comparable to the compared methods.

Author Contributions: conceptualization, N.Z., F.Z. and X.L.; methodology, N.Z., F.Z. and Y.Z.; software, N.Z.; validation, N.Z., F.Z. and H.X.; investigation, N.Z. and Y.Z.; data curation, N.Z. and F.Z.; writing original draft preparation, N.Z., G.N., and X.L.; writing review and editing, N.Z., Y.Z. G.N., and F.Z.; supervision, Y.Z. and H.X. All authors contributed to writing the final manuscript. All authors have read and agreed to the published version of the manuscript

Funding: This work is supported by the National Natural Science Foundation of China (Grant No. 62102466), the Fundamental Research Funds for the Central Universities, Sun Yat-sen University (Grant No. 24qnpy164), Guangdong Basic and Applied Basic Research Foundation Grant (No. 2022A1515140103 and No. 2023A1515011588), the Shenzhen Science and Technology Program (grant No. 202206193000001, 20220815171723002 and grant No. JCYJ20240813151238049), and the Open Fund of the State Key Laboratory of Satellite Navigation Systems and Equipment Technology (Grant No. CEPNT2023B08).

Data Availability Statement: All data that support the findings of this study are included within the article.

Acknowledgments: We would like to extend our sincere gratitude to Shiyou Xu for his invaluable contributions to the exploration and verification of this research, and Xiaoyan Lv for her helpful discussions and comments.

Conflicts of Interest: The authors declare no conflicts of interest.

References

1. Wen, L.; Zhen, Z.; Tao, C.; Ding, J. Distributed Cooperative Strategy of UAV Swarm Without Speed Measurement Under Saturation Attack Mission. *IEEE Trans. Aerosp. Electron. Syst.* **2024**, *60*, 4518–4529.
2. Lei, Y.; Tong, X.; Qiu, C.; Sun, Y.; Tang, J.; Guo, C.; Li, H. Onboard Data Management Approach Based on a Discrete Grid System for Multi-UAV Cooperative Image Localization. *IEEE Trans. Geosci. Remote Sens.* **2023**, *61*, 1–17.
3. Ye, X.; Xue, W.; Chen, X.; Zhang, Y.; Wang, X.; Guan, J. Cauchy Kernel-Based AEKF for UAV Target Tracking via Digital Ubiquitous Radar Under the Sea–Air Background. *IEEE Geosci. Remote Sens. Lett.* **2024**, *21*, 1–5.
4. Sun, Z.; Hu, C.; Cui, K.; Wang, R.; Ding, M.; Yan, Z.; Wu, D. Extracting Bird and Insect Migration Echoes From Single-Polarization Weather Radar Data Using Semi-Supervised Learning. *IEEE Trans. Geosci. Remote Sens.* **2024**, *62*, 1–12.
5. Sayed, A.N.; Ramahi, O.M.; Shaker, G. In the Realm of Aerial Deception: UAV Classification via ISAR Images and Radar Digital Twins for Enhanced Security. *IEEE Sens. Lett.* **2024**, *8*, 1–4.
6. Yuan, D.; Zhang, H.; Shu, X.; Liu, Q.; Chang, X.; He, Z.; Shi, G. Thermal Infrared Target Tracking: A Comprehensive Review. *IEEE Trans. Instrum. Meas.* **2024**, *73*, 1–19.
7. Chen, Y.; Li, Z.; Wu, L.; Chen, P. An Efficient Multi-Object Tracking Guided by Spatial Clustering on Vision Sensors. *IEEE Sens. J.* **2024**, *24*, 19344–19351.
8. Zhu, N.; Xi, Z.; Wu, C.; Zhong, F.; Qi, R.; Chen, H. Inductive Conformal Prediction Enhanced LSTM-SNN Network: Applications to Birds and UAVs Recognition. *IEEE Geosci. Remote Sens. Lett.* **2024**, *21*, 1–5.
9. Liu, Z.; Wang, Z.K.; Yang, Y.B.; Lu, Y. A Data-Driven Maneuvering Target Tracking Method Aided with Partial Models. *IEEE Trans. Veh. Technol.* **2024**, *73*, 414–425.
10. Ming, R.; Zhou, Z.; Luo, X.; Liu, W.; Le, Z.; Song, C.; Jiang, R.; Zang, Y. Optical Tracking System for Multi-UAV Clustering. *IEEE Sens. J.* **2021**, *21*, 19382–19394.
11. Liang, T.; Zhang, T.; Yang, J.; Feng, D.; Zhang, Q. UAV-Aided Positioning Systems for Ground Devices: Fundamental Limits and Algorithms. *IEEE Internet Things J.* **2022**, *9*, 13470–13485. [CrossRef]
12. Ye, T.; Qin, W.; Li, Y.; Wang, S.; Zhang, J.; Zhao, Z. Dense and Small Object Detection in UAV-Vision Based on a Global-Local Feature Enhanced Network. *IEEE Trans. Instrum. Meas.* **2022**, *71*, 1–13. [CrossRef]
13. Zhu, N.; Xu, S.; Li, C.; Hu, J.; Fan, X.; Wu, W.; Chen, Z. An Improved Phase-Derived Range Method Based on High-Order Multi-Frame Track-Before-Detect for Warhead Detection. *Remote Sens.* **2022**, *14*, 29. [CrossRef]
14. Song, Q.; Huang, S.; Zhang, Y.; Chen, X.; Chen, Z.; Zhou, X.; Deng, Z. Radar Target Classification Using Enhanced Doppler Spectrograms with ResNet34_CA in Ubiquitous Radar. *Remote Sens.* **2024**, *16*, 2860. [CrossRef]
15. Chen, X.; Zhang, H.; Song, J.; Guan, J.; Li, J.; He, Z. Micro-Motion Classification of Flying Bird and Rotor Drones via Data Augmentation and Modified Multi-Scale CNN. *Remote Sens.* **2022**, *14*, 1107. [CrossRef]
16. Zhu, N.; Xi, Z.; Chen, H.; Xu, S.; Wang, Y.; Zhong, F. Mondrian Conformal Prediction Enhanced LSTM for Birds and Drones Recognition. In *IET Conference Proceedings CP874 2023*; The Institution of Engineering and Technology: Stevenage, UK, 2023; pp. 760–765.
17. Zhao, Y.; Su, Y. Synchrosqueezing Phase Analysis on Micro-Doppler Parameters for Small UAVs Identification with Multichannel Radar. *IEEE Geosci. Remote. Sens. Lett.* **2020**, *17*, 411–415. [CrossRef]
18. Yu, X.; Wei, S.; Fang, Y.; Sheng, J.; Zhang, L. Low-Altitude Slow Small Target Threat Assessment Algorithm by Exploiting Sequential Multi-Feature with Long-Short-Term-Memory. *IEEE Sens. J.* **2023**, *23*, 21524–21533. [CrossRef]
19. Kang, K.B.; Choi, J.H.; Cho, B.L.; Lee, J.S.; Kim, K.T. Analysis of Micro-Doppler Signatures of Small UAVs Based on Doppler Spectrum. *IEEE Trans. Aerosp. Electron. Syst.* **2021**, *57*, 3252–3267. [CrossRef]
20. Chen, X.; Guan, J.; Chen, W.; Zhang, L.; Yu, X. Sparse Long-Time Coherent Integration-Based Detection Method for Radar Low-Observable Manoeuvring Target. *IET Radar Sonar Navig.* **2020**, *14*, 538–546. [CrossRef]
21. White, D.; Jahangir, M.; Baker, C.J.; Antoniou, M. Urban Bird-Drone Classification with Synthetic Micro-Doppler Spectrograms. *IEEE Trans. Radar Syst.* **2024**, *2*, 167–179. [CrossRef]

22. Chen, Z.; Ji, H.; Zhang, Y.; Zhu, Z.; Li, Y. High-Resolution Feature Pyramid Network for Small Object Detection on Drone View. *IEEE Trans. Circuits Syst. Video Technol.* **2023**, *34*, 475–489. [CrossRef]
23. Zheng, Y.; Chen, Z.; Lv, D.; Li, Z.; Lan, Z.; Zhao, S. Air-to-Air Visual Detection of Micro-UAVs: An Experimental Evaluation of Deep Learning. *IEEE Robot. Autom. Lett.* **2021**, *6*, 1020–1027.
24. Zhao, Y.; Su, Y. The Extraction of Micro-Doppler Signal with EMD Algorithm for Radar-Based Small UAVs' Detection. *IEEE Trans. Instrum. Meas.* **2020**, *69*, 929–940.
25. Dugger, K. Techniques for Wildlife Investigations and Management. *Condor* **2007**, *109*, 981–983.
26. Zhu, N.; Hu, J.; Xu, S.; Wu, W.; Zhang, Y.; Chen, Z. Micro-Motion Parameter Extraction for Ballistic Missile with Wideband Radar Using Improved Ensemble EMD Method. *Remote Sens.* **2021**, *13*, 3545. [CrossRef]
27. Drummond, O.E.; Society of Photo-Optical Instrumentation Engineers; CREOL (Research Center); SPIE Technical Symposium on Optical Engineering and Photonics in Aerospace Sensing. *Signal and Data Processing of Small Targets 1990 : 16–18 April 1990, Orlando, Florida / Oliver E. Drummond, Chair/Editor; Sponsored by SPIE–The International Society for Optical Engineering; Cooperative Organization, CREOL/University of Central Florida*; Society of Photo-Optical Instrumentation Engineers: Bellingham, WA, USA, 1990; Volume 1305.
28. Koch, J.W. Bayesian Approach to Extended Object and Cluster Tracking Using Random Matrices. *IEEE Trans. Aerosp. Electron. Syst.* **2008**, *44*, 1042–1059.
29. Gupta, A.K.; Nagar, D.K. *Matrix Variate Distributions*; CRC Press: Boca Raton, FL, USA, 2020; pp. 13–26.
30. Feldmann, M.; Fränken, D.; Koch, W. Tracking of Extended Objects and Group Targets Using Random Matrices. *IEEE Trans. Signal Process.* **2010**, *59*, 1409–1420.
31. Orguner, U. A Variational Measurement Update for Extended Target Tracking with Random Matrices. *IEEE Trans. Signal Process.* **2012**, *60*, 3827–3834. [CrossRef]
32. Bishop, C.M. *Pattern Recognition and Machine Learning*; Springer: Berlin/Heidelberg, Germany, 2006.
33. Lan, J.; Li, X.R. Tracking of Extended Object or Target Group Using Random Matrix: New Model and Approach. *IEEE Trans. Aerosp. Electron. Syst.* **2016**, *52*, 2973–2989.
34. Lan, J.; Li, X.R. Extended-object or Group-Target Tracking Using Random Matrix with Nonlinear Measurements. *IEEE Trans. Signal Process.* **2019**, *67*, 5130–5142.
35. Mo, L.; Song, X.; Zhou, Y.; Sun, Z.K. Unbiased Converted Measurements for Tracking. *IEEE Trans. Aerosp. Electron. Syst.* **1998**, *34*, 1023–1027.
36. Duan, Z.; Han, C.; Li, X.R. Comments on Unbiased Converted Measurements for Tracking. *IEEE Trans. Aerosp. Electron. Syst.* **2004**, *40*, 1374.
37. Granström, K.; Natale, A.; Braca, P.; Ludeno, G.; Serafino, F. Gamma Gaussian Inverse Wishart Probability Hypothesis Density for Extended Target Tracking Using X-Band Marine Radar Data. *IEEE Trans. Geosci. Remote Sens.* **2015**, *53*, 6617–6631.
38. Doucet, A.; Godsill, S.; Andrieu, C. On Sequential Monte Carlo Sampling Methods for Bayesian Filtering. *Stat. Comput.* **2000**, *10*, 197–208.
39. Mustiere, F.; Bolic, M.; Bouchard, M. Rao-Blackwellised Particle Filters: Examples of Applications. In Proceedings of the 2006 Canadian Conference on Electrical and Computer Engineering, Ottawa, ON, Canada, 7–10 May 2006; pp. 1196–1200.
40. Cheng, L.; Sengupta, A.; Cao, S. Deep Learning-Based Robust Multi-Object Tracking via Fusion of mmWave Radar and Camera Sensors. *IEEE Trans. Intell. Transp. Syst.* **2024**, *25*, 17218-17233.
41. Huang, Y.; Guo, R.; Zhang, Y.; Chen, Z. Deep-Reinforcement-Learning-Based Radar Parameter Adaptation for Multiple-Target Tracking. *IEEE Trans. Aerosp. Electron. Syst.* **2024**, *60*, 7125–7141.
42. Zhang, H.; Chen, H.; Zhang, W.; Zhang, X. Trajectory Planning for Airborne Radar in Extended Target Tracking Based on Deep Reinforcement Learning. *Digit. Signal Process.* **2024**, *153*, 104603
43. Cramér, H. *Mathematical Methods of Statistics (PMS-9), Volume 9*; Princeton University Press: Princeton, NJ, USA, 2016.
44. Chen, R.; Liu, J.S. Mixture Kalman Filters. *J. R. Stat. Soc. B* **2000**, *62*, 493–508.
45. Arulampalam, M.S.; Maskell, S.; Gordon, N.; Clapp, T. A Tutorial on Particle Filters for Online Nonlinear/Non-Gaussian Bayesian Tracking. *IEEE Trans. Signal Process.* **2002**, *50*, 2.

remote sensing

MDPI

Technical Note

Two-Dimensional Space-Variant Motion Compensation Algorithm for Multi-Hydrophone Synthetic Aperture Sonar Based on Sub-Beam Compensation

Haoran Wu [1,*], Fanyu Zhou [1], Zhimin Xie [2,3], Jingsong Tang [1], Heping Zhong [1] and Jiafeng Zhang [1]

[1] Naval Institute of Underwater Acoustic Technology, Naval University of Engineering, Wuhan 430033, China; trench2001@163.com (F.Z.); jinsongtangwh@163.com (J.T.); zheping525@sohu.com.cn (H.Z.); zhangjiafeng202211@outlook.com (J.Z.)

[2] College of Underwater Acoustic Engineering, Harbin Engineering University, Harbin 150001, China; xie1981minyue@sina.com

[3] Military Marine Environment Construction Office, Beijing 100161, China

* Correspondence: wuhaoran_wh@163.com; Tel.: +86-156-0863-9727

Abstract: For a multi-hydrophone synthetic aperture sonar (SAS), the instability of the platform and underwater turbulence easily lead to two-dimensional (2-D) space-variant (SV) motion errors. Such errors can cause serious imaging problems and are very difficult to compensate for. In this study, we propose a 2-D SV motion compensation algorithm for a multi-hydrophone SAS based on sub-beam compensation. The proposed algorithm is implemented using the following four-step process: (1) The motion error of each sub-beam is obtained by substituting the sonar's motion parameters measured in the exact motion error model established in this study. (2) The sub-beam's targets of all targets are compensated for motion error by implementing two-phase multiplications on the raw data of the multiple-hydrophone SAS in the order of hydrophone by hydrophone. (3) The data of the sub-beam's target compensated motion error are extracted from the raw data by utilizing the mapping relationship between the azimuth angle and the Doppler frequency. (4) The imaging result of each sub-beam is obtained by performing a monostatic imaging algorithm on each sub-beam's data and coherently added to obtain high-resolution imaging results. Finally, the validity of the proposed algorithm was tested using simulation and real data.

Keywords: motion compensation; aperture dependent; sub-beam; multi-hydrophone synthetic aperture sonar

1. Introduction

Synthetic aperture sonar (SAS) [1–3] is a high-resolution acoustic imaging system mounted on an autonomous underwater vehicle (AUV), remote-operated vehicle (ROV), and unmanned undersea vehicle (UUV) platforms. It is widely applied in underwater topography mapping, small-target detection, and buried object detection [4–7]. However, the instability of the platform, underwater turbulence, and other factors in the changeable underwater environment easily cause 2-D SV motion errors in the SAS, which vary with the range and azimuth dimensions. Such errors can result in several imaging problems, such as the loss of geometric resolution, reduction in image contrast, increase in sidelobes, and strong phase distortion.

To account for such errors, the SAS positions require an accurate measurement with inertial navigation units (INUs) or an accurate estimate, and the motion error requires accurate compensation with the motion compensation (MOCO) algorithm. Generally, this algorithm can be divided into two types: time-domain MOCO and line-by-line MOCO algorithms. The time-domain MOCO algorithm is the most accurate but requires very high computational effort [8]. The line-by-line MOCO algorithm has a much higher computational efficiency; however, the existing line-by-line MOCO algorithm for SAS only considers the range dependence of the 2-D SV motion error. Moreover, the azimuth dependence of the 2-D SV motion error has the ability to cause a dramatic phase error and an unfocused imaging result in high-resolution and complicated motion error cases. To the best of our knowledge, many 2-D SV MOCO algorithms for the airborne synthetic aperture radar (SAR) have been applied widely, such as in precise topography and aperture-dependent motion compensation algorithms (PTA) [9], subaperture topography and aperture-dependent motion compensation algorithms (SATA) [10], and frequency division motion algorithms (FD) [11–13]. Although SAS technology is derived from the SAR community, these algorithms cannot be directly applied to SAS. This is mainly due to the difference between SAS and SAR. Because of the low speed of sound in water, an SAS requires a hydrophone array to obtain a useful mapping rate [1], which is referred to as multi-hydrophone SAS. Owing to the multi-hydrophone configuration, the displacement of each sample from the ideal trajectory should take the offset displacement caused by the rotation of the hydrophone array into account, and this additional offset varies with the hydrophone position in the array. For example, the additional offset is up to 1.75 cm when the yaw angle is 1°, and the length of the hydrophone array is 1 m. This offset is not negligible for the SAS because its wavelength is usually less than 2 cm for the high-resolution SAS. Thus, the 2-D SV motion error for multi-hydrophone SAS is also hydrophone-dependent and cannot be compensated by the existing MOCO algorithms.

To compensate for the 2-D SV error for the multi-hydrophone SAS and overcome the shortcomings of the above algorithms, we propose a 2-D space-variant motion compensation algorithm for the multi-hydrophone SAS based on sub-beam compensation. The proposed algorithm is designed using the following four steps: (1) The exact motion error model for multi-hydrophone SAS, which includes five degrees of freedom (roll, pitch, yaw, sway, and heave), is established for the first time. In addition, the motion error of each target observed through a short observation aperture, viewed as a narrow beam or a sub-beam, is obtained by substituting the sonar's motion parameters measured or estimated into the exact motion error model. (2) Owing to the short observation aperture of the sub-beam, the motion error of all targets observed by the sub-beam can be replaced by the motion error of the targets at the beam center of the sub-beam. Moreover, because the motion error in the sub-beam is weakly dependent on the range, its delay can be considered the same as the delay along the range. Thus, by implementing two-phase multiplications on the raw data of the multiple-hydrophone SAS in the order of hydrophone by hydrophone, targets illuminated by the sub-beam of all targets illuminated by sonar are compensated for the phase error and delay error coming from motion error, respectively. (3) To extract the sub-beam data from raw data, the mapping relationship between the azimuth angle and Doppler frequency is utilized to split the equivalent single-hydrophone data, which is a result of implementing the Doppler spectrum extension and coherent superposition on the multi-hydrophone raw data. (4) To obtain the imaging result with high resolution, the imaging result of each sub-beam, obtained by performing the monostatic imaging algorithm on each sub-beam data, is coherently added. Finally, the validity of the proposed algorithm is tested using simulation and real data.

The remainder of this paper is organized as follows: In Section 2, a range history model is presented. In Section 3, the signal mode is introduced. In Section 4, the proposed algorithm is discussed in detail. Section 5 presents the experimental results based on the simulated and real measured data to verify the effectiveness of the proposed algorithm. The conclusions are presented in Section 6.

2. Range History Model

2.1. Exact Range History

As shown in Figure 1a, an imaging coordinate system $oxyz$ is established according to the right-hand rule, where the z-axis is pointed upward, and the x-axis and y-axis are in a horizontal plane and perpendicular to each other. An acoustic array, which contains an M + N + 1 hydrophone and one transponder, moves in the $oxyz$ coordinate system. The velocity of the acoustic array along the x-axis is v, and the offsets along the y-axis and z-axis are sway y_s and heave z_h, respectively. To describe the gestures of the acoustic array, a moving coordinate system $ox_ay_az_a$ is built as shown in Figure 1b, where the origin and x_a-axis are the phase center of the transponder and the phase center line of the hydrophones, respectively. The roll, pitch, and yaw of the acoustic array are denoted by θ_r, θ_p, and θ_y, respectively.

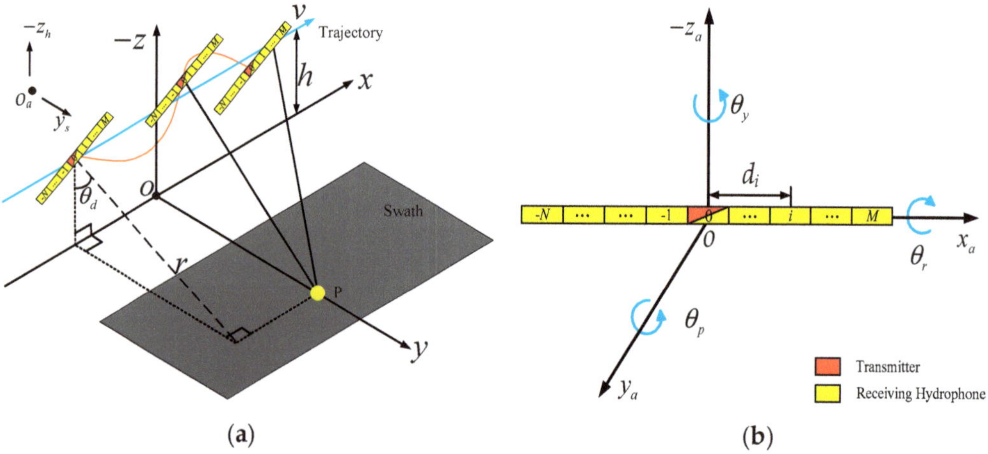

Figure 1. Geometric model of the multi-hydrophone SAS. (**a**) Model of range history; (**b**) array of hydrophones.

In Figure 1a, when the transponder transmits the signal at position $(vt, y_s, -h - z_h)$, the distance from the transponder to $P(x_0, r\sin\theta_d, 0)$ is as follows:

$$R_T^*(t; x_0, r) = \sqrt{(vt - x_0)^2 + (y_s - r\sin\theta_d)^2 + (h + z_h)^2} \tag{1}$$

where t is the slow time, h is the height of the array beyond the seafloor, r is the closest distance between the target and trajectory, and θ_d is the depression angle between r and the z-axis.

Owing to the low speed of underwater sound, the moving distance of the SAS between the transmission and reception cannot be ignored. Assuming that τ_{mi}^* is the delay between the moment of the ith hydrophone receiving the echo and the moment of the transponder transmitting signal, the position of the ith hydrophone at the receiving moment is given by the following:

$$\begin{bmatrix} x_i \\ y_i \\ z_i \end{bmatrix} = \begin{bmatrix} vt + v\tau_{mi}^* \\ y_s \\ -h - z_h \end{bmatrix} + M_r^T \begin{bmatrix} d_i \\ 0 \\ 0 \end{bmatrix} = \begin{bmatrix} v\tau_{mi}^* + vt + d_i\cos\theta_y\cos\theta_p \\ y_s - d_i\sin\theta_y\cos\theta_p \\ -h - z_h - d_i\sin\theta_p \end{bmatrix}, \tag{2}$$

where d_i is the distance from the ith hydrophone to the transponder and M_r is a rotation matrix given in [14]. According to the location of P and the ith hydrophone at the receiving moment, the distance from P to the ith hydrophone is as follows:

$$R_{Ri}^*(t; x_0, r) = \sqrt{\begin{aligned}&(vt + v\tau_{mi}^* + d_i\cos\theta_y\cos\theta_p - x_0)^2 \\ &+(r\sin\theta_d + d_i\sin\theta_y\cos\theta_p - y_s)^2 + (h + z_h + d_i\sin\theta_p)^2\end{aligned}}, \tag{3}$$

The exact propagation distance of sound from the transmission to reception is the sum of $R_T^*(t;x_0,r)$ and $R_{Ri}^*(t;x_0,r)$, which is expressed as follows:

$$R_{mi}^*(t;x_0,r) = R_T^*(t;x_0,r) + R_{Ri}^*(t;x_0,r). \tag{4}$$

This is referred to as the exact range history. However, because the echo signal received by the ith hydrophone is propagated for τ_{mi}^*, the exact range history can be expressed as follows:

$$R_{mi}^*(t;x_0,r) = c\tau_{mi}^*, \tag{5}$$

where c is the speed of underwater sound. By combining (4) and (5), the solution for τ_{mi}^* is given as follows:

$$\tau_{mi}^* = \frac{B_{mi} + \sqrt{B_{mi}^2 + C_{mi}}}{A}, \tag{6}$$

where A, B_{mi}, and C_{mi}, respectively, are represented as follows:

$$A = c^2 - v^2, \tag{7}$$

$$B_{mi} = vd_i\cos\theta_y\cos\theta_p + v^2\left(t - \frac{x_0}{v}\right) + c\sqrt{v^2\left(t - \frac{x_0}{v}\right)^2 + (r\sin\theta_d - y_s)^2 + (h + z_h)^2}, \tag{8}$$

$$C_{mi} = 2v\left(t - \frac{x_0}{v}\right)d_i\cos\theta_y\cos\theta_p + 2(h + z_h)d_i\sin\theta_p + d_i^2 + 2(r\sin\theta_d - y_s)d_i\sin\theta_y\cos\theta_p. \tag{9}$$

2.2. Exact Motion Error

Because yaw and pitch, particularly yaw, can change the direction of the SAS beam and cause the squint phenomenon, we selected the squint model as the ideal model for the multi-hydrophone SAS without loss of generality. Assuming that the mean of yaw and pitch are θ_{y0} and θ_{p0}, respectively, the ideal range history is given in [15] and is expressed as follows:

$$R_i^*(t;x_0,r) = \frac{B_i + \sqrt{B_i^2 + C_i}}{A}c, \tag{10}$$

where

$$B_i = vd_i\cos\theta_{y0}\cos\theta_{p0} + v^2\left(t - \frac{x_0}{v}\right) + c\sqrt{v^2\left(t - \frac{x_0}{v}\right)^2 + r^2}, \tag{11}$$

$$C_i = 2v\left(t - \frac{x_0}{v}\right)d_i\cos\theta_{s0}\cos\theta_{p0} + 2hd_i\sin\theta_{p0} + 2d_ir\sin\theta_d\sin\theta_{y0}\cos\theta_{p0} + d_i^2. \tag{12}$$

The exact motion error is the difference between the exact range history and the ideal range history and is obtained by subtracting (10) from (5) as follows:

$$\varepsilon_i^*(t;x_0,r) = \frac{B_{mi} + \sqrt{B_{mi}^2 + AC_{mi}}}{A}c - \frac{B_i + \sqrt{B_i^2 + AC_i}}{A}c, \tag{13}$$

It can be noted from (13) that this motion error is a 2-D SV and hydrophone-dependent term because its size is related to the target position at the beam and the hydrophone position in the acoustic array.

2.3. Sub-Beam Range History

The SAS echo can be viewed as the sum of all sub-beam echoes, as shown in Figure 2. Assuming that the number of sub-beams is K and the squint angle of the kth sub-beam is θ_k, the moment is $x_0/v + r\tan\theta_k/v$ when the beam center of the kth sub-beam crosses target P. Because the sub-beam beamwidth is very narrow, the motion error of all targets illuminated by the sub-beam can be replaced by the motion error of the targets at the beam center of the sub-beam. By substituting t_k into (13), the approximation motion error of the kth sub-beam can be obtained as follows:

$$\varepsilon_{i,k}(r) = \frac{B'_{mi} + \sqrt{{B'_{mi}}^2 + AC'_{mi}}}{A}c - \frac{B'_i + \sqrt{{B'_i}^2 + AC'_i}}{A}c \tag{14}$$

where

$$B'_{mi} = vd_i \cos\theta_y \cos\theta_p + vr\tan\theta_k + c\sqrt{r^2\tan^2\theta_k + (r\sin\theta_d - y_s)^2 + (h+z_h)^2}, \tag{15}$$

$$C'_{mi} = 2d_i r\tan\theta_k \cos\theta_y \cos\theta_p + 2(h+z_h)d_i \sin\theta_p + d_i^2 + 2(r\sin\theta_d - y_s)d_i \sin\theta_y \cos\theta_p, \tag{16}$$

$$B'_i = vd_i \cos\theta_{y0} \cos\theta_{p0} + vr\tan\theta_k + c\sqrt{r^2\tan^2\theta_k + r^2}, \tag{17}$$

$$C'_i = 2rd_i \tan\theta_k \cos\theta_{s0} \cos\theta_{p0} + 2hd_i \sin\theta_{p0} + 2rd_i \sin\theta_d \sin\theta_{y0} \cos\theta_{p0} + d_i^2. \tag{18}$$

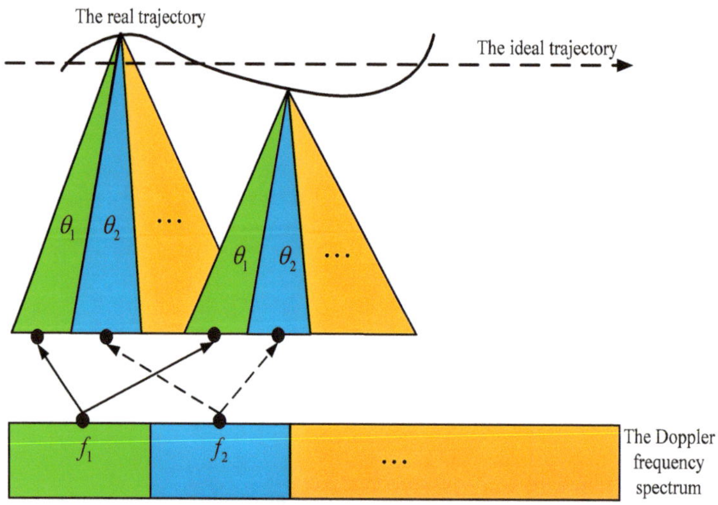

Figure 2. Relationship between the Doppler frequency and sub-beam.

The range history of the sub-beam can be viewed as the sum of the ideal range history and motion error; thus, the range history of the kth sub-beam can be expressed as follows:

$$R_{mi,k}(t;r) = R_i^*(t;r) + \varepsilon_{i,k}(r). \tag{19}$$

Considering that the expression of $R_i^*(t;r)$ is very complex, it is necessary to simplify it. According to the description of the displaced phase center antenna (DPCA) technology [16], the multi-hydrophone SAS can be viewed as a monostatic hydrophone transmit signal and receive echo at the phase center of the bistatic transponder/hydrophone pair. Thus, the range history $R_i^*(t;r)$ can be considered as the sum of a single root term and a range-dependent term and is rewritten as follows:

$$R_i^*(t;r) = R_i(t;r) + \Delta R_i(r), \tag{20}$$

where $R_i(t;r)$ is the single root term, which is equivalent to the range history of the monostatic hydrophone, and $\Delta R_i(r)$ is the range-dependent offset term, which is the offset distance attained by transforming a bistatic transponder/hydrophone pair into a monostatic hydrophone. $R_i(t;r)$ and $\Delta R_i(r)$ are given as follows:

$$R_i(t;r) = 2\sqrt{r^2 + v^2\left(t + \frac{r}{c} + \frac{d_i}{2v\cos\theta_{sq0}}\right)^2}, \tag{21}$$

$$\Delta R_i(r) = \frac{v^2}{4r}\left(\frac{2r}{c} + \frac{d_i}{v}\right)^2 - \frac{r}{2} + \frac{(d_i\sin\theta_{p0} + h)^2}{2r} + \frac{(d_i\sin\theta_{y0}\cos\theta_{p0} + r\sin\theta_d)^2}{2r}, \tag{22}$$

where θ_{sq0} is the squint angle.

It can be noted from (22) and (14) that $\Delta R_i(r)$ and $\varepsilon_{i,k}(r)$ are both range-dependent terms; thus, they can be integrated into one term for convenience as expressed below.

$$\zeta_{i,k}(r) = \Delta R_i(r) + \varepsilon_{i,k}(r), \tag{23}$$

referred to as the motion compensation term. Then, the range history of the kth sub-beam can be rewritten as follows:

$$R_i^*(t;r) = R_i(t;r) + \zeta_{i,k}(t;r). \tag{24}$$

To evaluate the size of the path error caused by approximating $R_i^*(t;r)$ into $R_{mi,k}(t;r)$, a simulation was performed. We assume that the acoustic array is shown in Figure 1b, where N and M are equal to 25 and 0, respectively. The system parameters of the SAS are shown in Table 1, and the motion parameters are shown in Figure 3, where θ_y, θ_p, y_s, and z_h vary randomly with the pulses in $[-1°, 1°]$, $[0°, 2°]$, $[-1.5, 1.5]$ m, and $[-0.1, 0.1]$ m, respectively. Considering that the beam width is a very important factor in determining the size of the 2-D SV motion error, this simulation was carried out for different beam widths. To compare the results of different beam widths under the same acoustic array and resolution, the SAS beamwidth was changed by adjusting the center frequency of the transmitted signal. Thus, there are three types of center frequencies, as listed in Table 1, whose wavelengths are λ_1(1.87 cm), λ_2(3.75 cm), and λ_3(7.5 cm) with beamwidths of 6°, 15°, and 24°, respectively.

Table 1. System parameters.

Carrier Frequency (kHz)	80,32,20	Pulse Width (ms)	20	PRF (Hz)	2.5
Bandwidth (kHz)	20	Speed (m/s)	2.5	Hydrophone (m)	0.08
Altitude (m)	30	Transmitter (m)	0.16	Hydrophone number	25

Without a loss of generality, the hydrophone at the farthest distance from the transponder was selected, and the maxima of θ_y, θ_p, y_s, and z_h were used to calculate the motion error. In the proposed method, the total number of sub-beams for beamwidths of 6°, 15°, and 24° are 20, 40, and 55, respectively. In the conventional algorithm, the motion errors of all targets illuminated by the SAS beam are replaced by the motion errors of the targets on the center line of the SAS beam. The results of the two algorithms were measured in terms of the wavelength (λ), as shown in Figure 4.

First, as shown in Figure 4a–c, the beamwidth has a significant influence on the size of the path error in the conventional algorithm. By using the threshold of the path error as 0.0625 (1/16) wavelengths, we measured the maximum path error in Figure 4a–c. The results were $0.3058\lambda_1$, $0.5878\lambda_2$, and $0.8919\lambda_3$, respectively, which are significantly larger than 0.0625 wavelengths. Therefore, it can be concluded that the size of the path error in the conventional algorithm has a significant effect on the imaging results after MOCO.

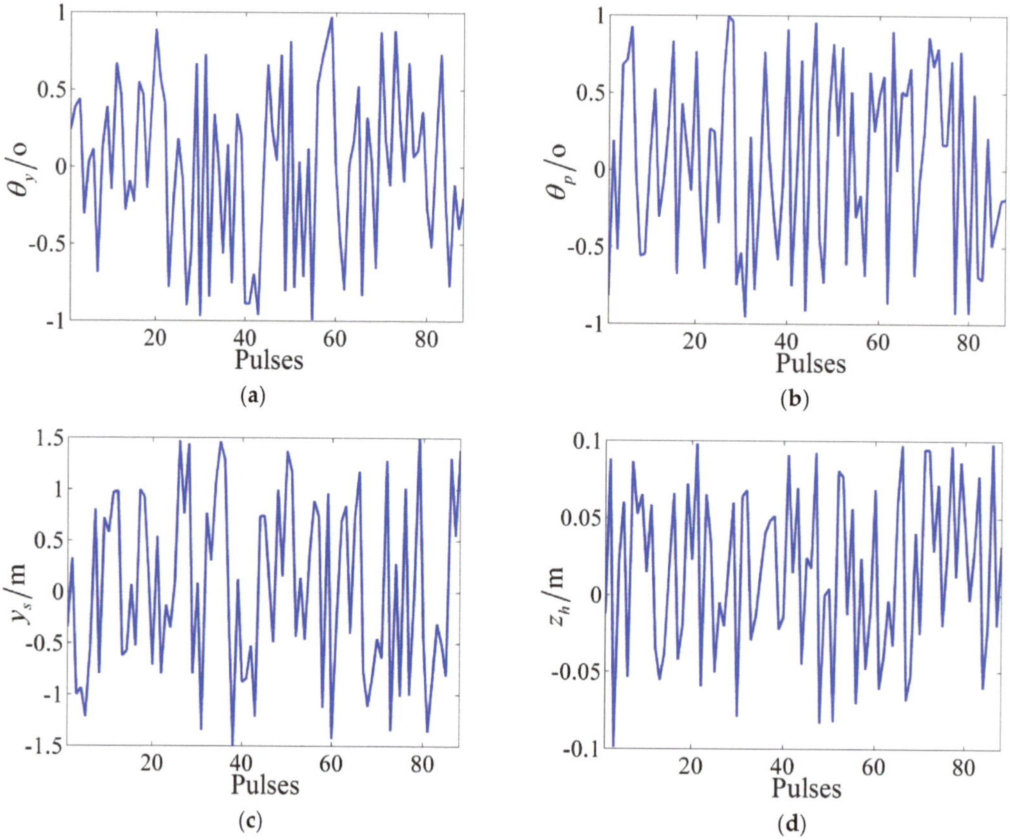

Figure 3. Motion error: (**a**) θ_y; (**b**) θ_p; (**c**) y_s; (**d**) z_h.

Next, it can be observed from Figure 4d–f that the beamwidth has little influence on the size of the path error. The main reason for this beneficial result is that the beamwidth of the central sub-beam is sufficiently narrow such that the size of the path error is small when the motion errors of all targets observed by the sub-beam are replaced by the motion errors of the targets at the beam center of the sub-beam. Theoretically, the farther the sub-beam is from the beam center of the SAS, the greater the size of the path error of the sub-beam. Thus, without loss of generality, we selected the sub-beam at the SAS beam edge for analysis. In Figure 4, the images of the second row and the third row are the path error results of the central and edge sub-beams at different beamwidths, respectively. By comparing the images of the two rows, it is clear that the edge sub-beams have a greater path error than the central sub-beams. To evaluate the size of the path error of these sub-beams quantitatively, we measured the maximum path error in Figure 4d–i. The results were $0.0104\lambda_1$, $0.0052\lambda_2$, $0.0059\lambda_3$, $0.0461\lambda_1$, $0.0527\lambda_2$, and $0.0538\lambda_3$, respectively. Although the results show that the path errors of the edge sub-beams are evidently larger than that of the central sub-beams, they are all less than the threshold of 0.0625 wavelengths. Therefore, it can be concluded that the size of the path error in the proposed algorithm has no effect on the imaging result after MOCO.

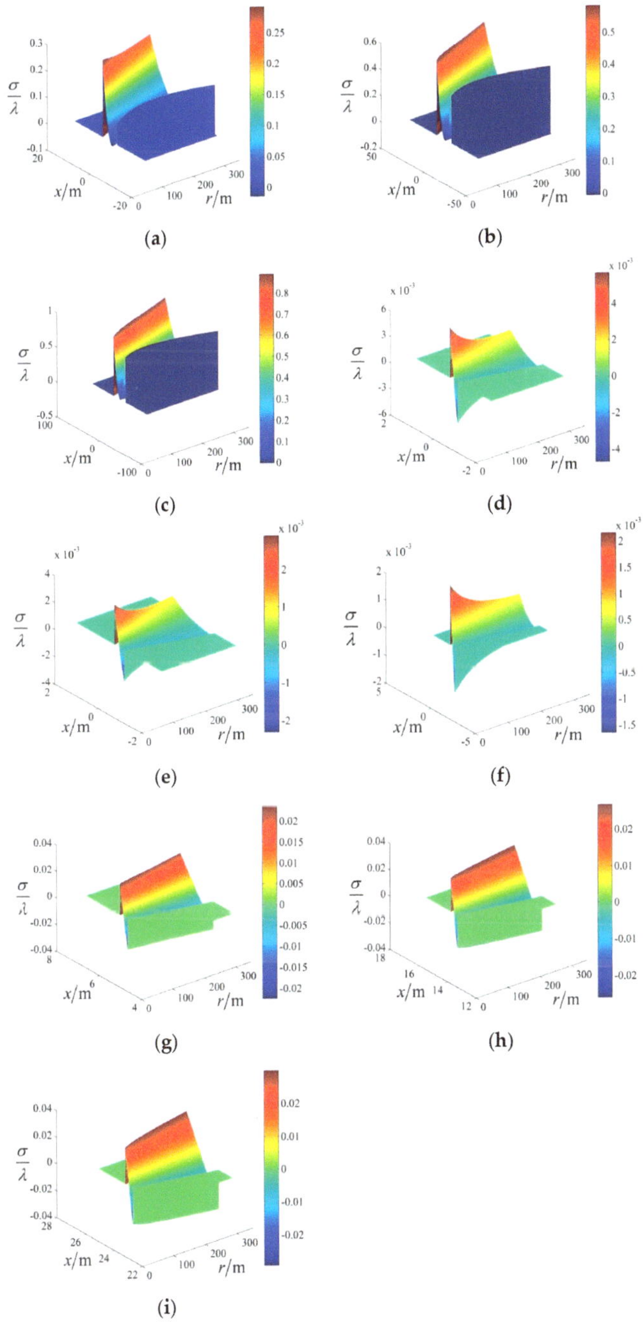

Figure 4. (**a–c**) are the path error results caused by the conventional algorithm for the beamwidths of 6°, 15°, and 24°, respectively; (**d–f**) are the path error results of the central sub-beam caused by the proposed algorithm for the beamwidths of 6°, 15°, and 24°, respectively; (**g–i**) are the path error results of the edge sub-beam caused by the proposed algorithm for the beamwidths of 6°, 15°, and 24°, respectively.

3. Signal Model

The echo signal of the *i*th hydrophone can be considered the sum of the echo signals collected by each sub-beam and is expressed as follows:

$$s_i(t,\tau;r) = \sum_{k=1}^{K} s_{i,k}(t,\tau;r) \tag{25}$$

$s_{i,k}(\tau,t;r)$ is the echo signal collected by the *k*th sub-beam of the *i*th hydrophone after demodulation and is expressed as follows:

$$\begin{aligned} s_{i,k}(t,\tau;r) &= w_r\left(\tau - \frac{R_i(t;r)+\zeta_{i,k}(t;r)}{c}\right)\omega_a(t)\exp\left\{-j\frac{2\pi f_0}{c}[R_i(t;r)+\zeta_{i,k}(t;r)]\right\} \\ &\times \exp\left\{j\pi K_r\left(\tau - \frac{R_i(t;r)+\zeta_{i,k}(t;r)}{c}\right)^2\right\} \end{aligned} \tag{26}$$

where $\omega_r(\Delta)$ represents the envelope of the transmitted signal, $\omega_{az}(\Delta)$ represents the beam pattern of a transponder and a hydrophone, τ is the fast time, K_r is the FM rate of the transmitted signal, and f_0 is the center frequency of the transmitted signal.

4. Development of the Motion Compensation

The proposed algorithm contains the range Fourier transformation, phase and delay correction for the sub-beam, azimuth Fourier transformation, azimuth spectrum replication, coherent addition for each hydrophone, Doppler spectrum division, monostatic imaging, azimuth inverse Fourier transformation, and coherent addition for the imaging result of each sub-beam, as shown in Figure 5.

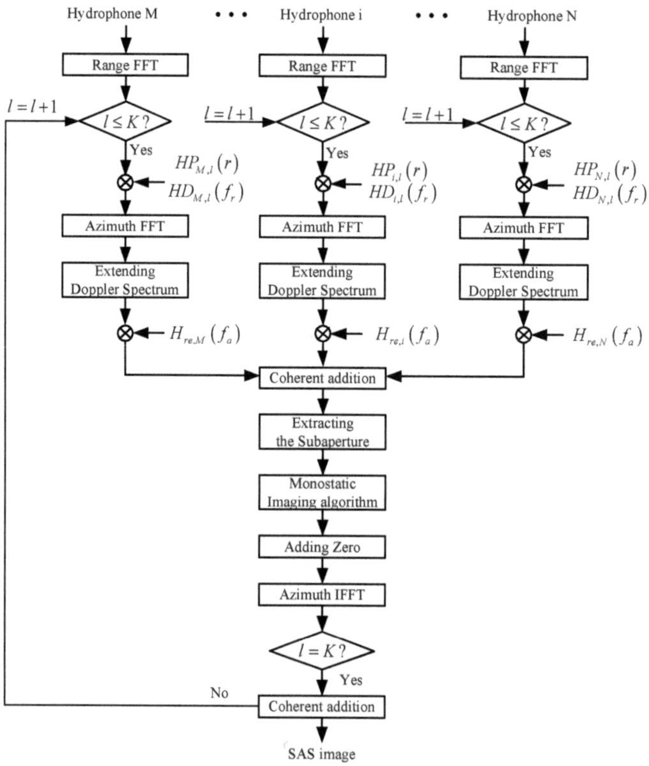

Figure 5. Flow chart combining the proposed algorithm.

The first step is to transform $s_i(\tau,t;r)$ into the range frequency domain. Here, the principle of the stationary phase (POSP) [17] is used to perform the range Fourier transformation on $s_i(\tau,t;r)$. The results of the range Fourier transformation are as follows:

$$\begin{aligned} S_i(f_r,t;r) &= A_0 W_r(f_r)\omega_a(t)\exp\left(-j\frac{\pi f_r^2}{K_r}\right)\exp\left[-j2\pi\frac{R_i(t;r)}{c}(f_0+f_r)\right] \\ &\times\sum_{k=1}^{K}\left\{\exp\left[-j2\pi\frac{\zeta_{i,k}(r)}{c}f_0\right]\times\exp\left[-j2\pi\frac{\zeta_{i,k}(r)}{c}f_r\right]\right\} \end{aligned} \tag{27}$$

where $W_r(.)$ represents the spectral envelope of the transmitted signal, and f_r is the range frequency.

The second step is to compensate for the sub-beam motion error by the phase correction and delay correction in the order of one-by-one hydrophone. The phase correction is performed on $S_i(f_r,t;r)$ with phase multiplication, where the factor for the phase multiplication is given by the following:

$$HP_{i,l}(r) = \exp\left[j\frac{2\pi f_0}{c}\zeta_{i,l}(r)\right] \tag{28}$$

where the subscript l represents the lth sub-beam. Because $\zeta_{i,l}(t;r)$ is weakly dependent on the range, the delay caused by $\zeta_{i,l}(t;r)$ can be considered the same as the delay along the range, which can be replaced by the delay at reference range. According to the Fourier transform shifting properties, the constant delay can be corrected by performing a phase multiplication on $S_i(f_r,t;r)$, where the factor for this phase multiplication is given as follows:

$$HD_{i,l}(t;f_r) = \exp\left[j\frac{2\pi f_r}{c}\zeta_{i,l}\left(r_{ref}\right)\right] \tag{29}$$

After multiplying (27) with (28) and (29) separately, the signal of the ith hydrophone is expressed as follows:

$$\begin{aligned} S_{i,l}(f_r,t;r) &= A_0 W_r(f_r)\omega_a(t)\exp\left(-j\frac{\pi f_r^2}{K_r}\right)\exp\left[-j2\pi\frac{R_i(t;r)}{c}(f_0+f_r)\right] \\ &\times\sum_{k=1}^{K}\left\{\exp\left[-j2\pi\frac{\zeta_{i,k}(r)-\zeta_{i,l}(r)}{c}f_0\right]\times\exp\left[-j2\pi\frac{\zeta_{i,k}(r)-\zeta_{i,l}\left(r_{ref}\right)}{c}f_r\right]\right\} \end{aligned} \tag{30}$$

The third step is the azimuth Fourier transformation. Here, we again utilize the POSP to perform the azimuth Fourier transform on $S_{i,l}(f_r,t;r)$. The results of the range Fourier transform are as follows:

$$\begin{aligned} S_{i,l}(f_r,f_a;r) &= A_0 W_r\left(\frac{f_r}{K_r}\right)W_a(f_a)\exp\left(-j\frac{\pi f_r^2}{K_r}\right)\exp\left(-j\frac{4\pi r f_0}{c}\sqrt{\frac{f_r^2}{f_0^2}+\frac{2f_r}{f_0}+D^2}\right) \\ &\times\exp\left(-j\frac{2\pi r f_a}{c}\right)\exp\left(-j\frac{2\pi d_i f_a}{2v\cos\theta_{sq0}}\right) \\ &\times\sum_{k=1}^{K}\left\{\exp\left[-j2\pi\frac{\zeta_{i,k}(r)-\zeta_{i,l}(r)}{c}f_0\right]\times\exp\left[-j2\pi\frac{\zeta_{i,k}(r)-\zeta_{i,l}\left(r_{ref}\right)}{c}f_r\right]\right\} \end{aligned} \tag{31}$$

where D is the range migration factor and is represented as follows:

$$D = \sqrt{1-\frac{c^2 f_a^2}{4v^2 f_0^2}} \tag{32}$$

The fourth step is the azimuth spectrum replication. The 2-D spectrum of each hydrophone in (31) suffers a serious problem of azimuth undersampling owing to the principle of multi-hydrophone SAS. Thus, it becomes difficult for the multi-hydrophone SAS to obtain a 2-D spectrum without azimuth aliasing before extracting the 2-D spectrum of the sub-beam from (31). The method of azimuth spectrum replication was proposed to solve this problem. This method includes two main steps. The first step is to replicate the undersampled 2-D spectrum of each hydrophone $N + M + 1$ times. The second step is to rank the replicated 2-D spectrum end-to-end in azimuth. Although the method of azimuth spectrum replication cannot suppress azimuth aliasing, it provides an effective method to deal with the undersampled 2-D spectrum in azimuth.

The fifth step is to coherently add the signal of each hydrophone. To remove the azimuth aliasing caused by the azimuth undersampling of each hydrophone, a method that coherently adds the corrected signal of each hydrophone is utilized as follows:

$$S_l(f_r, f_a; r) = \begin{bmatrix} H_{re,N}(f_a) & \cdots & H_{re,i}(f_a) & \cdots & H_{re,M}(f_a) \end{bmatrix} \begin{bmatrix} S_{N,l}(f_r, f_a; r) \\ \vdots \\ S_{i,l}(f_r, f_a; r) \\ \vdots \\ S_{M,l}(f_r, f_a; r) \end{bmatrix} \tag{33}$$

where

$$H_{re,i}(f_a) = \exp\left(j\frac{\pi d_i f_a}{v\cos\theta_{sq0}}\right) \tag{34}$$

The result is given as follows:

$$\begin{aligned} S_l(f_r, f_a; r) &= A_0 W_r\left(\frac{f_r}{K_r}\right) W_a(f_a)\exp\left(-j\frac{\pi f_r^2}{K_r}\right) \\ &\times \exp\left(-j\frac{2\pi r f_a}{c}\right)\exp\left(-j\frac{4\pi r f_0}{c}\sqrt{\frac{f_r^2}{f_0^2} + \frac{2f_r}{f_0} + D^2}\right) \\ &\times \begin{cases} 1, f_{a,l} - \dfrac{B_{a,l}}{2} \le f_a \le f_{a,l} + \dfrac{B_{a,l}}{2} \\ \sum\limits_{i=N}^{M}\sum\limits_{k=1,k\ne l}^{K}\left\{\begin{array}{l} \exp\left[-j2\pi\dfrac{\zeta_{i,k}(r) - \zeta_{i,l}(r)}{c}f_0\right] \\ \times\exp\left[-j2\pi\dfrac{\zeta_{i,k}(r) - \zeta_{i,l}\left(r_{ref}\right)}{c}f_r\right]\end{array}\right\}, else \end{cases} \end{aligned} \tag{35}$$

where $B_{a,l}$ and $f_{a,l}$ define the Doppler bandwidth and Doppler center frequency of the lth sub-beam, respectively. From (35), we can see that only the motion error corresponding to the lth sub-beam is effectively compensated.

The sixth step is to extract the sub-beam's signal from the 2-D spectrum of the SAS. $B_{a,l}$ and $f_{a,l}$ of the lth sub-beam are calculated by utilizing the mapping relationship between the azimuth view angle and Doppler frequency. According to the values of $B_{a,l}$ and $f_{a,l}$, the signal of the lth sub-beam can be obtained from (35) and is extracted as follows.

$$\begin{aligned} S_l(f_r, f_a; r) &= A_0 W_r\left(\frac{f_r}{K_r}\right) W_a(f_a) \\ &\times \exp\left(-j\frac{\pi f_r^2}{K_r}\right)\exp\left(-j\frac{2\pi r f_a}{c}\right) \\ &\times \exp\left(-j\frac{4\pi r f_0}{c}\sqrt{\frac{f_r^2}{f_0^2} + \frac{2f_r}{f_0} + D^2}\right), f_{a,l} - \frac{B_{a,l}}{2} \le f_a \le f_{a,l} + \frac{B_{a,l}}{2} \end{aligned} \tag{36}$$

The seventh step is to perform the imaging algorithm on the sub-beam signal. The fifth step is essentially a process in which a multi-hydrophone signal is transformed into a monostatic signal; thus, all monostatic imaging algorithms, such as the range–Doppler algorithm [18], chirp scaling algorithm [19], ωK algorithm (ωKA) [20], and their modifications, can be performed on the sub-beam signal to obtain the imaging result of each sub-beam. The imaging algorithm mainly includes range cell migration correction, range compression, second range compression, azimuth matched filter, and azimuth inverse

Fourier transformation. However, to meet the pixel points for the scene, the data in (36) must be padded with zero in the azimuth before the azimuth inverse Fourier transformation. In addition, to obtain the imaging results of the other sub-beams except for the lth sub-beam, steps 2 to 7 need to be executed repeatedly.

The eighth step is to coherently add the imaging results of all the sub-beams. Because the sub-beam's Doppler bandwidth is narrow, its imaging result has low resolution. To obtain a high-resolution imaging result, it is necessary to coherently add every sub-beam's imaging result.

5. Validation Test

5.1. Exact Range History Imaging Results of Simulation Data

To evaluate the performance of the proposed MOCO algorithm, a simulation was performed. The system parameters are shown in Table 1, the SAS motion error is shown in Figure 3, and the scene illuminated by sonar has five idea point targets, which are assumed to be located at positions P1 (−3 m, 152 m, 30 m), P2 (−3 m, 158 m, 30 m), P3 (3 m, 158 m, 30 m), P4 (3 m, 152 m, 30 m), and P5 (0 m, 155 m, 30 m), respectively. In the simulation, the exact range history of each target is given by (6), and to avoid the new path error caused by the imaging algorithm, a ωKA is utilized as the monostatic imaging algorithm in the seventh step. The results of the proposed algorithm and the conventional algorithm are shown in Figure 6.

From Figure 6a–c, it can be observed that the sidelobe energy and azimuth mainlobe width of the imaging results of the conventional algorithm become stronger and wider with the increase in beam width, respectively. Although the imaging results of the proposed algorithm have problems similar to those of the conventional algorithm, the increasing extent of sidelobe level and mainlobe width are less than those of the conventional algorithm, as shown in Figure 6d–f. Next, to compare the imaging results in more detail, point P5 was extracted from the subfigures of Figure 6. Theoretically, the range image is obtained using match filter technology, and the motion has a smaller effect on the range image than the azimuth image. Thus, for convenience, we only made the azimuth slice of point P5, as shown in Figure 7. Then, the impulse response width was measured and compared with the theoretical resolution of 8 cm. From Table 2, it can be observed from the comparison results that when the beam width is 6°, 15°, and 24°, the mainlobe broadening of the conventional algorithms is 8.9%, 24.9%, and 87.3% at beamwidths of 6°, 15°, and 24°, respectively, while the mainlobe broadening of the proposed algorithm is only 1.2%, 3.6%, and 11.8%. These results show that the imaging result of the proposed algorithm is better than that of the conventional algorithm and is close to the theoretical result. Therefore, the validity of the proposed algorithm was tested through simulation.

Table 2. The imaging quality of the results of the conventional algorithm and the proposed algorithm.

Algorithm	Beamwidth	IRW/cm	PSLR/dB	ISLR/dB
The conventional algorithm	6°	8.71	−17.80	−11.32
The proposed algorithm	6°	8.10	−18.05	−14.05
The conventional algorithm	15°	9.99	−14.80	−9.08
The proposed algorithm	15°	8.29	−19.89	−13.62
The conventional algorithm	24°	14.98	−16.01	−7.95
The proposed algorithm	24°	8.94	−18.93	−11.64

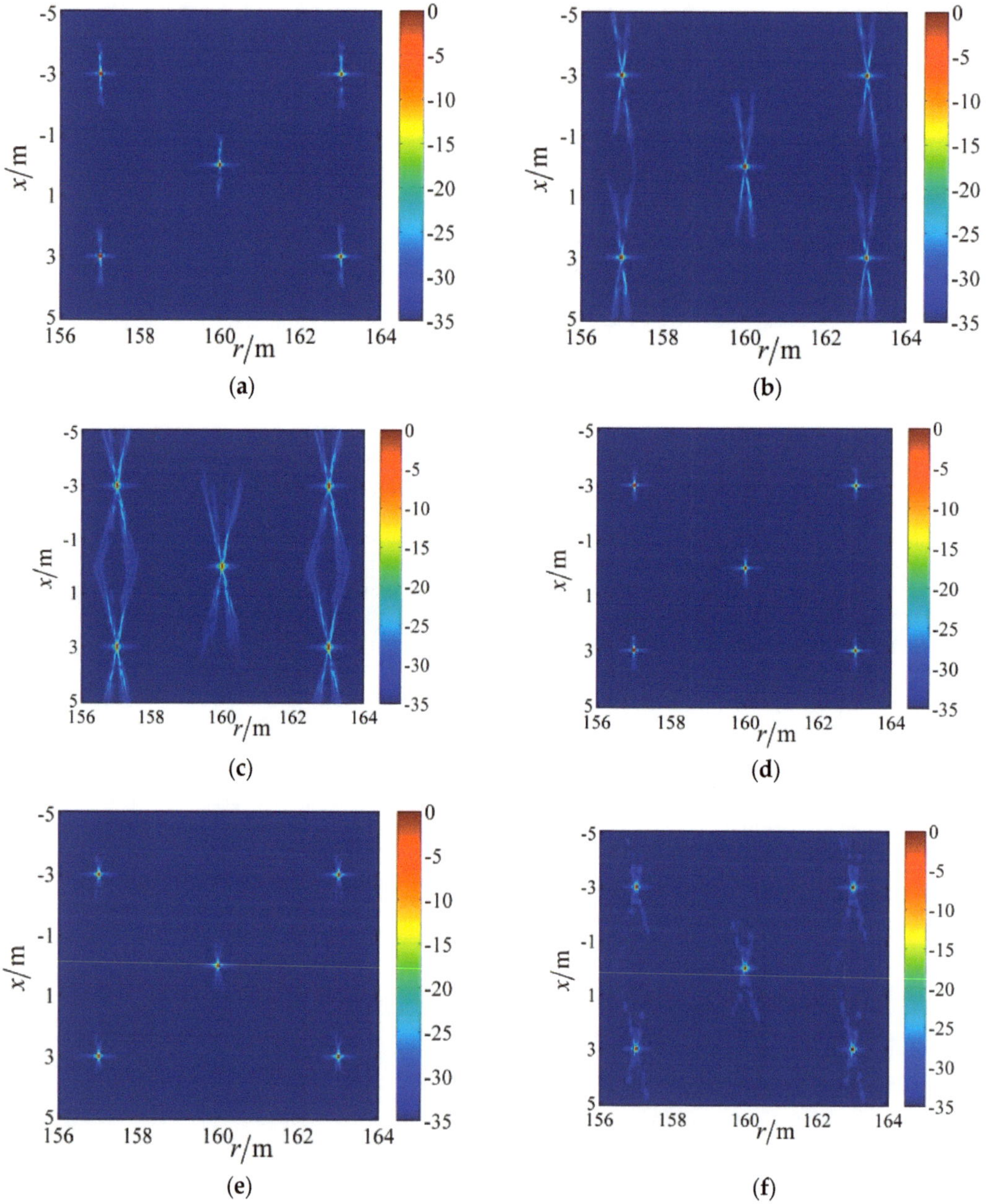

Figure 6. Imaging results of the conventional algorithm and the proposed algorithm at the different beamwidths. Conventional algorithm: (**a**) 6°, (**b**) 15°, and (**c**) 24°; Proposed algorithm: (**d**) 6°, (**e**) 15°, and (**f**) 24°.

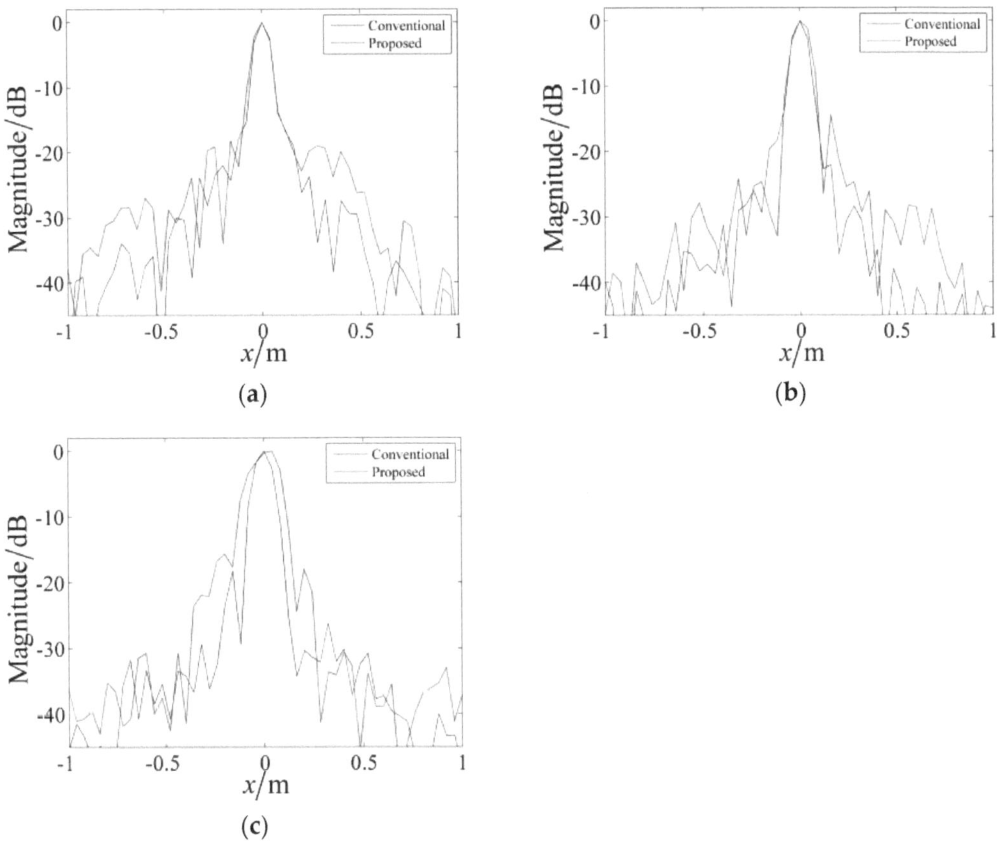

Figure 7. Azimuth slice of the imaging results of the conventional algorithm and the proposed algorithm at the different beamwidths. (**a**) 6°, (**b**) 15°, and (**c**) 24°.

5.2. Imaging Results of Real Data

In this section, the performance of the proposed algorithm is evaluated by comparing the imaging results of different algorithms. The real data were sourced from the South China Sea and were collected using a multi-hydrophone SAS in 2017. The imaging results with no motion compensation, the conventional algorithm, and the proposed algorithm are shown in Figure 8a–c, respectively, where the number of sub-beams in the proposed algorithm is seven.

It can be observed that the image in Figure 8b is clearer and shows more detailed information than the image in Figure 8a; thus, the existing algorithm is effective. However, the image result of the far scene in Figure 8b is worse than that of the near scene. This result indicates that there is still a large path error that is not compensated for by the conventional algorithm. By comparing Figure 8b with Figure 8c, it can be seen that the image result in Figure 8c is better than that in Figure 8b, both in the far scene and in the near scene. Therefore, the validity and superiority of the proposed algorithm are verified.

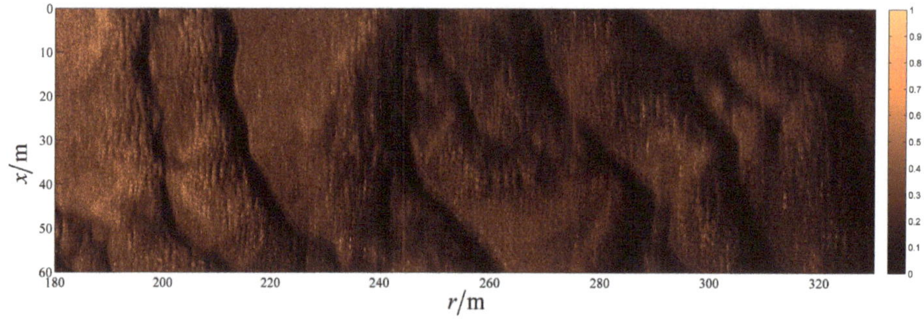

(a)

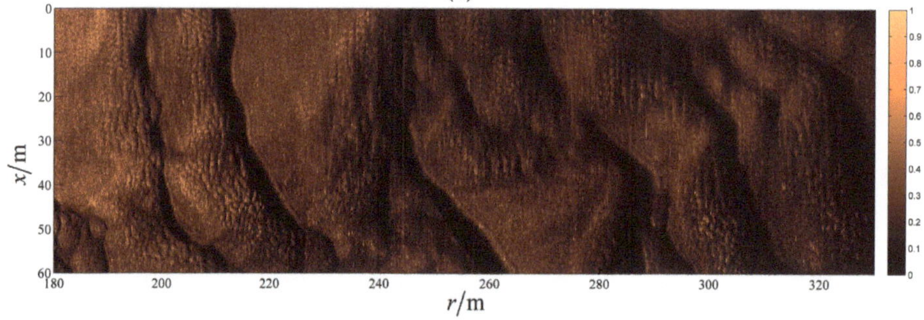

(b)

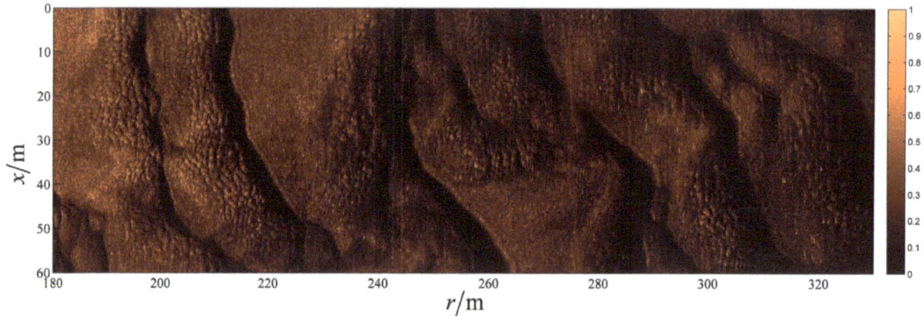

(c)

Figure 8. The imaging results. (**a**) No motion compensation; (**b**) the conventional algorithm; (**c**) the proposed algorithm.

6. Conclusions

A 2-D SV motion compensation algorithm for a multi-hydrophone SAS based on sub-beam compensation was proposed in this study. To the best of our knowledge, such an algorithm had not yet been reported. The main contributions of this study are as follows: (1) An exact motion error model for the multi-hydrophone SAS, which includes five degrees of freedom, namely roll, pitch, yaw, sway, and heave, is newly established. (2) The ability of the proposed algorithm to compensate for the 2-D SV motion error with hydrophone dependency for the multi-hydrophone SAS is demonstrated. (3) The ability of the proposed algorithm to compensate for low-frequency and high-frequency motion errors simultaneously is also shown. Moreover, the algorithm can be applied to multi-channel SARs, such as the spaceborne SAR. However, the premise of the proposed algorithm is to determine the motion parameters of the platform measured.

Author Contributions: Conceptualization, H.W.; Methodology, H.W. and J.Z.; Software, H.W. and H.Z.; Validation, Z.X.; Formal analysis, J.T.; Writing—original draft, H.W.; Writing—review & editing, F.Z.; Project administration, H.W. All authors have read and agreed to the published version of the manuscript.

Funding: This research was funded by the National Natural Science Foundation of China (Grant No. 41906162).

Data Availability Statement: The original contributions presented in the study are included in the article, further inquiries can be directed to the corresponding author.

Conflicts of Interest: The authors declare no conflict of interest.

References

1. Hayes, M.P.; Gough, P.T. Synthetic Aperture Sonar: A Review of Current Status. *IEEE J. Ocean. Eng.* **2009**, *34*, 207–224. [CrossRef]
2. Zhang, X.; Wu, H.; Sun, H.; Ying, W. Multireceiver SAS imagery based on monostatic conversion. *IEEE J. Sel. Top. Appl. Earth Obs. Remote Sens.* **2021**, *14*, 10835. [CrossRef]
3. Zhang, X. An efficient method for the simulation of multireceiver SAS raw signal. *Multimed. Tools Appl.* **2023**, *83*, 37351–37368. [CrossRef]
4. Scott, B.T.; Brett, B.; Sowmya, R.; Ethan, L.; Bradley, M.; Patrick, R. Environmentally adaptive automated recognition of underwater mines with synthetic aperture sonar imagery. *J. Acoust. Soc. Am.* **2021**, *150*, 851–863.
5. Angeliki, X.; Yan, P. Compressive synthetic aperture sonar imaging with distributed optimization. *J. Acoust. Soc. Am.* **2019**, *146*, 1839–1850.
6. Nadimi, N.; Javidan, R.; Layeghi, K. An Efficient Acoustic Scattering Model Based on Target Surface Statistical Descriptors for Synthetic Aperture Sonar Systems. *J. Mar. Sci. Appl.* **2020**, *19*, 494–507. [CrossRef]
7. Piper, J.E.; Lim, R.; Thorsos, E.I.; Williams, K.L. Buried Sphere Detection Using a Synthetic Aperture Sonar. *IEEE J. Ocean. Eng.* **2009**, *34*, 485–494. [CrossRef]
8. Callow, H.J. Signal Processing for Synthetic Aperture Sonar Image Enhancement. Ph.D. Thesis, Electrical and Electronic Engineering, University of Canterbury, Christchurch, New Zealand, 2003.
9. Macedo, K.A.C.d.; Scheiber, R. Precise topography- and aperture-dependent motion compensation for airborne SAR. *IEEE Geosci. Remote Sens. Lett.* **2005**, *2*, 172–176.
10. Prats, P.; Reigber, A.; Mallorqui, J.J. Topography-dependent motion compensation for repeat-pass interferometric SAR systems. *IEEE Geosci. Remote Sens. Lett.* **2005**, *2*, 206–210. [CrossRef]
11. Zheng, X.; Yu, W.; Li, Z. A Novel Algorithm for Wide Beam SAR Motion Compensation Based on Frequency Division. In Proceedings of the 2006 IEEE International Symposium on Geoscience and Remote Sensing, Denver, CO, USA, 31 July–4 August 2006; pp. 3160–3163.
12. Lu, Q.; Huang, P.; Gao, Y.; Liu, X. Precise frequency division algorithm for residual aperture-variant motion compensation in synthetic aperture radar. *Electron. Lett.* **2019**, *55*, 51–53. [CrossRef]
13. Lu, Q.; Gao, Y.; Huang, P.; Liu, X. Range and Aperture-Dependent Motion Compensation Based on Precise Frequency Division and Chirp Scaling for Synthetic Aperture Radar. *IEEE Sens. J.* **2019**, *19*, 1435–1442. [CrossRef]
14. Wu, H.; Tang, J.; Zhong, H. A moderate squint imaging algorithm for the multiple-hydrophone SAS with receiving hydrophone dependence. *IET Radar Sonar Navig.* **2019**, *13*, 139–147. [CrossRef]
15. Wu, H.; Tang, J.; Zhong, H. A Correction Approach for the Inclined Array of Hydrophones in Synthetic Aperture. *Sensors* **2018**, *18*, 2000. [CrossRef] [PubMed]
16. Zhang, X.; Ying, W.; Dai, X. High-resolution imaging for the multireceiver SAS. *J. Eng.* **2019**, *2019*, 6057–6062. [CrossRef]
17. Cumming, I.G.; Wong, F.H. *Digital Processing of Synthetic Aperture Radar Data: Algorithms and Implementation*; Artech House: Norwood, MA, USA, 2005.
18. Zhang, X.; Yang, P.; Wang, Y. A Novel Multireceiver SAS RD Processor. *IEEE Trans. Geosci. Remote Sens.* **2023**, *62*, 4203611. [CrossRef]
19. Zhang, X.; Yang, P.; Wang, Y. LBF-based CS algorithm for multireceiver SAS. *IEEE Geosci. Remote Sens. Lett.* **2024**, *21*, 1502505. [CrossRef]
20. Cafforio, C.; Prati, C.; Rocca, F. SAR data focusing using seismic migration techniques. *IEEE Trans. Aerosp. Electron. Syst.* **1991**, *27*, 194–207. [CrossRef]

remote sensing

MDPI

Technical Note

Two-Dimensional Direction Finding for L-Shaped Coprime Array via Minimization of the Ratio of the Nuclear Norm and the Frobenius Norm

Lang Zhou [1], Kun Ye [2] and Xuebo Zhang [3,*]

[1] School of Electronic Science and Engineering, Xiamen University, Xiamen 361005, China; zhoulang@stu.xmu.edu.cn
[2] School of Informatics, Xiamen University, Xiamen 361005, China; yekun@stu.xmu.edu.cn
[3] College of Physics and Electronic Engineering, Northwest Normal University, Lanzhou 730070, China
* Correspondence: xby_zhang@nwnu.edu.cn

Abstract: More recently, the ability of the coprime array to yield large array apertures and high degrees of freedom in comparison with the uniform linear array (ULA) has drawn an enormous amount of attention. In light of this, we propose a low-rank matrix completion algorithm via minimization of the ratio of the nuclear norm and the Frobenius norm (N/F) to solve the two-dimensional (2D) direction finding problem for the L-shaped coprime array (LsCA). Specifically, we first interpolate the virtual co-array signal related to the cross-correlation matrix (CCM) and utilize the interpolated virtual signal for Toeplitz matrix reconstruction. Then, the N/F method is employed to perform low-rank matrix completion on the reconstructed matrix. Finally, exploiting the conjugate symmetry characteristics of the completed matrix, we further develop a direction-finding algorithm that enables 2D angle estimation. Remarkably, the 2D angles are able to be automatically paired by the proposed algorithm. Numerical simulation findings demonstrate that the proposed N/F algorithm generates excellent angular resolution and computational complexity. Furthermore, this algorithm yields better estimation accuracy compared to the competing algorithms.

Keywords: low-rank matrix completion; minimization of the ratio of the nuclear norm and the Frobenius norm; 2D direction finding; L-shaped coprime array; automatically paired

1. Introduction

Direction finding has been a popular subject in array signal processing, which has received an enormous amount of interest in applications including navigation, sonar, and radar [1–10]. Practically, two-dimensional (2D) direction finding is more in line with the requirements of practical applications than one-dimensional (1D) direction finding. In recent decades, 2D array structures, including the parallel-shaped array (PsA) [11–13], L-shaped array (LsA) [14–26], uniform circular array (UCA) [27–32], and uniform rectangular array (URA) [33–38], have been employed in 2D direction finding algorithms. In addition, since the LsA has a simple structure and easy implementation, the 2D direction finding for the LsA is the most popular research topic.

To enable the 2D direction finding for the LsA, various methods have been proposed. In [17], a generalized propagator method (PM) that only needs matrix linear operations is proposed. Gu et al. [18] proposed a joint SVD (JSVD) to enhance the 2D direction finding performance by utilizing the cross-correlation matrix (CCM) that has no noise term. In [19], an effective array aperture extension technique (EAET) is proposed by taking into account the conjugate symmetry characteristic of the CCM, which in turn enhances the estimation performance. To simultaneously enlarge the array apertures and virtual snapshots, Dong et al. [20] constructed a large array-received-data-like matrix, taking into account the conjugate symmetry characteristic of the CCM. To extend the number of resolvable source signals, generalized conjugate symmetry is utilized in [21] to further enlarge the array aperture and virtual snapshots, followed by building a QCCQP problem to boost the direction finding accuracy. In [22], an iterative tensor-based 2D angle estimation is developed, which can effectively remove the cross-term associated with the noise term. Jin et al. [23] investigate an LsA-based localization method for achieving underdetermined 2D angle estimation of incoherently distributed sources, which generates high-accuracy estimation performance in the absence of any prior information regarding the angular distribution. However, the majority of the above methods are only applicable to the LsA, whose subarrays are uniform linear arrays (ULAs), thereby not directly utilizing the advantage of large virtual difference co-arrays of sparse arrays.

This paper proposes a 2D direction finding algorithm via minimization of the ratio of the nuclear norm and the Frobenius norm (N/F) for the L-shaped coprime array (LsCA), motivated by the larger array apertures and higher degrees of freedom of the LsCA. This algorithm first interpolates the virtual co-array signal corresponding to the CCM and performs a Toeplitz matrix reconstruction. Then, the N/F method is utilized to perform low-rank matrix completion on the reconstructed matrix. Finally, the decoupled estimation of the 2D angles is achieved by exploiting the conjugate symmetry characteristics of the completed matrix. Numerical simulation findings illustrate that the proposed N/F algorithm generates superior performance both with regard to estimation accuracy, angular resolution, and computational complexity.

This paper's major contributions are summarized below:

(1) We derive the virtual co-array signal model corresponding to the CCM of LsCA and perform Toeplitz matrix reconstruction utilizing the interpolated virtual co-array signal.
(2) Considering the zero regions of the reconstructed Toeplitz matrix, we utilize the N/F method for low-rank matrix completion. Taking into account the conjugate symmetry characteristics of the completed matrix, a direction-finding algorithm that enables 2D angle estimation is developed.
(3) It can be observed from numerical simulation findings that the proposed N/F algorithm generates excellent performance with respect to angular resolution and computational complexity. In addition, this algorithm yields superior estimation accuracy in comparison with the competing algorithms.

The remaining sections herein are outlined below: Section 2 describes the LsCA signal model in preparation for the subsequent direction finding. In Section 3, we propose a 2D direction finding algorithm for the LsCA based on the N/F approach. Section 4 gives numerical simulations to discuss the proposed algorithm's performance with respect to estimation accuracy, angular resolution, and computational complexity. Finally, the conclusion is presented in Section 5.

Notations: The explanations of the symbols mentioned all over the paper are given in Table 1.

Table 1. Symbols and associated explanations.

Symbol	Explanation
a, $\mathbf{a}$, $\mathbf{A}$	Scalars, vectors, matrices
$\mathbf{I}_n$, $\mathbf{0}_{p\times q}$, $\mathbf{J}_n$	n-dimensional identity matrix, $p \times q$ zero matrix, n-dimensional permutation matrix with 1 on the antidiagonal and 0 elsewhere
$(\cdot)^*$, $(\cdot)^T$, $(\cdot)^H$	Conjugate, transpose, Hermitian transpose
$diag(\cdot)$	Diagonalization: Convert vectors into diagonal matrices
$arg(\cdot)$	The phase of the argument
$E[\cdot]$	Mathematical expectation
$vec(\cdot)$, $\mathcal{T}(\cdot)$	Vectorization: Convert matrices into column vectors, Toeplitz matrix operator
$\|\cdot\|_F$	Frobenius norm
$\|\cdot\|_*$	Nuclear norm
$\mathbb{A}^{(i)}$	The i-th element of the set $\mathbb{A}$

Table 1. *Cont.*

Symbol	Explanation
$\lvert\mathbb{A}\rvert$	Cardinality of set $\mathbb{A}$
$\mathbf{A}^{-1}$, $\mathbf{A}^{\dagger}$, $Tr(\mathbf{A})$	Inverse of matrix $\mathbf{A}$, pseudo-inverse of matrix $\mathbf{A}$, trace of matrix $\mathbf{A}$
$\otimes, \odot$	Kronecker product, Khatri-Rao product
$[\mathbf{A}]_{i,j}$, $[\mathbf{a}]_i$	The (i,j)-th element of the matrix $\mathbf{A}$, the ith element of the vector $\mathbf{a}$
$\langle a_1, a_2\rangle$	Integer set $\{a \in \mathbb{Z} \vert a_1 \leq a \leq a_2\}$

2. L-Shaped Coprime Array Signal Model

As illustrated in Figure 1, the LsCA located in the $x-z$ plane is composed of two coprime arrays $\mathbb{S}_i, i = x, z$ placed orthogonally along the x and z axes, respectively. Every coprime array $\mathbb{S}_i$ has $|\mathbb{S}_i| = 2M_i + N_i - 1$ physical sensors, consisting of two sparse ULAs [39]. Note that M_i and N_i are coprime integers, and the two sparse ULAs contain $2M_i$ and N_i physical sensors with corresponding spacings of $N_i d$ and $M_i d$, respectively, where $d = \frac{\lambda}{2}$ and λ represents the carrier wavelength. Therefore, the sensor positions of each coprime array $\mathbb{S}_i$ can be denoted as $\mathbb{S}_i d$, where $\mathbb{S}_i = \{\mathbb{S}_i^{(l)}, l \in \langle 1, |\mathbb{S}_i|\rangle\} = \{1, N_i + 1, 2N_i + 1, \cdots, (2M_i - 1)N_i + 1, M_i + 1, 2M_i + 1, \cdots, (N_i - 1)M_i + 1\}$.

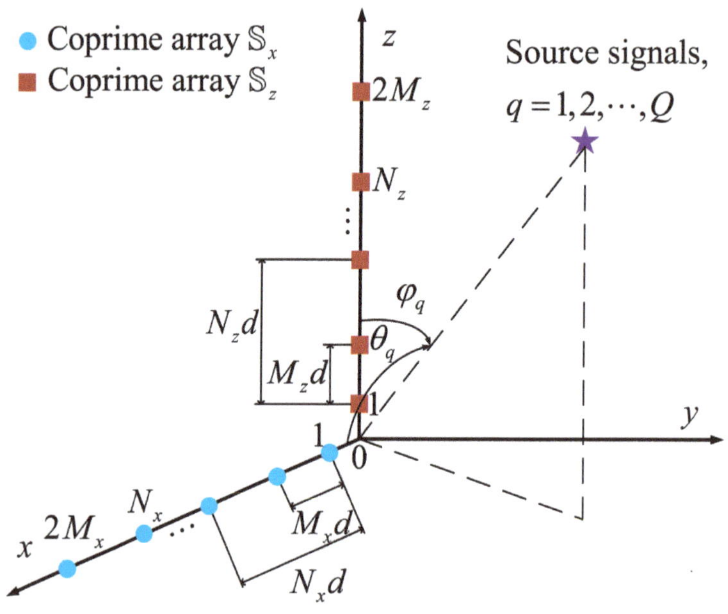

Figure 1. The LsCA structure.

Assume that Q uncorrelated far-field narrowband source signals from directions $\{(\theta_q, \varphi_q), q \in \langle 1, Q\rangle\}$ are incident on the LsCA, where $\theta_q \in [0, \pi]$ and $\varphi_q \in [0, \pi]$ represent the azimuth and elevation angles of the qth source signal, respectively. Let $\mathbf{x}(t)$ and $\mathbf{z}(t)$ denote the data vectors received at the time t by the coprime array $\mathbb{S}_x$ and the coprime array $\mathbb{S}_z$, respectively, which are represented as

$$\mathbf{x}(t) = \mathbf{A_x}\mathbf{s}(t) + \mathbf{n_x}(t), \tag{1}$$

$$\mathbf{z}(t) = \mathbf{A_z}\mathbf{s}(t) + \mathbf{n_z}(t), \tag{2}$$

where $\mathbf{s}(t)$ indicates the source signal vector. $\mathbf{A_x}$ and $\mathbf{A_z}$ are the array manifold matrices related to the $\mathbb{S}_x$ and the $\mathbb{S}_z$, which are given as follows:

$$\mathbf{A_x} = \left[\mathbf{a_x}(\theta_1), \mathbf{a_x}(\theta_2), \cdots, \mathbf{a_x}(\theta_Q)\right], \tag{3}$$

$$\mathbf{a_x}(\theta_q) = \left[e^{-j\pi\mathbb{S}_x^{(1)}\cos\theta_q}, \ldots, e^{-j\pi\mathbb{S}_x^{(|\mathbb{S}_x|)}\cos\theta_q}\right]^T, \tag{4}$$

$$\mathbf{A_z} = \left[\mathbf{a_z}(\varphi_1), \mathbf{a_z}(\varphi_2), \cdots, \mathbf{a_z}(\varphi_Q)\right], \tag{5}$$

$$\mathbf{a_z}(\varphi_q) = \left[e^{-j\pi\mathbb{S}_z^{(1)}\cos\varphi_q}, \ldots, e^{-j\pi\mathbb{S}_z^{(|\mathbb{S}_z|)}\cos\varphi_q}\right]^T. \tag{6}$$

Furthermore, $\mathbf{n_x}(t)$ and $\mathbf{n_z}(t)$ are additive Gaussian white noises satisfying $\mathcal{CN}(\mathbf{0}, \sigma_n^2\mathbf{I})$, which are independent.

Next, the CCM of the LsCA can be presented as

$$\begin{aligned}\mathbf{R_{xz}} &= E[\mathbf{x}(t)\mathbf{z}^H(t)] \\ &= \mathbf{A_x}\mathbf{R_{ss}}\mathbf{A_z}^H,\end{aligned} \tag{7}$$

where $\mathbf{R_{ss}} = E[\mathbf{s}(t)\mathbf{s}^H(t)] = diag(\mathbf{p})$, and $\mathbf{p} = [\sigma_1^2, \sigma_2^2, \cdots, \sigma_Q^2]^T$ is the source signal power vector.

Vectorizing $\mathbf{R_{xz}}$, the following virtual co-array signal is obtained:

$$\begin{aligned}\mathbf{y} &= vec(\mathbf{R_{xz}}) \\ &= (\mathbf{A_z^*} \otimes \mathbf{A_x})vec(\mathbf{R_{ss}}) \\ &= (\mathbf{A_z^*} \odot \mathbf{A_x})\mathbf{p} \\ &= \mathbf{Bp}.\end{aligned} \tag{8}$$

Then, eliminating the duplicate rows in $\mathbf{y}$ and sorting the rows, we obtain

$$\bar{\mathbf{y}} = \bar{\mathbf{B}}\mathbf{p}, \tag{9}$$

where $\bar{\mathbf{B}} = [\bar{\mathbf{b}}(\theta_1, \varphi_1), \bar{\mathbf{b}}(\theta_2, \varphi_2), \cdots, \bar{\mathbf{b}}(\theta_Q, \varphi_Q)]$ is the virtual array manifold matrix. Here, the vector $\bar{\mathbf{b}}(\theta_q, \varphi_q)$ can be represented as

$$\bar{\mathbf{b}}(\theta_q, \varphi_q) = \bar{\mathbf{b}}_\mathbf{z}^*(\varphi_q) \otimes \bar{\mathbf{b}}_\mathbf{x}(\theta_q), \tag{10}$$

$$\bar{\mathbf{b}}_\mathbf{x}(\theta_q) = \left[e^{-j\pi\mathbb{D}_x^{(1)}\cos\theta_q}, \ldots, e^{-j\pi\mathbb{D}_x^{(|\mathbb{D}_x|)}\cos\theta_q}\right]^T, \tag{11}$$

$$\bar{\mathbf{b}}_\mathbf{z}(\varphi_q) = \left[e^{-j\pi\mathbb{D}_z^{(1)}\cos\varphi_q}, \ldots, e^{-j\pi\mathbb{D}_z^{(|\mathbb{D}_z|)}\cos\varphi_q}\right]^T, \tag{12}$$

where $\mathbb{D}_x = \mathbb{D}_z$. Furthermore, $\mathbb{D}_x$ and $\mathbb{D}_z$ represent the difference co-array of $\mathbb{S}_x$ and $\mathbb{S}_z$, respectively.

3. The Proposed Algorithm

This section first utilizes the N/F method to implement the Toeplitz matrix completion associated with the virtual co-array signal $\bar{\mathbf{y}}$. Then, exploiting the conjugate symmetry characteristics of the completed matrix, the estimated 2D angles are obtained.

3.1. Low-Rank Matrix Completion

According to [39], one can deduce that $\mathbb{D}_x$ and $\mathbb{D}_z$ are not hole-free. Then, to enable the virtual co-array corresponding to the co-array signal $\bar{\mathbf{y}}$ to be consecutive, we perform the interpolation as follows:

$$\mathbb{D}_{\mathcal{I}} = \min(\mathbb{D}_x) : 1 : \max(\mathbb{D}_x). \tag{13}$$

These holes can be identified as missing sensors, with their associated virtual signals set to zero. Consequently, the interpolated signal $\hat{\mathbf{y}}_{\mathcal{I}}$ is calculated as

$$[\hat{\mathbf{y}}_{\mathcal{I}}]_i = \begin{cases} [\bar{\mathbf{y}}]_i & ,i \in \mathbb{D}_x \\ 0 & ,i \in \mathbb{D}_{\mathcal{I}} - \mathbb{D}_x \end{cases}, \tag{14}$$

Based on the interpolated signal $\hat{\mathbf{y}}_{\mathcal{I}}$, a Toeplitz matrix $\tilde{\mathbf{R}}_{\mathbf{v}}$ is constructed of the form

$$\tilde{\mathbf{R}}_{\mathbf{v}} = \begin{bmatrix} [\hat{\mathbf{y}}_{\mathcal{I}}]_{L_v} & [\hat{\mathbf{y}}_{\mathcal{I}}]_{L_v-1} & \cdots & [\hat{\mathbf{y}}_{\mathcal{I}}]_1 \\ [\hat{\mathbf{y}}_{\mathcal{I}}]_{L_v+1} & [\hat{\mathbf{y}}_{\mathcal{I}}]_{L_v} & \cdots & [\hat{\mathbf{y}}_{\mathcal{I}}]_2 \\ \vdots & \vdots & \ddots & \vdots \\ [\hat{\mathbf{y}}_{\mathcal{I}}]_{2L_v-1} & [\hat{\mathbf{y}}_{\mathcal{I}}]_{2L_v-2} & \cdots & [\hat{\mathbf{y}}_{\mathcal{I}}]_{L_v} \end{bmatrix}, \tag{15}$$

where $L_v = \frac{|\mathbb{D}_{\mathcal{I}}|+1}{2}$.

Since $\hat{\mathbf{y}}_{\mathcal{I}}$ is obtained by zero interpolation, there are zero regions in $\tilde{\mathbf{R}}_{\mathbf{v}}$, i.e., $\tilde{\mathbf{R}}_{\mathbf{v}}$ exists with missing elements. To complete $\tilde{\mathbf{R}}_{\mathbf{v}}$, low-rank regularization can be applied to $\tilde{\mathbf{R}}_{\mathbf{v}}$. Then, in light of the rank relaxation principle [40], the low-rank matrix completion can be modeled as a matrix nuclear norm minimization (NNM) optimization problem

$$\begin{aligned} &\min_{\hat{\mathbf{R}}_{\mathbf{v}}} \|\hat{\mathbf{R}}_{\mathbf{v}}\|_* \\ &\quad s.t.\ \hat{\mathbf{R}}_{\mathbf{v}} \circ \mathbf{G} = \tilde{\mathbf{R}}_{\mathbf{v}}, \end{aligned} \tag{16}$$

where the optimized variable $\hat{\mathbf{R}}_{\mathbf{v}}$ is the completed matrix. $\mathbf{G}$ is a binary matrix, with the elements in the positions related to the zero elements of $\tilde{\mathbf{R}}_{\mathbf{v}}$ being 0 and the remaining elements being 1. While the NNM problem is tractable, it is extremely likely to yield suboptimal performance because the nuclear norm is essentially determined by large singular values. As such, we introduce the N/F [41] as an alternative to the nuclear norm to address the problem of unbalanced penalization for all singular values that are involved in the nuclear norm. Then, Equation (16) can be reformulated as

$$\min_{\hat{\mathbf{R}}_{\mathbf{v}}} \lambda \frac{\|\hat{\mathbf{R}}_{\mathbf{v}}\|_*}{\|\hat{\mathbf{R}}_{\mathbf{v}}\|_F} + \frac{1}{2}\|\hat{\mathbf{R}}_{\mathbf{v}} \circ \mathbf{G} - \tilde{\mathbf{R}}_{\mathbf{v}}\|_2^2, \tag{17}$$

where λ represents the regularization parameter. Taking into account that $\tilde{\mathbf{R}}_{\mathbf{v}}$ has the Toeplitz property, Equation (17) can be further reformulated as

$$\begin{aligned} &\min_{\mathbf{X},\mathbf{y},\hat{\mathbf{R}}_{\mathbf{v}}} \lambda \frac{\|\hat{\mathbf{R}}_{\mathbf{v}}\|_*}{\|\hat{\mathbf{R}}_{\mathbf{v}}\|_F} + \frac{1}{2}\|\hat{\mathbf{R}}_{\mathbf{v}} \circ \mathbf{G} - \tilde{\mathbf{R}}_{\mathbf{v}}\|_2^2 + \frac{\gamma}{2}[Tr(\mathbf{X}) + Tr(\mathcal{T}(\mathbf{y}))] \\ &s.t. \begin{bmatrix} \mathcal{T}(\mathbf{y}) & \hat{\mathbf{R}}_{\mathbf{v}} \\ \hat{\mathbf{R}}_{\mathbf{v}}^H & \mathbf{X} \end{bmatrix} \succeq 0, \end{aligned} \tag{18}$$

where γ and $\mathbf{X}$ represent the regularization parameter and the regularization matrix of the free variables for preventing trivial solutions, respectively. $\hat{\mathbf{R}}_{\mathbf{v}}$ and $\mathbf{y}$ represent the optimized variables that need to be solved. As the optimization Problem (18) is convex, we prefer to solve it employing ADMM [42]. First, by introducing the constraint matrices $\mathbf{W}$ and $\mathbf{U}$, the augmented Lagrangian function for Equation (18) is defined as

$$\begin{aligned}\mathcal{L}(\hat{\mathbf{R}}_{\mathbf{v}},\mathbf{W},\mathbf{X},\mathbf{y},\mathbf{U},\mathbf{S},\mathbf{\Lambda}) = \lambda\frac{\|\hat{\mathbf{R}}_{\mathbf{v}}\|_*}{\|\hat{\mathbf{R}}_{\mathbf{v}}\|_F} + \frac{1}{2}\|\hat{\mathbf{R}}_{\mathbf{v}}\circ\mathbf{G}-\tilde{\mathbf{R}}_{\mathbf{v}}\|_2^2 + \langle\mathbf{S},\hat{\mathbf{R}}_{\mathbf{v}}-\mathbf{W}\rangle \\ +\frac{\mu_1}{2}\|\hat{\mathbf{R}}_{\mathbf{v}}-\mathbf{W}\|_F^2 + \frac{\gamma}{2}[Tr(\mathbf{X})+Tr(\mathcal{T}(\mathbf{y}))] \\ +\left\langle\mathbf{\Lambda},\mathbf{U}-\begin{bmatrix}\mathcal{T}(\mathbf{y}) & \hat{\mathbf{R}}_{\mathbf{v}}\\ \hat{\mathbf{R}}_{\mathbf{v}}^H & \mathbf{X}\end{bmatrix}\right\rangle + \frac{\mu_2}{2}\left\|\mathbf{U}-\begin{bmatrix}\mathcal{T}(\mathbf{y}) & \hat{\mathbf{R}}_{\mathbf{v}}\\ \hat{\mathbf{R}}_{\mathbf{v}}^H & \mathbf{X}\end{bmatrix}\right\|_F^2,\end{aligned} \tag{19}$$

where μ_1 and μ_2 represent the penalty parameters. $\mathbf{S}$ and $\mathbf{\Lambda}$ represent the Lagrange multipliers. Based on the ADMM framework, one can obtain the solutions to (19) by iteratively updating $\hat{\mathbf{R}}_{\mathbf{v}}$, $\mathbf{W}$, $\mathbf{X}$, $\mathbf{y}$, $\mathbf{U}$, $\mathbf{S}$, and $\mathbf{\Lambda}$. Specifically, the updating process is given by

$$\hat{\mathbf{R}}_{\mathbf{v}}^{(n+1)} = \arg\min_{\hat{\mathbf{R}}_{\mathbf{v}}} \mathcal{L}\left(\hat{\mathbf{R}}_{\mathbf{v}},\mathbf{W}^{(n)},\mathbf{X}^{(n)},\mathbf{y}^{(n)},\mathbf{U}^{(n)},\mathbf{S}^{(n)},\mathbf{\Lambda}^{(n)}\right), \tag{20}$$

$$\mathbf{W}^{(n+1)} = \arg\min_{\mathbf{W}} \mathcal{L}\left(\hat{\mathbf{R}}_{\mathbf{v}}^{(n+1)},\mathbf{W},\mathbf{X}^{(n)},\mathbf{y}^{(n)},\mathbf{U}^{(n)},\mathbf{S}^{(n)},\mathbf{\Lambda}^{(n)}\right), \tag{21}$$

$$\mathbf{X}^{(n+1)} = \arg\min_{\mathbf{X}} \mathcal{L}\left(\hat{\mathbf{R}}_{\mathbf{v}}^{(n+1)},\mathbf{W}^{(n+1)},\mathbf{X},\mathbf{y}^{(n)},\mathbf{U}^{(n)},\mathbf{S}^{(n)},\mathbf{\Lambda}^{(n)}\right), \tag{22}$$

$$\mathbf{y}^{(n+1)} = \arg\min_{\mathbf{y}} \mathcal{L}\left(\hat{\mathbf{R}}_{\mathbf{v}}^{(n+1)},\mathbf{W}^{(n+1)},\mathbf{X}^{(n+1)},\mathbf{y},\mathbf{U}^{(n)},\mathbf{S}^{(n)},\mathbf{\Lambda}^{(n)}\right), \tag{23}$$

$$\mathbf{U}^{(n+1)} = \arg\min_{\mathbf{U}} \mathcal{L}\left(\hat{\mathbf{R}}_{\mathbf{v}}^{(n+1)},\mathbf{W}^{(n+1)},\mathbf{X}^{(n+1)},\mathbf{y}^{(n+1)},\mathbf{U},\mathbf{S}^{(n)},\mathbf{\Lambda}^{(n)}\right), \tag{24}$$

$$\mathbf{S}^{(n+1)} = \mathbf{S}^{(n)} + \mu_1\left(\mathbf{S},\hat{\mathbf{R}}_{\mathbf{v}}^{(n+1)} - \mathbf{W}^{(n+1)}\right), \tag{25}$$

$$\mathbf{\Lambda}^{(n+1)} = \mathbf{\Lambda}^{(n)} + \mu_2\left(\mathbf{U}^{(n+1)} - \begin{bmatrix}\mathcal{T}\left(\mathbf{y}^{(n+1)}\right) & \hat{\mathbf{R}}_{\mathbf{v}}^{(n+1)}\\ \left[\hat{\mathbf{R}}_{\mathbf{v}}^{(n+1)}\right]^H & \mathbf{X}^{(n+1)}\end{bmatrix}\right). \tag{26}$$

The iteration converges if and only if the relative error of the completion matrix between two contiguous iterations is less than the convergence threshold $\xi_{ADMM} > 0$, i.e.,

$$\frac{\left\|\hat{\mathbf{R}}_{\mathbf{v}}^{(n+1)} - \hat{\mathbf{R}}_{\mathbf{v}}^{(n)}\right\|_F}{\left\|\hat{\mathbf{R}}_{\mathbf{v}}^{(n)}\right\|_F} \leqslant \xi_{ADMM}. \tag{27}$$

3.2. Angles Estimation

The completed matrix $\hat{\mathbf{R}}_{\mathbf{v}}$ is equivalent to the CCM of LsA, i.e.,

$$\hat{\mathbf{R}}_{\mathbf{v}} = \hat{\mathbf{A}}_{\mathbf{x}}\mathbf{R}_{\mathbf{ss}}\hat{\mathbf{A}}_{\mathbf{z}}^H, \tag{28}$$

where $\hat{\mathbf{A}}_{\mathbf{x}} = [\hat{\mathbf{a}}_{\mathbf{x}}(\theta_1),\hat{\mathbf{a}}_{\mathbf{x}}(\theta_2),\cdots,\hat{\mathbf{a}}_{\mathbf{x}}(\theta_Q)]$, $\hat{\mathbf{A}}_{\mathbf{z}} = [\hat{\mathbf{a}}_{\mathbf{z}}(\varphi_1),\hat{\mathbf{a}}_{\mathbf{z}}(\varphi_2),\cdots,\hat{\mathbf{a}}_{\mathbf{z}}(\varphi_Q)]$. Here, $\hat{\mathbf{a}}_{\mathbf{x}}(\theta_q)$ and $\hat{\mathbf{a}}_{\mathbf{z}}(\varphi_q)$ can respectively be represented as

$$\hat{\mathbf{a}}_{\mathbf{x}}(\theta_q) = \left[1,e^{-j\pi\cos\theta_q},\cdots,e^{-j\pi(L_v-1)\cos\theta_q}\right]^T, \tag{29}$$

$$\hat{\mathbf{a}}_{\mathbf{z}}(\varphi_q) = \left[1,e^{-j\pi\cos\varphi_q},\cdots,e^{-j\pi(L_v-1)\cos\varphi_q}\right]^T. \tag{30}$$

To facilitate the selection of the first and last $L_v - 1$ rows of the matrix, the selection matrixs can be defined as

$$\mathbf{\Phi}_i = \left[\mathbf{0}_{(L_v-1)\times(i-1)},\mathbf{I}_{L_v-1},\mathbf{0}_{(L_v-1)\times(2-i)}\right], \quad i = 1,2. \tag{31}$$

Based on $\hat{\mathbf{A}}_{\mathbf{x}}$, the following relationship is satisfied:

$$\mathbf{\Phi}_2\hat{\mathbf{A}}_{\mathbf{x}} = \mathbf{\Phi}_1\hat{\mathbf{A}}_{\mathbf{x}}\mathbf{\Psi}_{\mathbf{x}}, \tag{32}$$

where $\mathbf{\Psi}_{\mathbf{x}} = diag\left(\left[e^{-j\pi\cos\theta_1}, \cdots, e^{-j\pi\cos\theta_Q}\right]\right)$. Next, applying eigenvalue decomposition to $\hat{\mathbf{R}}_{\mathbf{v}}$ yields the signal subspace $\mathbf{E}_{\mathbf{s}}$. Because $\mathbf{E}_{\mathbf{s}}$ spans the column space of $\hat{\mathbf{A}}_{\mathbf{x}}$, we have

$$\mathbf{E}_{\mathbf{s}} = \hat{\mathbf{A}}_{\mathbf{x}}\mathbf{T}. \tag{33}$$

Combining Equations (32) and (33), we can derive

$$\mathbf{\Phi}_2\mathbf{E}_{\mathbf{s}} = \mathbf{\Phi}_1\mathbf{E}_{\mathbf{s}}\mathbf{\Theta}, \tag{34}$$

where $\mathbf{\Theta} = \mathbf{T}^{-1}\mathbf{\Psi}_{\mathbf{x}}\mathbf{T}$. Obviously, the azimuth angle information is included in the eigenvalues of $\mathbf{\Theta}$. Therefore, $\mathbf{\Theta}$ can be estimated as

$$\hat{\mathbf{\Theta}} = (\mathbf{\Phi}_1\mathbf{E}_{\mathbf{s}})^{\dagger}(\mathbf{\Phi}_2\mathbf{E}_{\mathbf{s}}). \tag{35}$$

Let ω_q be the qth eigenvalue of $\hat{\mathbf{\Theta}}$, then the azimuth angles can be estimated as

$$\hat{\theta}_q = \cos^{-1}\left(-\frac{\arg(\omega_q)}{\pi}\right), \quad q = 1, 2, \cdots, Q. \tag{36}$$

Since $\hat{\mathbf{A}}_{\mathbf{x}}$ and $\hat{\mathbf{A}}_{\mathbf{z}}$ are Vandermonde matrices, they satisfy the conjugate symmetry characteristics [19–21], i.e.,

$$\mathbf{J}_{L_v}\hat{\mathbf{A}}_{\mathbf{x}}^{*} = \hat{\mathbf{A}}_{\mathbf{x}}\mathbf{\Psi}_{\mathbf{x}}{}^{-(L_v-1)}, \tag{37}$$

$$\hat{\mathbf{A}}_{\mathbf{z}}^{T}\mathbf{J}_{L_v} = \mathbf{\Psi}_{\mathbf{z}}{}^{(L_v-1)}\hat{\mathbf{A}}_{\mathbf{z}}^{H}, \tag{38}$$

where $\mathbf{\Psi}_{\mathbf{z}} = diag\left(\left[e^{-j\pi\cos\varphi_1}, \cdots, e^{-j\pi\cos\varphi_Q}\right]\right)$. Combining Equations (28), (37) and (38), we can derive

$$\begin{aligned}\mathbf{Y} &= \mathbf{J}_{L_v}\hat{\mathbf{R}}_{\mathbf{v}}^{*}\mathbf{J}_{L_v} \\ &= \hat{\mathbf{A}}_{\mathbf{x}}\mathbf{\Psi}_{\mathbf{x}}{}^{-(L_v-1)}\mathbf{R}_{\mathbf{ss}}\mathbf{\Psi}_{z}{}^{(L_v-1)}\hat{\mathbf{A}}_{z}^{H}.\end{aligned} \tag{39}$$

Next, applying Equation (31) to $\mathbf{Y}$ yields four maximally overlapping submatrices, i.e.,

$$\begin{aligned}\mathbf{Y}_1 &= \mathbf{\Phi}_1\mathbf{Y}\mathbf{\Phi}_1^{T} \\ &= \hat{\mathbf{A}}_{\mathbf{x}1}\mathbf{\Psi}_{\mathbf{x}}{}^{-(L_v-1)}\mathbf{R}_{\mathbf{ss}}\mathbf{\Psi}_{z}{}^{(L_v-1)}\hat{\mathbf{A}}_{\mathbf{z}1}^{H},\end{aligned} \tag{40}$$

$$\begin{aligned}\mathbf{Y}_2 &= \mathbf{\Phi}_1\mathbf{Y}\mathbf{\Phi}_2^{T} \\ &= \hat{\mathbf{A}}_{\mathbf{x}1}\mathbf{\Psi}_{\mathbf{x}}{}^{-(L_v-1)}\mathbf{R}_{\mathbf{ss}}\mathbf{\Psi}_{z}{}^{(L_v-2)}\hat{\mathbf{A}}_{\mathbf{z}1}^{H},\end{aligned} \tag{41}$$

$$\begin{aligned}\mathbf{Y}_3 &= \mathbf{\Phi}_2\mathbf{Y}\mathbf{\Phi}_1^{T} \\ &= \hat{\mathbf{A}}_{\mathbf{x}1}\mathbf{\Psi}_{\mathbf{x}}{}^{-(L_v-2)}\mathbf{R}_{\mathbf{ss}}\mathbf{\Psi}_{z}{}^{(L_v-1)}\hat{\mathbf{A}}_{\mathbf{z}1}^{H},\end{aligned} \tag{42}$$

$$\begin{aligned}\mathbf{Y}_4 &= \mathbf{\Phi}_2\mathbf{Y}\mathbf{\Phi}_2^{T} \\ &= \hat{\mathbf{A}}_{\mathbf{x}1}\mathbf{\Psi}_{\mathbf{x}}{}^{-(L_v-2)}\mathbf{R}_{\mathbf{ss}}\mathbf{\Psi}_{z}{}^{(L_v-2)}\hat{\mathbf{A}}_{\mathbf{z}1}^{H},\end{aligned} \tag{43}$$

where $\hat{\mathbf{A}}_{\mathbf{x}1} = \mathbf{\Phi}_1\hat{\mathbf{A}}_{\mathbf{x}}$, $\hat{\mathbf{A}}_{\mathbf{z}1} = \mathbf{\Phi}_1\hat{\mathbf{A}}_{\mathbf{z}}$. To extend the virtual array apertures and snapshots, a generalized received data matrix is built of the form

$$\hat{\mathbf{Y}} = \begin{bmatrix} \mathbf{Y}_1 & \mathbf{Y}_3 \\ \mathbf{Y}_2 & \mathbf{Y}_4 \end{bmatrix} = \hat{\mathbf{A}}_{\mathbf{ex}}(\theta,\varphi)\hat{\mathbf{S}}_{\mathbf{ex}}(\theta,\varphi), \tag{44}$$

where

$$\mathbf{A}_{\mathbf{ex}}(\theta,\varphi) = \left[\hat{\mathbf{A}}_{\mathbf{x}1}\mathbf{\Psi}_{\mathbf{z}}{}^{(L_v-1)}, \hat{\mathbf{A}}_{\mathbf{x}1}\mathbf{\Psi}_{\mathbf{z}}{}^{(L_v-2)}\right]^T, \tag{45}$$

$$\mathbf{S}_{\mathbf{ex}}(\theta,\varphi) = \left[\mathbf{R}_{\mathbf{ss}}\mathbf{\Psi}_{\mathbf{x}}{}^{-(L_v-1)}\hat{\mathbf{A}}_{\mathbf{z}1}^H, \mathbf{R}_{\mathbf{ss}}\mathbf{\Psi}_{\mathbf{x}}{}^{-(L_v-2)}\hat{\mathbf{A}}_{\mathbf{z}1}^H\right]. \tag{46}$$

Taking into account the matrix $\hat{\mathbf{Y}}$, the generalized covariance matrix is denoted as $\mathbf{R}_{\mathbf{ex}} = E\left[\hat{\mathbf{Y}}\hat{\mathbf{Y}}^H\right]$. Setting $\mathbf{U}_{\mathbf{n}}$ to be the noise subspace of $\mathbf{R}_{\mathbf{ex}}$, we obtain

$$\mathbf{a}_{\mathbf{ex}}^H(\theta,\varphi)\mathbf{U}_{\mathbf{n}}\mathbf{U}_{\mathbf{n}}^H\mathbf{a}_{\mathbf{ex}}(\theta,\varphi) = 0, \tag{47}$$

where $\mathbf{a}_{\mathbf{ex}}(\theta,\varphi) = \left[\hat{\mathbf{a}}_{\mathbf{x}1}(\theta)e^{-j\pi(L_v-1)\cos\varphi}, \hat{\mathbf{a}}_{\mathbf{x}1}(\theta)e^{-j\pi(L_v-2)\cos\varphi}\right]^T = \mathbf{v}(\varphi)\otimes\hat{\mathbf{a}}_{\mathbf{x}1}(\theta)$, $\mathbf{v}(\varphi) = \left[e^{-j\pi(L_v-1)\cos\varphi}, e^{-j\pi(L_v-2)\cos\varphi}\right]^T$.

Given $(\mathbf{AB})\otimes(\mathbf{CD}) = (\mathbf{A}\otimes\mathbf{C})(\mathbf{B}\otimes\mathbf{D})$, we have

$$\begin{aligned} \mathbf{a}_{\mathbf{ex}}(\theta,\varphi) &= (\mathbf{I}_2\mathbf{v}(\varphi))\otimes(\hat{\mathbf{a}}_{\mathbf{x}1}(\theta)\cdot 1) \\ &= (\mathbf{I}_2\otimes\hat{\mathbf{a}}_{\mathbf{x}1}(\theta))\mathbf{v}(\varphi), \end{aligned} \tag{48}$$

Substituting Equation (48) into (47) yields

$$\mathbf{v}^H(\varphi)\mathbf{M}_{\mathbf{v}}(\theta)\mathbf{v}(\varphi) = 0, \tag{49}$$

where

$$\mathbf{M}_{\mathbf{v}}(\theta) = (\mathbf{I}_2\otimes\hat{\mathbf{a}}_{\mathbf{x}1}(\theta))^H\mathbf{U}_{\mathbf{n}}\mathbf{U}_{\mathbf{n}}^H(\mathbf{I}_2\otimes\hat{\mathbf{a}}_{\mathbf{x}1}(\theta)). \tag{50}$$

Considering all constraints, (49) can be modeled as an optimization problem as below:

$$\begin{aligned} \min_{\mathbf{v}(\varphi),\mathbf{M}_{\mathbf{v}}(\theta)} \quad & \mathbf{v}^H(\varphi)\mathbf{M}_{\mathbf{v}}(\theta)\mathbf{v}(\varphi) \\ s.t.\ & \mathbf{v}^H(\varphi)\mathbf{v}(\varphi) = 2. \end{aligned} \tag{51}$$

Next, substituting Equation (36) into (51), the elevation angles can be calculated as [20]

$$\hat{\varphi}_q = \cos^{-1}\left(\frac{\arg\left(-\frac{\left[e_{\min}\left(\mathbf{M}_{\mathbf{v}}(\hat{\theta}_q)\right)\right]_1}{\left[e_{\min}\left(\mathbf{M}_{\mathbf{v}}(\hat{\theta}_q)\right)\right]_2}\right)}{\pi}\right), \quad q = 1,2,\cdots,Q, \tag{52}$$

where $\left[e_{\min}\left(\mathbf{M}_{\mathbf{v}}(\hat{\theta}_q)\right)\right]_1$ and $\left[e_{\min}\left(\mathbf{M}_{\mathbf{v}}(\hat{\theta}_q)\right)\right]_2$ are the first and second elements of the eigenvector corresponding to the smallest eigenvalue of $\mathbf{M}_{\mathbf{v}}(\hat{\theta}_q)$, respectively. Remarkably, the 2D angles are able to be automatically paired by the proposed algorithm.

The specific implementation of the proposed N/F algorithm is outlined as presented in Algorithm 1.

Algorithm 1 Proposed N/F algorithm for LsCA structure.

Input: The signal vectors $\mathbf{x}(t)$ and $\mathbf{z}(t)$ received by the LsCA structure, number of source signals Q.
Output: Paired angles $(\hat{\theta}_q, \hat{\varphi}_q), q \in \langle 1, Q \rangle$.
1: $\mathbf{R}_{\mathbf{xz}} \leftarrow E[\mathbf{x}(t)\mathbf{z}^H(t)]$.
2: $\bar{\mathbf{y}} \leftarrow$ (8) and (9).
Low-rank Matrix Completion:
3: $\hat{\mathbf{y}}_{\mathcal{I}} \leftarrow$ utilize (14) for virtual interpolation.
4: $\tilde{\mathbf{R}}_{\mathbf{v}} \leftarrow$ Toeplitz matrix reconstruction based on the interpolated signal $\hat{\mathbf{y}}_{\mathcal{I}}$.
5: The N/F optimization problem is modeled based on (18) for low-rank matrix completion.
6: $\mathcal{L}(\hat{\mathbf{R}}_{\mathbf{v}}, \mathbf{W}, \mathbf{X}, \mathbf{y}, \mathbf{U}, \mathbf{S}, \mathbf{\Lambda}) \leftarrow$ utilize (19) to obtain the augmented Lagrangian function.
7: $\hat{\mathbf{R}}_{\mathbf{v}} \leftarrow$ based on the ADMM framework, $\hat{\mathbf{R}}_{\mathbf{v}}$, $\mathbf{W}$, $\mathbf{X}$, $\mathbf{y}$, $\mathbf{U}$, $\mathbf{S}$, and $\mathbf{\Lambda}$ are updated utilizing (20)–(26) until the iteration converges to obtain the optimal completion matrix $\hat{\mathbf{R}}_{\mathbf{v}}$.
Angles estimation:
8: $\hat{\mathbf{\Theta}} \leftarrow$ (32)–(35).
9: $\hat{\theta}_q \leftarrow$ perform eigenvalue decomposition to $\hat{\mathbf{\Theta}}$ and utilize (36) to obtain the estimated azimuth angles.
10: $\hat{\mathbf{Y}} \leftarrow$ based on the conjugate symmetry characteristics, utilize (44) to construct the generalized received data matrix.
11: Utilize (51) to model the optimization problem for solving the estimated elevation angles.
12: $\hat{\varphi}_q \leftarrow$ utilize (52) to obtain the estimated elevation angles.

4. Numerical Simulations

This section gives numerical simulations for discussing the proposed N/F algorithm's performance, both regarding estimation accuracy, angular resolution, and computational complexity. In addition, the results of the 2D-MUSIC [14], EAET [19], CSAP [20], QC-CQP [21], and 2D-DFT [8] algorithms are included for comparison. For 2D-MUSIC, 2D-DFT, and proposed N/F algorithms, the LsCA structure consisting of the coprime array with 10 physical sensors as subarrays in the x-axis and z-axis, respectively, is chosen as the array structure, and $\mathbb{S}_x = \mathbb{S}_z = \{1, 4, 6, 7, 10, 11, 13, 16, 21, 26\}$. For EAET, CSAP, and QCCQP algorithms, the LsA structure consisting of ULAs with 20 physical sensors as subarrays in the x-axis and z-axis, respectively, is chosen as the array structure. On top of that, for the proposed N/F algorithm, the parameters λ, γ, μ_1, μ_2, and ξ_{ADMM} are configured as 10, 0.01, 0.1, 0.1, and 10^{-5}, respectively.

4.1. Estimation Accuracy Comparison

Firstly, the estimation accuracy of the 2D direction finding is assessed by the root mean square error (RMSE), of which the definition is

$$RMSE = \sqrt{\frac{1}{KQ}\sum_{k=1}^{K}\sum_{q=1}^{Q}\left(\left(\hat{\theta}_q^k - \theta_q\right)^2 + \left(\hat{\varphi}_q^k - \varphi_q\right)^2\right)}, \tag{53}$$

where $K = 300$ represents the number of Monte Carlo trials. $\hat{\theta}_q^k$ and $\hat{\varphi}_q^k$ represent the qth estimated azimuth and elevation angles for the kth trial, respectively.

We consider three uncorrelated source signals from directions $(\theta_1, \varphi_1) = (80^\circ, 100^\circ)$, $(\theta_2, \varphi_2) = (120^\circ, 80^\circ)$ and $(\theta_3, \varphi_3) = (100^\circ, 120^\circ)$ incident on the array structure. In addition, the CRB [43] curve is chosen as a reference for analyzing the estimation performance. The RMSE of the estimated 2D angles at the number of snapshots $T = 500$ with respect to the SNR when the SNR has a variation range of -10 dB to 20 dB is presented in Figure 2. It is apparent from the figure that the estimation results of all the algorithms tend to be more precise and stable when the SNR grows. Moreover, the proposed N/F algorithm is preferable to the 2D-MUSIC, EAET, CSAP, QCCQP, and 2D-DFT algorithms throughout the region of SNR variation. Contributing to this fact is that the proposed N/F algorithm completes the reconstructed Toplitz matrix and provides more useful information. The RMSE of the estimated 2D angles at SNR $= 5$ dB with respect to the T when the T ranges from 100 to 1000 is shown in Figure 3. For all algorithms, the corresponding RMSE decreases with increasing T. In addition, by comparison with the 2D-MUSIC, EAET, CSAP, QCCQP, and 2D-DFT algorithms, the proposed N/F algorithm generates the highest estimation accuracy throughout a wide region of T.

On top of that, for the purpose of discussing the proposed N/F algorithm's estimation accuracy in the cases with a larger amount of source signals, we consider $Q = 12$ and $Q = 20$ uncorrelated source signals uniformly located over $[30°, 150°]$ directions impinging on the LsCA structure. In these simulations, SNR and T are configured as 0 dB and 500, respectively. From the estimation results presented in Figures 4 and 5, it can be found that the proposed N/F algorithm accurately detects all source signals in the two scenarios, demonstrating that the proposed N/F algorithm still possesses excellent estimation accuracy under the condition of the greater amount of source signals.

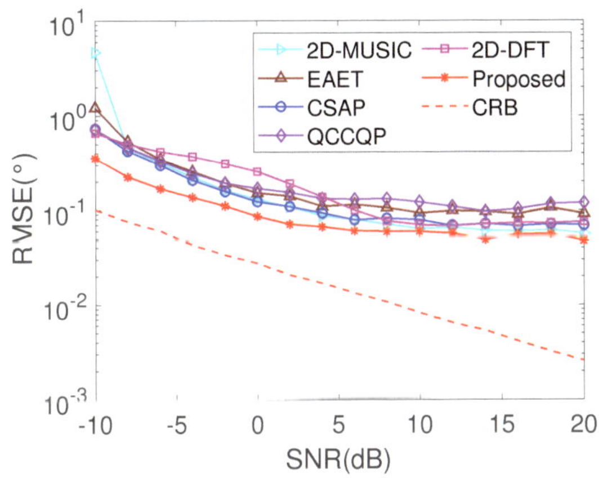

Figure 2. RMSE versus SNR with $T = 500$.

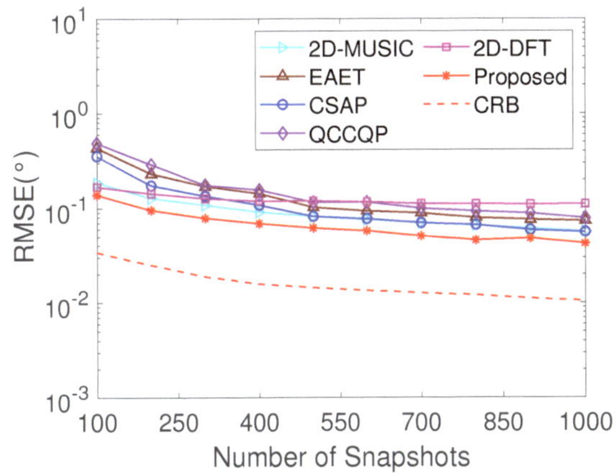

Figure 3. RMSE versus T with SNR $= 5$ dB.

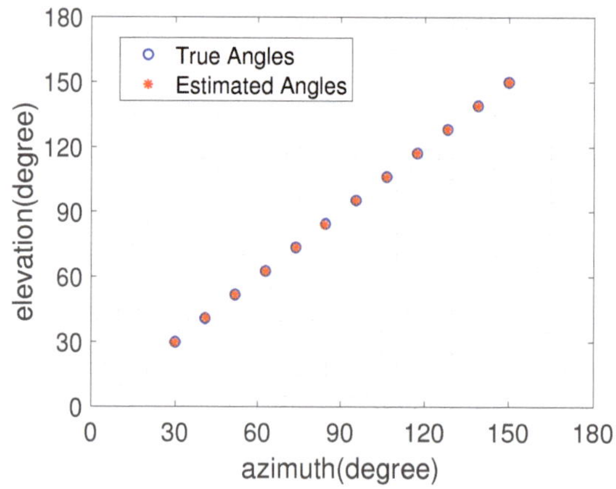

Figure 4. The proposed N/F algorithm's estimation accuracy with $Q = 12$.

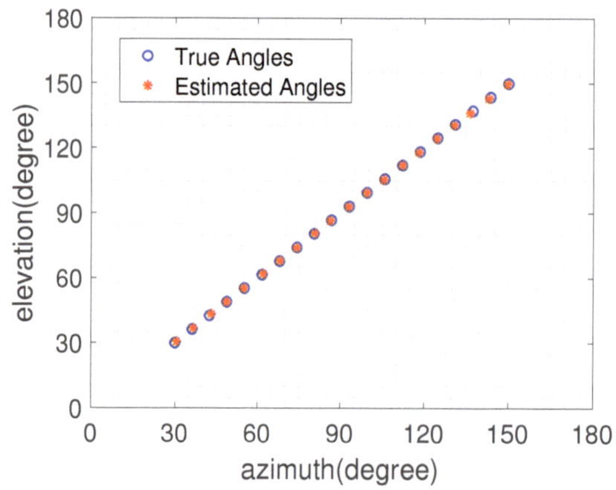

Figure 5. The proposed N/F algorithm's estimation accuracy with $Q = 20$.

4.2. Angular Resolution Comparison

To further evaluate the proposed N/F algorithms' performance, we utilize resolution probability to examine the angular resolution of the various algorithms. We consider two uncorrelated source signals from directions $(\theta_1, \varphi_1) = (80°, 82°)$ and $(\theta_2, \varphi_2) = (82°, 80°)$ incident on the array structure. Here, the angular separations are $\Delta\theta = \Delta\varphi = 2°$ for azimuth and elevation angles. We regard the two source signals as successfully localized if

$$\begin{aligned} &\left\| \hat{\theta}_q - \theta_q \right\| \leqslant 0.5° \\ &\left\| \hat{\varphi}_q - \varphi_q \right\| \leqslant 0.5°, \quad q = 1, 2 \end{aligned} \tag{54}$$

holds true. This example involves 300 Monte Carlo trials.

The angular resolution probabilities for various algorithms with SNR ranging from -10 dB to 20 dB and $T = 500$ are presented in Figure 6, from which one can observe that the proposed N/F algorithm's angular resolution is preferable to that of the 2D-MUSIC, EAET, CSAP, QCCQP, and 2D-DFT algorithms when SNR < 5 dB, and its resolution probability reaches 1 when the SNR equals -4 dB. In contrast, the resolution probabilities of the 2D-MUSIC, EAET, CSAP, QCCQP, and 2D-DFT algorithms reach 1 at SNRs of 6 dB, 6 dB, 6 dB, -2 dB, and 2 dB, respectively. Remarkably, the 2D-MUSIC, EAET, CSAP, and 2D-DFT algorithms are completely unsuccessful in localizing the two source signals when

SNR < −5 dB. The angular resolution probabilities for various algorithms with T ranging from 50 to 1000 and SNR = 5 dB are presented in Figure 7, from which one can observe that only the proposed N/F algorithm succeeds in identifying the two source signals when $T < 150$ and has a resolution probability of 1 when $T = 100$. In comparison, the 2D-MUSIC, EAET, CSAP, QCCQP and 2D-DFT algorithms achieve an angular resolution probability of 1 at the number of snapshots T of 500, 350, 400, 150, and 500, respectively. In summary, the proposed N/F algorithm is preferable to the 2D-MUSIC, EAET, CSAP, QCCQP, and 2D-DFT algorithms in terms of angular resolution.

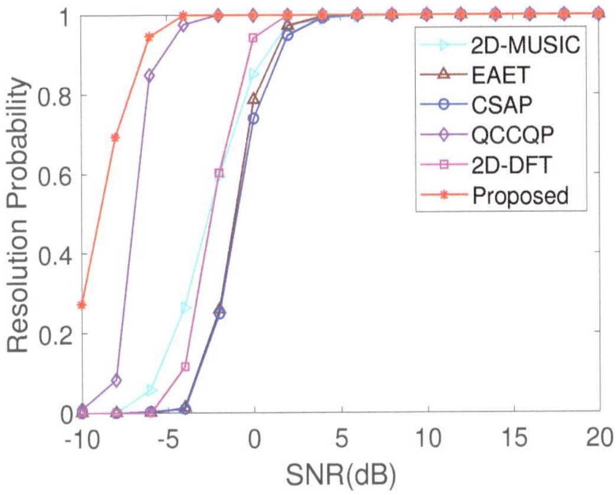

Figure 6. Resolution probability versus SNR with $T = 500$.

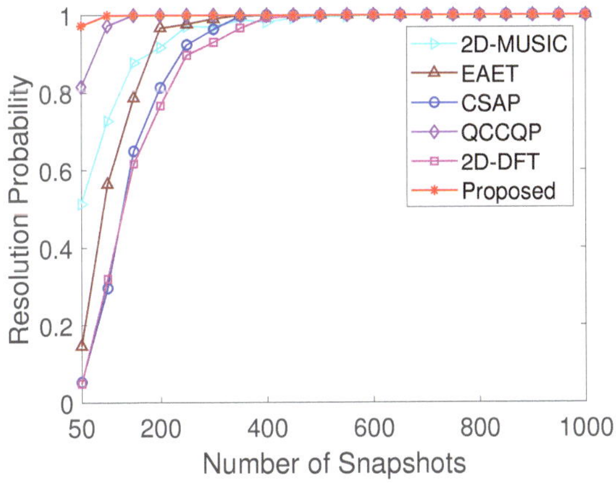

Figure 7. Resolution probability versus T with SNR = 5 dB.

4.3. Computational Complexity Comparison

We also compare the computational complexity of various algorithms. In this example, we consider three uncorrelated source signals impinging on the array structure. On top of that, SNR and T are configured as 5 dB and 500, respectively. Figure 8 presents the computational complexity of various algorithms that perform 300 Monte Carlo trials. One can observe that the QCCQP algorithm has the highest computational complexity, followed by the CSAP and 2D-MUSIC algorithms. In addition, the computational complexity of EAET, 2D-DFT, and the proposed N/F algorithms is lower and close to the same. To summarize, the proposed N/F algorithm realizes moderate computational complexity.

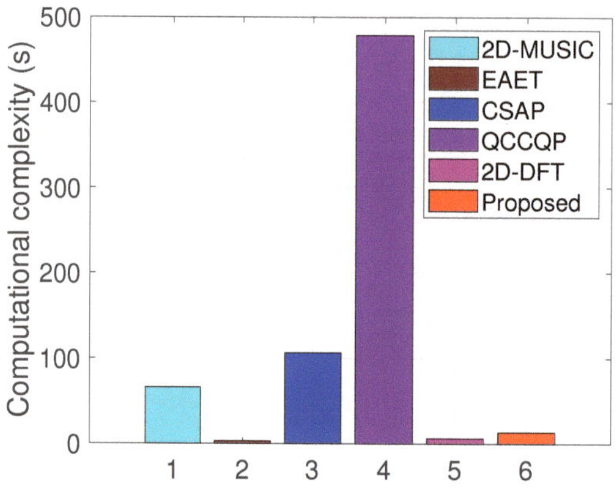

Figure 8. Comparison of computational complexity among different algorithms.

5. Conclusions

In this paper, a 2D direction-finding algorithm for the LsCA is proposed via the N/F method. This algorithm not only enhances the 2D direction finding performance but also has the capability to automatically pair 2D angles. Numerical simulation findings demonstrate that the proposed N/F algorithm generates superior performance regarding angular resolution and computational complexity. In addition, this algorithm yields better estimation accuracy in comparison to the existing algorithms. In future research, the N/F method should be considered and applied to matrix completion for sparse planar arrays.

Author Contributions: Writing—original draft preparation, conceptualization, L.Z.; methodology, software validation, L.Z. and K.Y.; supervision, funding acquisition, X.Z.; writing—review and editing, L.Z., K.Y., and X.Z. All authors have read and agree to the published version of the manuscript.

Funding: This work was supported by the National Key Laboratory Foundation under Grant 9140C290401150C29132.

Conflicts of Interest: The authors declare no conflicts of interest.

References

1. Krim, H.; Viberg, M. Two decades of array signal processing research: The parametric approach. *IEEE Signal Process. Mag.* **1996**, *13*, 67–94. [CrossRef]
2. Liu, D.; Zhao, Y.; Zhang, T. Sparsity-Based Two-Dimensional DOA Estimation for Co-Prime Planar Array via Enhanced Matrix Completion. *Remote Sens.* **2022**, *14*, 4690. [CrossRef]
3. Wen, F.; Shi, J.; Zhang, Z. Joint 2D-DOD, 2D-DOA, and polarization angles estimation for bistatic EMVS-MIMO radar via PARAFAC analysis. *IEEE Trans. Veh. Technol.* **2019**, *69*, 1626–1638. [CrossRef]
4. Cheng, L.; Wu, Y.C.; Zhang, J.; Liu, L. Subspace identification for DOA estimation in massive/full-dimension MIMO systems: Bad data mitigation and automatic source enumeration. *IEEE Trans. Signal Process.* **2015**, *63*, 5897–5909. [CrossRef]
5. Wen, F.; Shi, J.; He, J.; Truong, T.K. 2D-DOD and 2D-DOA estimation using sparse L-shaped EMVS-MIMO radar. *IEEE Trans. Aerosp. Electron. Syst.* **2022**, *59*, 2077–2084. [CrossRef]
6. Zhou, L.; Ye, K.; Qi, J.; Sun, H. DOA estimation based on pseudo-noise subspace for relocating enhanced nested array. *IEEE Signal Process. Lett.* **2022**, *29*, 1858–1862. [CrossRef]
7. Zhou, L.; Qi, J.; Hong, S. Enhanced Dilated Nested Arrays With Reduced Mutual Coupling for DOA Estimation. *IEEE Sens. J.* **2024**, *24*, 615–623. [CrossRef]
8. Zhang, Q.; Li, J.; Li, Y.; Li, P. A DOA tracking method based on offset compensation using nested array. *IEEE Trans. Circuits Syst. II Exp. Briefs* **2021**, *69*, 1917–1921. [CrossRef]
9. Xu, T.; Wang, X.; Huang, M.; Lan, X.; Sun, L. Tensor-Based Reduced-Dimension MUSIC Method for Parameter Estimation in Monostatic FDA-MIMO Radar. *Remote Sens.* **2021**, *13*, 3772. [CrossRef]
10. Ge, S.; Fan, C.; Wang, J.; Huang, X. Low-complexity one-bit DOA estimation for massive ULA with a single snapshot. *Remote Sens.* **2022**, *14*, 3436. [CrossRef]

11. Li, J.; Li, Y.; Zhang, X. Two-dimensional off-grid DOA estimation using unfolded parallel coprime array. *IEEE Commun. Lett.* **2018**, *22*, 2495–2498. [CrossRef]
12. Zhang, Y.; Xu, X.; Sheikh, Y.A.; Ye, Z. A rank-reduction based 2D DOA estimation algorithm for three parallel uniform linear arrays. *Signal Process.* **2016**, *120*, 305–310. [CrossRef]
13. Chen, H.; Hou, C.; Liu, W.; Zhu, W.P.; Swamy, M. Efficient two-dimensional direction-of-arrival estimation for a mixture of circular and noncircular sources. *IEEE Sens. J.* **2016**, *16*, 2527–2536. [CrossRef]
14. Zheng, Z.; Mu, S. 2D Direction FindingWith Pair-Matching Operation for L-Shaped Nested Array. *IEEE Commun. Lett.* **2020**, *25*, 975–979. [CrossRef]
15. Tayem, N.; Kwon, H.M. L-shape 2Dimensional arrival angle estimation with propagator method. *IEEE Trans. Antennas Propag.* **2005**, *53*, 1622–1630. [CrossRef]
16. Kikuchi, S.; Tsuji, H.; Sano, A. Pair-matching method for estimating 2D angle of arrival with a cross-correlation matrix. *IEEE Antennas Wirel. Propag. Lett.* **2006**, *5*, 35–40. [CrossRef]
17. Tayem, N.; Majeed, K.; Hussain, A.A. Two-dimensional DOA estimation using cross-correlation matrix with L-shaped array. *IEEE Antennas Wirel. Propag. Lett.* **2015**, *15*, 1077–1080. [CrossRef]
18. Gu, J.F.; Wei, P. Joint SVD of two cross-correlation matrices to achieve automatic pairing in 2D angle estimation problems. *IEEE Antennas Wirel. Propag. Lett.* **2007**, *6*, 553–556. [CrossRef]
19. Dong, Y.Y.; Dong, C.x.; Xu, J.; Zhao, G.q. Computationally efficient 2D DOA estimation for L-shaped array with automatic pairing. *IEEE Antennas Wirel. Propag. Lett.* **2016**, *15*, 1669–1672. [CrossRef]
20. Dong, Y.Y.; Dong, C.x.; Liu, W.; Chen, H.; Zhao, G.q. 2D DOA estimation for L-shaped array with array aperture and snapshots extension techniques. *IEEE Signal Process. Lett.* **2017**, *24*, 495–499. [CrossRef]
21. Wu, R.; Zhang, Z. Convex optimization-based 2D DOA estimation with enhanced virtual aperture and virtual snapshots extension for L-shaped array. *IEEE Trans. Veh. Technol.* **2020**, *69*, 6473–6484. [CrossRef]
22. Xu, F.; Zheng, H.; Vorobyov, S.A. Tensor-based 2D DOA estimation for L-shaped nested array. *IEEE Trans. Aerosp. Electron. Syst.* **2024**, *60*, 604–618. [CrossRef]
23. Jin, L.; Fang, J.; Liu, W.; Yu, Z.; Xu, G.; Zhu, M.; Chen, H. Underdetermined two-dimensional angles parameters estimation for incoherent distributed sources based on an L-shaped array. *IEEE Sens. J.* **2024**, *24*, 18479–18487. [CrossRef]
24. Rao, W.; Li, D.; Zhang, J.Q. A tensor-based approach to L-shaped arrays processing with enhanced degrees of freedom. *IEEE Signal Process. Lett.* **2017**, *25*, 1–5. [CrossRef]
25. Wu, X.; Zhu, W.P. On efficient gridless methods for 2D DOA estimation with uniform and sparse L-shaped arrays. *Signal Process.* **2022**, *191*, 108351. [CrossRef]
26. Xi, N.; Liping, L. A computationally efficient subspace algorithm for 2D DOA estimation with L-shaped array. *IEEE Signal Process. Lett.* **2014**, *21*, 971–974.
27. Zhang, T.; Lu, Y.; Hui, H. Compensation for the mutual coupling effect in uniform circular arrays for 2D DOA estimations employing the maximum likelihood technique. *IEEE Trans. Aerosp. Electron. Syst.* **2008**, *44*, 1215–1221. [CrossRef]
28. Wax, M.; Sheinvald, J. Direction finding of coherent signals via spatial smoothing for uniform circular arrays. *IEEE Trans. Antennas Propag.* **1994**, *42*, 613–620. [CrossRef]
29. Mathews, C.P.; Zoltowski, M.D. Eigenstructure techniques for 2D angle estimation with uniform circular arrays. *IEEE Trans. Signal Process.* **1994**, *42*, 2395–2407. [CrossRef]
30. Ding, X.; Xu, W.; Wang, Y.; Li, Y.; Huang, Y. Wideband 2D DoA Estimation with Uniform Circular Array. *IEEE Sens. J.* **2024**, *24*, 11585–11598. [CrossRef]
31. Zhou, T.; He, Z.; Shi, Q.; Lin, C.; Zhang, S. Multisnapshot High-Resolution Gridless DOA Estimation for Uniform Circular Arrays. *IEEE Signal Process. Lett.* **2024**, *31*, 1705–1709. [CrossRef]
32. Labbaf, N.; Oskouei, H.R.D.; Abedi, M.R. Robust DoA estimation in a uniform circular array antenna with errors and unknown parameters using Deep Learning. *IEEE Trans. Green Commun. Netw.* **2023**, *7*, 2143–2152. [CrossRef]
33. Zoltowski, M.D.; Haardt, M.; Mathews, C.P. Closed-form 2D angle estimation with rectangular arrays in element space or beamspace via unitary ESPRIT. *IEEE Trans. Signal Process.* **1996**, *44*, 316–328. [CrossRef]
34. Wu, H.; Hou, C.; Chen, H.; Liu, W.; Wang, Q. Direction finding and mutual coupling estimation for uniform rectangular arrays. *Signal Process.* **2015**, *117*, 61–68. [CrossRef]
35. Shen, F.F.; Liu, Y.M.; Zhao, G.H.; Chen, X.Y.; Li, X.P. Sparsity-based off-grid DOA estimation with uniform rectangular arrays. *IEEE Sens. J.* **2018**, *18*, 3384–3390. [CrossRef]
36. Ahmed, T.; Zhang, X.; Hassan, W.U. A higher-order propagator method for 2D-DOA estimation in massive MIMO systems. *IEEE Commun. Lett.* **2019**, *24*, 543–547. [CrossRef]
37. Zhang, Z.; Wen, F.; Shi, J.; He, J.; Truong, T.K. 2D-DOA estimation for coherent signals via a polarized uniform rectangular array. *IEEE Signal Process. Lett.* **2023**, *30*, 893–897. [CrossRef]
38. Heidenreich, P.; Zoubir, A.M.; Rubsamen, M. Joint 2D DOA estimation and phase calibration for uniform rectangular arrays. *IEEE Trans. Signal Process.* **2012**, *60*, 4683–4693. [CrossRef]
39. Vaidyanathan, P.P.; Pal, P. Sparse sensing with co-prime samplers and arrays. *IEEE Trans. Signal Process.* **2010**, *59*, 573–586. [CrossRef]

40. Recht, B.; Fazel, M.; Parrilo, P.A. Guaranteed minimum-rank solutions of linear matrix equations via nuclear norm minimization. *SIAM Rev.* **2010**, *52*, 471–501. [CrossRef]
41. Gao, K.; Huang, Z.H.; Guo, L. Low-rank matrix recovery problem minimizing a new ratio of two norms approximating the rank function then using an ADMM-type solver with applications. *J. Comput. Appl. Math.* **2024**, *438*, 115564. [CrossRef]
42. Boyd, S.; Parikh, N.; Chu, E.; Peleato, B.; Eckstein, J. Distributed optimization and statistical learning via the alternating direction method of multipliers. *Found. Trends Mach. Learn.* **2011**, *3*, 1–122. [CrossRef]
43. Xia, T.; Zheng, Y.; Wan, Q.; Wang, X. Decoupled estimation of 2D angles of arrival using two parallel uniform linear arrays. *IEEE Trans. Antennas Propag.* **2007**, *55*, 2627–2632. [CrossRef]

remote sensing

MDPI

Technical Note

The Time Difference of Arrival Estimation Method Utilizing an Inexact Reconstruction Within the Framework of Compressed Sensing

Shanhe Wang *, Yu Xiang, Yuanyuan Gao, Yu Hua, Changjiang Huang and Xian Zhao

National Time Service Center, Chinese Academy of Sciences, No. 3, Shuyuan East Road, Lintong District, Xi'an 710600, China; xy@ntsc.ac.cn (Y.X.); gaoyy@ntsc.ac.cn (Y.G.); hy@ntsc.ac.cn (Y.H.); huangcj@ntsc.ac.cn (C.H.); zhaoxian@ntsc.ac.cn (X.Z.)

* Correspondence: wangshanhe@ntsc.ac.cn

Abstract: The time difference of arrival (TDOA) estimation plays a crucial role in emitter localization and time synchronization applications. When time among multiple sensors is synchronized, the TDOAs between the sensors and the emitter can be measured to achieve hyperbolic positioning of the emitter. Conversely, if the positions of both the sensors and the emitter are known, TDOAs can be utilized to synchronize the clocks across the sensors. Given that compressed sensing (CS) can reduce both the sampling rate and data volume, thereby enhancing the efficiency of TDOA estimation, there has been growing interest among researchers in exploring TDOA estimation within the CS framework. In scenarios such as passive positioning, the signals received by sensors are often non-cooperative, and the underlying signal system is unknown, making it difficult to obtain a sparse representation of the signal. This paper introduces an incomplete reconstruction-based TDOA estimation method along with an improved variant. By selecting a partial Fourier transform matrix as the measurement matrix and a Fourier transform matrix as the projection matrix, the orthogonal matching pursuit (OMP) algorithm is employed to reconstruct the compressed measurement data. Through subsequent processing steps, such as conjugate mirroring, TDOA estimation between two signals can be performed. Although the reconstructed signal may substantially differ from the original, the accuracy of TDOA estimation remains reliable. Simulation results demonstrate that when the signal-to-noise ratio (SNR) of the received signal is at least 0 dB and the compressed sampling length exceeds one-tenth of the original signal length, the TDOA estimation error of the proposed method is nearly identical to that of the cross-correlation method.

Keywords: compressed sensing; time difference of arrival; inexact reconstruction

1. Introduction

In the field of signal processing, measuring the time shift between a signal and its delayed version is of considerable importance. For example, in passive localization [1,2] and indoor positioning [3–5], the critical measurement data in many scenarios comprise the time difference of arrival (TDOA) from the transmission or reflection point to two sensors. The more precise the TDOA estimation, the greater the accuracy of the positioning. In time synchronization applications, when the positions of the transmission or reflection point and the two sensors are accurately known, the TDOA measurement, after correcting for distance differences, reflects the clock difference between the local clocks of the two sensors [6]. Similarly, the more accurate the TDOA estimation, the higher the precision of time synchronization.

In practice, TDOA is typically calculated using the cross-correlation function [7,8]. When improving the signal-to-noise ratio (SNR) or bandwidth is not feasible—such as in passive positioning with opportunistic signals—enhancing the accuracy of TDOA estimation necessitates increasing the sampling rate or extending the sampling duration.

However, this approach significantly complicates data sampling, transmission, and storage, particularly when numerous sensors are involved, and the vast amount of data cannot be effectively and promptly transmitted. Consequently, many researchers have proposed the application of compressed sensing techniques to TDOA estimation methods to reduce the volume of sampling data [9–13].

For a signal of length N, denoted as $\boldsymbol{s}$, the measurement result in compressed sensing is given by [14,15]

$$\boldsymbol{y} = \boldsymbol{\Phi}(\boldsymbol{s} + \boldsymbol{n}) + \boldsymbol{e} \tag{1}$$

where $\boldsymbol{\Phi}$ represents the $M \times N$ measurement matrix with $M \ll N$, $\boldsymbol{n}$ represents the additive noise to the signal, and $\boldsymbol{e}$ is the measurement error (noise) at the measurement stage.

TDOA estimation methods based on compressed sensing determine the time difference using a set of measurements $\boldsymbol{y}$ and a measurement matrix $\boldsymbol{\Phi}$. Depending on whether the $\boldsymbol{y}$ is reconstructed, these methods can be broadly classified into two categories.

The first approach involves estimating the TDOA by reconstructing the original signals s_1 and s_2 from the measurement results $\boldsymbol{y}_1$ and $\boldsymbol{y}_2$ of two sensors, followed by calculating the cross-correlation function. This method requires the signal to have a sparse representation in a specific basis, meaning there must be a projection matrix that renders the signal sparse when projected. There are two primary strategies for designing the projection matrix: one involves constructing a set of orthogonal bases based on the mathematical expression of the signal, ensuring the signal has a sparse representation in this orthogonal basis, which necessitates prior knowledge of the received signal system; the other strategy is to use time-shifted versions of the signal as row vectors to construct a projection matrix. In theory, the received signal's sparsity in this projection matrix is 1, but the constructed matrix is typically not orthogonal. Therefore, this approach demands sufficient prior knowledge of the signal and a high SNR for the received signal; otherwise, significant errors in signal reconstruction can lead to an inaccurate TDOA estimation.

The second approach estimates the TDOA without reconstructing the measurement results $\boldsymbol{y}$. In most cases, since $\boldsymbol{\Phi}$ projects the signal from a high-dimensional space to a lower dimension, the measurement results $\boldsymbol{y}$ may disrupt the time–delay relationship between the original signals $\boldsymbol{s}$, rendering the cross-correlation method ineffective for TDOA estimation from $\boldsymbol{y}$. Cao, H. proposed estimating the TDOA by leveraging the phase relationship between the product of the discrete Fourier transform (DFT) coefficients from each sensor and the unknown TDOA, thereby avoiding reconstruction [15]. The phase of the product of the DFT coefficients is proportional to the unknown TDOA, which can then be estimated. When the TDOA between the two signals is small, accurate estimation can be achieved by progressively narrowing the search range and step size; however, larger TDOAs may introduce significant deviations.

Salari, S. highlighted that when the measurement matrix $\boldsymbol{\Phi}$ is a Hadamard matrix, the signal projection preserves the time–domain relationship of the signal [10], allowing the correlation of the measurement results $\boldsymbol{y}$ from two sensors to yield a TDOA estimate. However, when the TDOA is large, the estimate may suffer from periodic ambiguities, with a period corresponding to the product of the number of rows in the measurement matrix $\boldsymbol{\Phi}$ and the sampling rate.

Another method is that one sensor performs compressed sampling on the signal to obtain $\boldsymbol{y}_1$, and another reference sensor samples the signal at the Nyquist sampling rate to obtain $\boldsymbol{s}_2$. Then, after delaying $\boldsymbol{s}_2$ by a certain time τ, and projecting through the measurement matrix, the measurement result is obtained.

$$\boldsymbol{y}_2(\tau) = \boldsymbol{\Phi}(\boldsymbol{s}_2(\tau)) \tag{2}$$

By employing cross-correlation calculations between $\boldsymbol{y}_2(\tau)$ and $\boldsymbol{y}_1$, an estimate of the TDOA is obtained. A limitation of this approach is that one sensor must sample at the Nyquist rate, and the estimation of the TDOA requires a certain amount of search time [16].

This paper proposes an incomplete reconstruction-based TDOA estimation method and its improved variant, which estimate the TDOA without requiring an exact reconstruction of the original signal from the measurement results $\boldsymbol{y}$. The primary focus of this method lies in the accuracy of the TDOA estimation rather than the reconstruction fidelity, thereby eliminating the need for the signal to possess a sparse representation. Although the reconstructed signal may significantly deviate from the original, the TDOA estimation accuracy remains robust.

2. Methods

2.1. Cross-Correlation Method

Assuming that the TDOA between the signal from the transmission or reflection point to the two sensors is constant over the observation interval, the signal model received by Sensor 1 and Sensor 2 is represented as

$$\begin{cases} x_1(t) = a_1 s(t-\tau_1) + n_1(t) \\ x_2(t) = a_2 s(t-\tau_2) + n_2(t) \end{cases} \tag{3}$$

where $s(t)$ is the signal radiated by the transmission or reflection point, the transmission coefficients a_1 and a_2 represent the attenuation experienced by the signal during its propagation path. τ_1 and τ_2 represent the time delay of the signal from transmission to reception, and $n_1(t)$ and $n_2(t)$ are the additive noise received together with the signal, $x_1(t)$ and $x_2(t)$ represent the received signals of the two sensors, and $[0, T)$ represents the observation interval.

To simplify the signal model, taking the signal received by Sensor 1 as a reference, we have

$$\begin{cases} x_1(t) = s(t) + n_1(t) \\ x_2(t) = as(t-D) + n_2(t) \end{cases} \tag{4}$$

where $a = a_2/a_1$ is the ratio of transmission coefficients, and $D = \tau_2 - \tau_1$ is the time difference between signals. Using digital signal processing methods, Equation (4) is written in discrete form after sampling as

$$\begin{cases} x_1(n) = s(n) + n_1(n) \\ x_2(n) = as(n-D) + n_2(n) \end{cases} \tag{5}$$

where N denotes the number of sampling points. We assume that $s(n)$ is a stationary band-limited zero-mean random signal and the noises $n_1(n)$ and $n_2(n)$ are uncorrelated zero-mean white noise. Moreover, the signal is assumed to be independent of the noises, and the cross-correlation function between signals $x_1(n)$ and $x_2(n)$ can be expressed as

$$\begin{aligned} R_{x_1x_2}(\tau) &= E[x_1(n)x_2(n+\tau)] \\ &= aE[s(n)s(n+\tau-D)] + E[s(n)n_2(n+\tau)] \\ &\quad + aE[n_1(n)s(n+\tau-D)] + E[n_1(n)n_2(t+\tau)] \\ &= aR_{ss}(\tau - D) \end{aligned} \tag{6}$$

where $R_{ss}(\tau) = E[s(n)s(n+\tau)]$ represents the autocorrelation function of $s(t)$.

For the ergodic random process, the statistical cross-correlation function can be replaced by the time cross-correlation function; that is, the biased estimation expression of $R_{x_1x_2}(\tau)$ within the observation time is

$$\hat{R}_{x_1x_2}(\tau) = \frac{1}{N} \sum_{n=0}^{N-|\tau|-1} x_1(n)x_2(n+\tau) \tag{7}$$

According to the properties of the autocorrelation function of a stationary random process, when $s(n)$ is not a periodic signal, for any $\tau \neq 0$, we have

$$|R_{ss}(\tau)| \leq |R_{ss}(0)| \tag{8}$$

Thus, $|R_{x_1x_2}(\tau)| \leq |a| \cdot |R_{ss}(0)| = |R_{x_1x_2}(D)|$, and the parameter τ that maximizes $|\hat{R}_{x_1x_2}(\tau)|$ is an estimate of the D, denoted as $\hat{D}$; that is,

$$\hat{D} = \arg\{\max(|\hat{R}_{x_1x_2}(\tau)|)\} \tag{9}$$

Based on the above method, many improved algorithms have been derived, such as the generalized cross-correlation method [7,8,17], adaptive filtering algorithms [18,19], cross-power spectral phase method [20,21], wavelet transform method [22], and other methods.

For the above methods and improved methods, under the conditions of given signals, increasing the amount of integrated data and improving the SNR of the received signal can effectively improve the accuracy of the TDOA estimation. However, improving the SNR of the received signal requires increasing the antenna aperture or other hardware costs, and increasing the amount of integrated data requires increasing the sampling rate or sampling time, which will lead to the need for more data storage space and increase the pressure of data transmission.

2.2. IRCS Method

Write Equation (5) in vector form, and then after performing a compressed measurement with the measurement matrix $\mathbf{\Phi}$, we obtain

$$\begin{cases} \boldsymbol{y}_1 = \mathbf{\Phi}\boldsymbol{x}_1 = \mathbf{\Phi}(\boldsymbol{s}_1 + \boldsymbol{n}_1) \\ \boldsymbol{y}_2 = \mathbf{\Phi}\boldsymbol{x}_2 = \mathbf{\Phi}(a\boldsymbol{s}_2 + \boldsymbol{n}_2) \end{cases} \tag{10}$$

where $\boldsymbol{x}_i = [x_i(1), x_i(2), \cdots, x_i(N)]^{\mathrm{T}}$, $\boldsymbol{n}_i = [n_i(1), n_i(2), \cdots, n_i(N)]^{\mathrm{T}}$, $i = 1,2$, $\boldsymbol{s}_1 = [s(1), s(2), \cdots, s(N)]^{\mathrm{T}}$, and $\boldsymbol{s}_2$ is the vector $\boldsymbol{s}_1$ that is delayed or advanced by $|D|$ sampling points. The superscript T indicates the transpose of a vector.

Without loss of generality, assume that $a = 1$, $D \neq 0$, and denote $\boldsymbol{x} = \boldsymbol{s}_1 + \boldsymbol{n}_1$, $\Delta\boldsymbol{n}_{21} = \boldsymbol{n}_2 - \boldsymbol{n}_1$, with $\boldsymbol{x}^D$ being the vector $\boldsymbol{x}$ that is delayed or advanced by $|D|$ sampling points.

$$\begin{cases} \boldsymbol{y}_1 = \mathbf{\Phi}\boldsymbol{x} \\ \boldsymbol{y}_2 = \mathbf{\Phi}(\boldsymbol{x}^D + \Delta\boldsymbol{n}_{21}) \end{cases} \tag{11}$$

Let

$$\boldsymbol{\theta} = \mathbf{\Psi}\boldsymbol{x} \ or \ \boldsymbol{x} = \mathbf{\Psi}^{-1}\boldsymbol{\theta} \tag{12}$$

That is, the projection of $\boldsymbol{x}$ onto the projection matrix $\mathbf{\Psi}$ is denoted as $\boldsymbol{\theta}$. For opportunity signals or non-cooperative signals, such as non-cooperative signals radiated by satellites, the characteristics of the signal are unknown, and the received signal can be considered as a band-limited signal with a fixed frequency as the center frequency. Its power spectral density is relatively continuous with frequency. Therefore, finding a suitable projection matrix $\mathbf{\Psi}$ to make $\boldsymbol{x}$ have sparsity in its projection is very difficult; therefore, most of the time, $\boldsymbol{\theta}$ is not sparse.

Applying the same reconstruction algorithm to the two equations in Equation (11), since $\boldsymbol{\theta}$ is non-sparse. Theoretically, it is impossible to accurately reconstruct the signal, that is, the reconstructed values $\hat{\boldsymbol{\theta}}_1$ and $\hat{\boldsymbol{\theta}}_2$ differ significantly from $\boldsymbol{\theta}$.

However, if the projection matrix $\mathbf{\Psi}$ is the Fourier transform matrix, then $\boldsymbol{\theta}$ is the Fourier transform of $\boldsymbol{x}$, which represents its spectral information. Rewrite Equation (11) as

$$\begin{cases} \boldsymbol{y}_1 = \boldsymbol{A}\boldsymbol{\theta} \\ \boldsymbol{y}_2 = \boldsymbol{A}(\boldsymbol{\theta}e^{j\frac{2\pi D}{N}} + \Delta\boldsymbol{\theta}_{21}) \end{cases} \tag{13}$$

where, $\boldsymbol{A} = \mathbf{\Phi}\mathbf{\Psi}^{-1}$, $\Delta\boldsymbol{\theta}_{21} = \mathbf{\Psi}\Delta\boldsymbol{n}_{21}$.

When the measurement matrix $\mathbf{\Phi}$ is a partial Fourier transform matrix, it is composed of M rows that are randomly sampled from the Fourier transform matrix, assuming the row index set is denoted by κ. In that case, A is a matrix formed by extracting M rows corresponding to κ from an $N \times N$ identity matrix. In Equation (13), y_1 is a vector composed of M elements selected from θ corresponding to κ, which practically means that M components of x's spectrum have been randomly sampled.

$$y_1 = [\theta(\kappa(1)), \theta(\kappa(2)), \cdots, \theta(\kappa(M))]^T \tag{14}$$

The reconstruction algorithm selects orthogonal matching pursuit (OMP) algorithm [23], with the number of iterations set to M. Considering the characteristics of A and y_1, the k-th ($k \in \kappa$) element of the reconstructed signal $\hat{\theta}_1$ is identical to the k-th element of θ. All other elements of $\hat{\theta}_1$ are zero.

In Figure 1, x is a signal of length 8, θ is its DFT, $M = 4$, y_1 is the measurement result corresponding to $\kappa = \{1, 2, 4, 5\}$, which is composed of the frequency components corresponding to the red line with a dot in θ, and $\hat{\theta}_1$ is the reconstruction result obtained with the OMP algorithm.

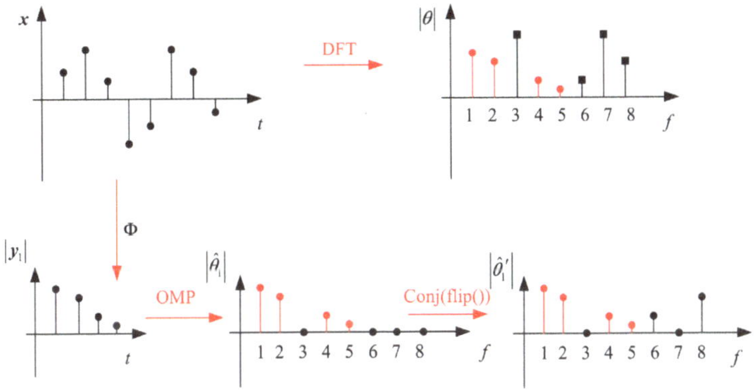

Figure 1. Compressed sampling and reconstruction results.

Similar to $\hat{\theta}_1$, $\hat{\theta}_2$ reconstructed from A and y_2 is

$$\hat{\theta}_2 = \hat{\theta}_1 e^{j\frac{2\pi D}{N}} + \Delta\hat{\theta}_{21} \tag{15}$$

where the k-th ($k \in \kappa$) element of $\Delta\hat{\theta}_{21}$ is identical to the k-th element of $\Delta\theta_{21}$. All other elements of $\Delta\hat{\theta}_{21}$ are zero.

Since the reconstructed $\hat{\theta}_1$ and $\hat{\theta}_2$ still retain some of the original signal's frequency and phase information, the cross-correlation function between signals x_1 and x_2 is

$$\begin{aligned} R_{x_1x_2}^{CS}(\tau) &= IFFT(\hat{\theta}_1 \hat{\theta}_2{}^* e^{j\frac{2\pi\tau}{N}}) \\ &= IFFT(\hat{\theta}_1 \hat{\theta}_1{}^* e^{j\frac{2\pi(\tau - D)}{N}}) + IFFT(\hat{\theta}_1 \Delta\hat{\theta}_{21}{}^* e^{j\frac{2\pi\tau}{N}}) \end{aligned} \tag{16}$$

where $IFFT()$ stands for inverse Fourier transform, and * denotes the complex conjugate.

When $\tau = D$, the first term on the right side of Equation (16) reaches its maximum value. The second term on the right side is the result of the correlation between some components of x_1 and some components of the noise $n_2 - n_1$. Under the assumption that the signal and noise are uncorrelated, the expectation of this part is approximately 0. Therefore, the TDOA estimate value is

$$\hat{D} = \arg\{\max(|R_{x_1x_2}^{CS}(\tau)|)\} \tag{17}$$

The aforementioned method is referred to as the TDOA estimation method, utilizing an inexact reconstruction within the framework of compressed sensing, abbreviated as IRCS.

2.3. EIRCS Method

According to the definition of the DFT, it can be seen that the first half and the second half of the data in $\boldsymbol{\theta}$ are conjugates of each other. When N is an even number, we have

$$\boldsymbol{\theta}(k) = \boldsymbol{\theta}^*(N-k+2),\ k \in [2, N/2] \tag{18}$$

This conclusion can also be drawn from $\boldsymbol{\theta}$ in Figure 1. If the received signal $\boldsymbol{x}$ undergoes a DFT before generating the measurement matrix $\boldsymbol{\Phi}$, and the index set $\boldsymbol{\kappa}_1$ is recorded for the $M-2$ largest elements among the 2nd to $N/2$ elements in $\boldsymbol{\theta}$, let the set $\boldsymbol{\kappa} = \left\{1, \boldsymbol{\kappa}_1, \frac{N}{2}+1\right\}$. Then, $\boldsymbol{y}_1$ retains a minimum of $M-2$ largest frequency components of the received signal. Based on the conjugate symmetry of the DFT, the final reconstruction result obtained from $\boldsymbol{y}_1$ is

$$\hat{\boldsymbol{\theta}}_1{}' = \begin{bmatrix} \hat{\boldsymbol{\theta}}_1 \\ conj(flip(\boldsymbol{\theta}^{\boldsymbol{\kappa}_1})) \end{bmatrix} \tag{19}$$

where $conj(flip(\boldsymbol{\theta}^{\boldsymbol{\kappa}_1}))$ is the vector obtained by reversing the order of elements in $\hat{\boldsymbol{\theta}}_1(\boldsymbol{\kappa}_1)$ and then taking the complex conjugate. As shown by the $\hat{\boldsymbol{\theta}}_1{}'$ curve in Figure 1, $\hat{\boldsymbol{\theta}}_2{}'$ is similar.

For example, in Figure 1, $\boldsymbol{\kappa} = \{1,2,3,5\}$ (at this time $\boldsymbol{\kappa}_1 = \{2,3\}$), which retains more signal energy compared to a randomly selected $\boldsymbol{\kappa} = \{1,2,4,5\}$, leading to a higher estimation accuracy for the TDOA. This method is called the enhanced TDOA estimation method utilizing inexact reconstruction within the framework of compressed sensing, abbreviated as EIRCS.

Algorithm Procedure:

- Initialization: set the signal sampling duration T or sampling length N, as well as the number of rows M of the measurement matrix $\boldsymbol{\Phi}$.
- Construct a partial Fourier transform matrix as the measurement matrix, where $\boldsymbol{\kappa}$ is random (IRCS) or specified (EIRCS), and extract the corresponding M rows from the $N \times N$ identity matrix to form the sensing matrix $\boldsymbol{A}$.
- Compress the signal according to Equation (10) to obtain the measurement values $\boldsymbol{y}_1$ and $\boldsymbol{y}_2$.
- Select the OMP reconstruction algorithm with M iterations to reconstruct the signal, and then obtain $\hat{\boldsymbol{\theta}}_1$ and $\hat{\boldsymbol{\theta}}_2$ (IRCS), or $\hat{\boldsymbol{\theta}}_1{}'$ and $\hat{\boldsymbol{\theta}}_2{}'$ from Equation (19) (EIRCS).
- Estimate the TDOA according to Equation (17).

2.4. Performance Analysis

This section analyzes the IRCS and EIRCS methods in terms of algorithm complexity and Cramér–Rao lower bound (CRLB).

2.4.1. Algorithm Complexity

Algorithm complexity is a critical metric for evaluating the performance of algorithms. In the case of the OMP algorithm, its computational complexity is expressed as $O(kMN)$, where k denotes the sparsity of the reconstructed signal, and M and N represent the number of rows and columns of matrix $\boldsymbol{A}$, respectively. Since the matrix $\boldsymbol{A}$ consists of M rows extracted from an N-dimensional identity matrix, this significantly simplifies many computations.

In the research presented in this paper, we set $k = M$. This allows us to implement the OMP algorithm using a very straightforward method. Taking the reconstruction of signal $\hat{\boldsymbol{\theta}}_1$ using matrix $\boldsymbol{A}$ and vector $\boldsymbol{y}_1$ as an example, we first derive $\boldsymbol{\kappa}$ based on the generation of $\boldsymbol{A}$ and generate an $N \times 1$ zero vector $\boldsymbol{b}$. We then initialize the counter $i = 0$ and perform the following steps:

- Compute $i = i + 1$.
- Assign the i-th element of $\boldsymbol{y}_1$ to the $\boldsymbol{\kappa}(i)$-th element of vector $\boldsymbol{b}$.
- Repeat the above steps until $i = M$.

At this point, the vector $\boldsymbol{b}$ becomes the reconstructed signal $\hat{\boldsymbol{\theta}}_1$. The complexity of this algorithm is only $O(M)$, indicating that the OMP algorithm exhibits a very high computational efficiency in the context of the IRCS or EIRCS.

This simplification not only reduces the computational burden but also enhances the algorithm's practical performance when dealing with high-dimensional data. This is particularly important for signal reconstruction problems that require real-time processing and efficient computation. Therefore, we can assert that the IRCS or EIRCS provides an efficient solution for practical applications.

2.4.2. CRLB

When using communication satellites or broadcasting satellites as radiation sources, the signals they emit are generally random non-periodic signals, which may occupy a wide frequency range and include various modulation methods. For example, in digital satellite TV signals, there are signals from multiple channels, with each channel including the amplitude-modulated signal for the TV image and the frequency-modulated signal for the accompanying audio [24]. After source coding and channel coding, these signals approximate random signals. These random signals are very similar to band-limited white noise (BLWN), which is non-sparse in the time domain and remains a non-sparse signal after a Fourier transformation.

Therefore, we select the signal $s(t)$ and noise $n_1(t)$, $n_2(t)$ as zero-mean independent Gaussian BLWN sequences, while denoting the noise variances, and

$$SNR = 10\log_{10}\frac{\sigma_s^2}{\sigma^2} \tag{20}$$

The CRLB of the TDOA estimation is [14,15]

$$CRLB = \frac{3(1+2R)}{N\pi^2 R^2} \tag{21}$$

where $R = 10^{SNR/10}$.

It is noteworthy that the CLRB of the IRCS or EIRCS is not only influenced by the signal's SNR but also closely related to the parameter M. Generally, under the same conditions, the CLRB of the EIRCS is often not greater than that of the IRCS, which provides the EIRCS with certain advantages in specific applications.

When the value of M is large, the reconstructed signal is relatively close to the original signal, leading to a higher degree of similarity between the two. In this case, the CLRB of both the IRCS and EIRCS can be effectively estimated using Equation (21).

Conversely, when the value of M is small, the energy contained in the reconstructed signal from the original signal is relatively low, resulting in a decrease in effective bandwidth. Under these circumstances, the CLRB of both the IRCS and EIRCS is often higher than the values calculated by Equation (21). This phenomenon indicates that a smaller M value may lead to a deterioration in reconstruction performance, adversely affecting the overall quality of the reconstruction.

3. Experiments

This section conducts simulation experiments on the TDOA estimation using cross-correlation (CC), IRCS, and EIRCS methods. It explores the mean squared error (MSE) of the TDOA estimation values over 100 independent trials under a different SNR (with both signal paths having the same SNR) and different compressed sampling numbers M (for the number of rows in the measurement matrix $\boldsymbol{\Phi}$ of the IRCS and EIRCS methods). The MSE is obtained from averaging 100 independent trials, so that

$$\text{MSE} = \frac{\sum_{i=1}^{100}\left(\hat{\tau}(i) - D\right)^2}{100} \tag{22}$$

where $\hat{\tau}(i)$ is the TDOA estimate of the i trial.

3.1. The MSE with Respect to the SNR

The signal length N is chosen to be 512, $D = 20$, and the SNR range ranges from −10 dB to 30 dB. We explore the MSE of various methods under different SNRs. Figure 2 presents the MSE of the three methods for the TDOA estimation and the CRLB at different SNRs. The horizontal axis represents the signal SNR in dB. For the ease of presentation, the horizontal axis is expressed as $10\log_{10}(\text{MSE})$, with the unit converted to dB (the same below), but this is still referred to as the MSE in this paper. In Figure 2, the values of M are taken in sequence as $M = N/32, N/16, N/8, N/4$. The black dashed line represents the CRLB, the blue line with triangles represents the CC method, the green line with circles represents the IRCS method, and the red line with stars represents the EIRCS method.

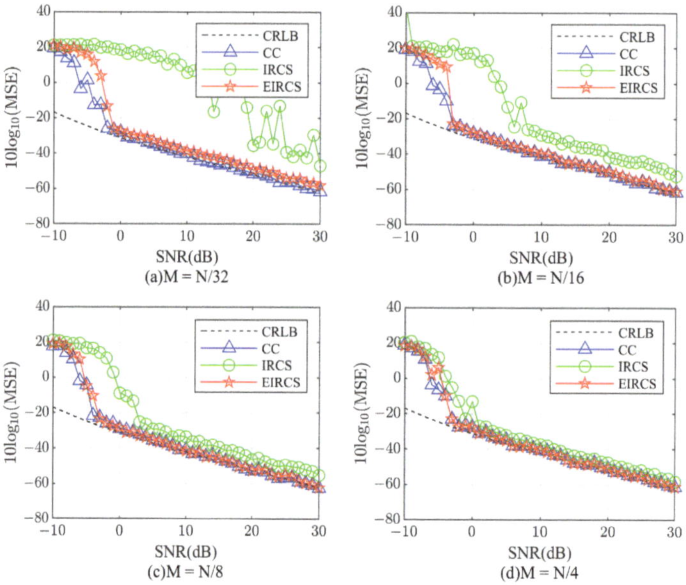

Figure 2. The MSE of the TDOA corresponding to different SNRs.

Figure 2 shows that the MSE of the TDOA estimated by the three methods decreases with the increase in the SNR, and all are above the CRLB curve.

Under the same conditions (SNR > −3 dB), the MSE of CC is generally approximately equal that of the EIRCS, and the EIRCS is less than that of the IRCS. However, as M increases, the IRCS gradually approaches the EIRCS, and the EIRCS gradually approaches CC.

For example, at an SNR of 20 dB, when M takes the values of $N/32$, $N/16$, $N/8$, and $N/4$, the difference in the MSE between the IRCS and EIRCS is 13.8 dB, 8.9 dB, 4.2 dB, and 3.54 dB, respectively, the difference in the MSE between the EIRCS and CC is 2.1 dB, 1.5 dB, 0.3 dB, and 0.01 dB, respectively. When $M = N/2$, the three methods are almost overlapping, which is because the increase in M makes the data used by the three methods to calculate TDOA almost identical.

3.2. The MSE with Respect to the M

At fixed SNRs of −10 dB, 0 dB, 10 dB, and 20 dB, Figure 3 presents the MSE of the TDOA estimation for CC and different M values corresponding to the IRCS and EIRCS methods.

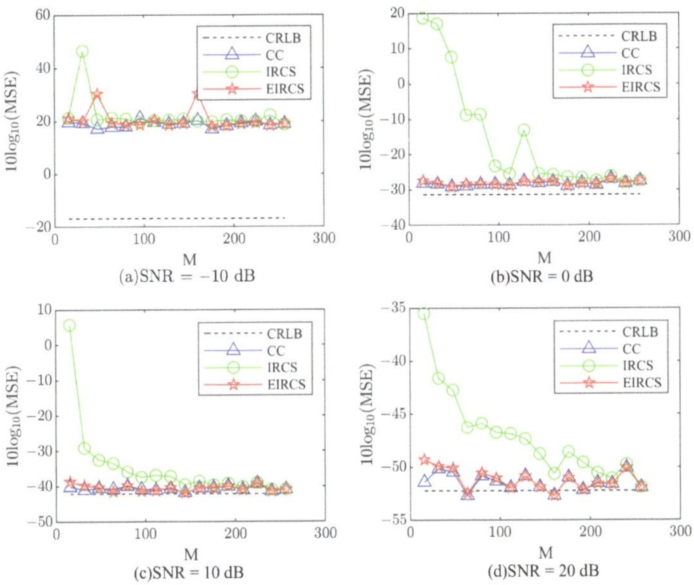

Figure 3. The MSE of the TDOA corresponding to different M values.

Figure 3a indicates that when the SNR is −10 dB, the TDOA estimated from a smaller amount of data has a significant error level. Even at some points, the MSE of the EIRCS exceeds that of the IRCS, such as $M = 48, 160$. The reason is that the noise energy is greater than the true signal. In some cases, the signal reconstructed by the EIRCS method contains more noise relative to the IRCS method, leading to larger time difference estimation errors.

Figure 3b–d show that under the same conditions, the MSE of EIRCS is less than that of IRCS, which also confirms the higher TDOA estimation accuracy of EIRCS mentioned in Section 2.3.

Figure 3 also indicates that when the SNR of the received signal is not lower than 0 dB and $M > 50$ (approximately $N/10$), the performance of the EIRCS is almost identical to CC. This means that despite compressed sensing reducing the sampling data, the estimation accuracy of the TDOA is hardly decreased. In some cases, such as when the SNR = 10 dB and $M = 96$, the MSE of the EIRCS is less than that of CC. This is because the noise is Gaussian white noise, which can be considered to have its spectrum evenly distributed across the signal frequency domain. After reconstruction, most elements of $\hat{\boldsymbol{\theta}}_1$ and $\hat{\boldsymbol{\theta}}_2$ are zero, which can, in some cases, improve the SNR and enhance the accuracy of the TDOA estimation.

Combining Figures 2 and 3, when the SNR is fixed at 0 dB, the CRLB for the TDOA estimation using the IRCS approaches its theoretical value when $M > 100$, and it is significantly greater than the theoretical value when $M < 100$. When the SNR is fixed at 10 dB, M being around 32 is the tipping point where the CRLB approaches and exceeds the theoretical value. Therefore, with different SNRs, the range of M for accurately estimating the CRLB using Equation (21) varies.

4. Conclusions

This paper proposes a TDOA estimation method based on an incomplete reconstruction IRCS and its improved method, EIRCS. Moreover, it provides a brief analysis of their algorithm complexity and the CRLB.

In the IRCS and EIRCS, the measurement matrix $\mathbf{\Phi}$ is a partial Fourier transform matrix, and the projection matrix $\mathbf{\Psi}$ is a Fourier transform matrix. The measurement of the signal is composed of some components of its DFT. After reconstruction using the OMP algorithm, partial frequency information of the signal can be obtained. Although the reconstructed signal differs significantly from the original signal, the TDOA estimation can still be performed. This method does not require a sparse representation of the signal; thus, it has a broader range of application scenarios.

If there is prior knowledge of the signal, by optimizing the generation method of the measurement matrix $\mathbf{\Phi}$, the EIRCS can achieve the same MSE as a traditional cross-correlation method. This is beneficial for reducing the cost of data sampling and storage, improving the efficiency of the TDOA estimation, and it has certain significance for the research of subsequent new methods.

Author Contributions: Conceptualization, S.W. and Y.H.; methodology, S.W.; software, S.W.; validation, Y.X., Y.G. and C.H.; formal analysis, X.Z.; investigation, Y.H.; resources, Y.X.; data curation, Y.G.; writing—original draft preparation, S.W.; writing—review and editing, Y.H.; visualization, Y.G.; supervision, Y.X.; project administration, S.W.; funding acquisition, S.W. All authors have read and agreed to the published version of the manuscript.

Funding: This research was funded by the Basic Research Program of Natural Sciences of Shaanxi Province, grant number 2023-JC-QN-0096).

Data Availability Statement: The original contributions presented in the study are included in the article, further inquiries can be directed to the corresponding author.

Acknowledgments: The authors would like to thank all editors and reviewers for their valuable comments and suggestions.

Conflicts of Interest: The authors declare no conflicts of interest.

References

1. Wang, Z.; Hu, D.; Zhao, Y.; Hu, Z.; Liu, Z. Real-Time Passive Localization of TDOA via Neural Networks. *IEEE Commun. Lett.* **2021**, *25*, 3320–3324. [CrossRef]
2. Pine, K.C.; Pine, S.; Cheney, M. The Geometry of Far-Field Passive Source Localization With TDOA and FDOA. *IEEE Trans. Aerosp. Electron. Syst.* **2021**, *57*, 3782–3790. [CrossRef]
3. Xin, J.; Ge, X.; Zhang, Y.; Liang, X.; Li, H.; Wu, L.; Wei, J.; Bu, X. High-Precision Time Difference of Arrival Estimation Method Based on Phase Measurement. *Remote Sens.* **2024**, *16*, 1197. [CrossRef]
4. Xiao, G.; Dong, Q.; Liao, G.; Li, S.; Xu, K.; Quan, Y. High-Precision Joint TDOA and FDOA Location System. *Remote Sens.* **2024**, *16*, 693. [CrossRef]
5. Khalafalla, M.; Jiang, K.; Tian, K.; Feng, H.; Xiong, Y.; Tang, B. Two-Dimensional Target Localization Approach via a Closed-Form Solution Using Range Difference Measurements Based on Pentagram Array. *Remote Sens.* **2024**, *16*, 1370. [CrossRef]
6. Zhao, K.; Zhao, T.; Zheng, Z.; Yu, C.; Ma, D.; Rabie, K.; Kharel, R. Optimization of Time Synchronization and Algorithms with TDOA Based Indoor Positioning Technique for Internet of Things. *Sensors* **2020**, *20*, 6513. [CrossRef] [PubMed]
7. Azaria, M.; Hertz, D. Time delay estimation by generalized cross correlation methods. *IEEE Trans. Acoust. Speech Signal Process.* **1984**, *32*, 280–285. [CrossRef]
8. Knapp, C.Y.; Carter, G.C. The generalized correlation method for estimation of time delay. *IEEE Trans. Acoust. Speech Signal Process.* **2003**, *24*, 320–327. [CrossRef]
9. Cao, H.; Chan, Y.T.; So, H.C. Maximum Likelihood TDOA Estimation From Compressed Sensing Samples Without Reconstruction. *IEEE Signal Process. Lett.* **2017**, *24*, 564–568. [CrossRef]
10. Salari, S.; Chan, F.; Chan, Y.T.; Read, W. TDOA Estimation With Compressive Sensing Measurements and Hadamard Matrix. *IEEE Trans. Aerosp. Electron. Syst.* **2018**, *54*, 3137–3142. [CrossRef]
11. Daponte, P.; De Vito, L.; Picariello, F.; Rapuano, S.; Remondini, E.; Schettini, C.; Tudosa, I. A Compressed-Sensing System for Radio Spectrum Monitoring and Localization of Noncooperative Sources. *IEEE Trans. Instrum. Meas.* **2024**, *73*, 1–11. [CrossRef]
12. Balestrieri, E.; Vito, L.D.; Picariello, F.; Rapuano, S.; Tudosa, I. A TDoA-based Measurement Method for RF Emitters Localization by Exploiting Wideband Compressive Sampling. In Proceedings of the 2020 IEEE International Instrumentation and Measurement Technology Conference (I2MTC), Dubrovnik, Croatia, 25–28 May 2020; pp. 1–6. [CrossRef]
13. Yang, D.; Li, H.; Peterson, G.D.; Fathy, A. Compressive sensing TDOA for UWB positioning systems. In Proceedings of the 2011 IEEE Radio and Wireless Symposium, Phoenix, AZ, USA, 16–19 January 2011; pp. 194–197. [CrossRef]
14. Cao, H.; So, H.; Chan, Y. Optimum time delay estimation for complex-valued stationary signals. *Signal Process.* **2017**, *131*, 434–440. [CrossRef]

15. Cao, H.; Chan, Y.T.; So, H.C. Compressive TDOA Estimation: Cramér–Rao Bound and Incoherent Processing. *IEEE Trans. Aerosp. Electron. Syst.* **2020**, *56*, 3326–3331. [CrossRef]
16. Chan, Y.; Chan, F.; Rajan, S.; Lee, B. Direct estimation of time difference of arrival from compressive sensing measurements. In Proceedings of the 2015 3rd International Workshop on Compressed Sensing Theory and its Applications to Radar, Sonar and Remote Sensing (CoSeRa), Pisa, Italy, 17–19 June 2015; pp. 273–276. [CrossRef]
17. Carter, G.C. Coherence and time delay estimation. *Proc. IEEE* **2005**, *75*, 236–255. [CrossRef]
18. Jelfs, B.; Sun, S.; Ghorbani, K.; Gilliam, C. An Adaptive All-Pass Filter for Time-Varying Delay Estimation. *IEEE Signal Process. Lett.* **2021**, *28*, 628–632. [CrossRef]
19. Liu, B. Generalized Cross-Correlation Delay Estimation Algorithm Based on Improved LMS Adaptive Filtering. In Proceedings of the 2023 3rd International Conference on Mobile Networks and Wireless Communications (ICMNWC), Tumkur, India, 4–5 December 2023; pp. 1–6. [CrossRef]
20. Zhao, Z.; Hou, Z. The generalized phase spectrum method for time delay estimation. *Chin. J. Acoust.* **1985**, *10*, 201–215.
21. Nikias, C.L.; Pan, R. Time delay estimation in unknown Gaussian spatially correlated noise. *Trans. Acoust. Speech Signal Process.* **1988**, *36*, 1706–1714. [CrossRef]
22. Zhang, L.; Deng, Z.; Han, K.; Li, X.; Zhang, C. TDOA Location Algorithm Based on Adaptive Threshold Function Wavelet Transform. In *Lecture Notes in Electrical Engineering, Proceedings of the China Satellite Navigation Conference (CSNC 2022) Proceedings, Beijing, China, 22–25 May 2022*; Springer: Singapore, 2022; pp. 442–451.
23. Tropp, J.; Gilbert, A. Signal recovery from random measurements via orthogonal matching pursuit. *Trans. Inf. Theory* **2007**, *53*, 4655–4666. [CrossRef]
24. Wang, S. Research on High Precision Time Transfer Method Based on Non-Cooperative Satellite. Ph.D. Thesis, National Time Service Center, Chinese Academy of Sciences, Beijing, China, 2021.

remote sensing

MDPI

Technical Note

Elevation Angle Estimation in a Multipath Environment Using MIMO-OFDM Signals

Yeo-Sun Yoon

Department of Defense System Engineering, Sejong University, 209 Neungdong-ro, Gwangjin-gu, Seoul 05006, Republic of Korea; ysyoon@sejong.ac.kr

Abstract: It is challenging to estimate the elevation angle of low-altitude targets due to the multipath effect. Various signal processing techniques have been proposed to mitigate these effects, including the use of multi-frequency signals as opposed to single narrowband signals. However, the optimal type of multi-frequency signals and their effective utilization have not been thoroughly explored. Compressive sensing was also proposed as a high-resolution angle estimation method. But, that was conducted with narrowband signals. In this paper, we employ MIMO-OFDM signals along with a block sparse Bayesian learning fast marginalized (BSBL-FM) method. This combination allows for the effective processing of multi-frequency signals and provides high resolution estimates. The MIMO-OFDM approach represents radar signals in a block-sparse matrix form, and the BSBL-FM method leverages this sparsity to achieve high-resolution angle estimates. Simulation results demonstrate that our method can accurately estimate angles at extremely low altitudes where the elevation angle is less than 1 degree.

Keywords: multipath; low-altitude targets; OFDM radar; compressive sensing

1. Introduction

Multi-path refers to the phenomenon whereby radio signals arrive at an antenna via more than one path. In wireless communications, this is common since the communication units are usually in a dense environment like an urban area. Although it compromises the communication quality, it is not a severe problem since there are many ways to mitigate the degradation due to the multi-path. However, in radar applications, this very simple physical phenomenon could disrupt the entire function of the radar. Radars generally assume that signals propagate directly from the target, estimating the angle based on this direct path. However, multi-path causes reflections from a single target to arrive from multiple directions. When the target flies at a low altitude, these angles of arrival are so close to each other that resolving them becomes difficult. Higher resolution typically requires either longer antenna arrays or high-resolution algorithms. However, using longer antenna arrays is often impractical, especially for low-frequency radars. High-resolution angle estimation methods such as MUSIC can be used for better estimates [1,2]. But, this subspace-based method performs poorly when signals are correlated. Although there are smoothing methods to address this correlation problem [3], it can only improve the estimation performance, and is not effective when two signals are fully correlated and very close in angle, as in low-altitude target scenarios.

Another approach to achieve high-resolution estimation is to apply compressive sensing (CS) [4]. CS is a linear optimization method which can reconstruct data from a very limited number of samples [5]. In the context of angle estimations, it is used to find angles from a dictionary which involves a set of angles discretized at intervals smaller than the Rayleigh resolution [6]. By applying CS, it is possible to resolve very close angles that conventional beamforming methods cannot resolve. While this approach outperforms previous methods, its effectiveness diminishes when the target's altitude is extremely low.

At such low altitudes, the difference between the direct path angle and the multipath angles becomes too small to distinguish. Even with MIMO radars [7], the CS cannot resolve very close signals. Recently, Chen, et al, proposed a beamspace method to estimate elevation angles [8]. While this method demonstrates improved performance over other techniques, it requires an initial value, and the estimation error increases when the elevation angle is less than one degree.

MIMO radars have been used to address the multipath effects in other approaches too [9,10]. Shi, et al. proposed a multi-static MIMO radar where widely separated transmitters emit orthogonal signals, which are received by a single receiver [9]. In another study [10], they focused primarily on the power efficiency and the tracking process with a single-frequency signal as a radar waveform. But, both of them exploited the divergence among the received signals obtained from multi-static MIMO radars and this is not suitable for mono-static radars.

The use of multiple frequencies, instead of a single-frequency signal, was another method to improve angle estimation [11]. In their approach, the radar changes the frequency of the signal according to the target's elevation angle estimated in the previous frame in order to improve the results of the refined maximum likelihood algorithm. However, this method does not fully exploit the advantages of multi-frequency signals as it uses a single frequency for each transmission and process.

In this paper, a multi-frequency signal with CS methodology is proposed to improve elevation angle estimation for extremely low-altitude targets. The recent proliferation of advanced drones has increased the importance of detecting and tracking very low-flying targets. MIMO-OFDM signals are employed as a multi-frequency signal. While previous approaches took advantages of OFDM signals in radar applications, in general, cases [12,13], we chose this waveform to estimate two very close elevation angles in multi-path scenarios. For the angle estimation method, the block sparse Bayesian learning fast marginalized (BSBL-FM) algorithm was used. This method was originally proposed to solve the sparse approximation problem [14]. Then, it was employed to estimate physiological signals in telemonitoring systems [15]. In a recent paper [16], this algorithm was proposed to estimate target location in JCR (joint communication and radar) systems. The BSBL-FM algorithm was used to find the row-sparse matrix [14]. It converts the optimization problem into a Bayesian term and then applies a maximum a posteriori probability (MAP) algorithm. The multi-frequency components of the MIMO-OFDM signal enable the received signal to be represented by a product of an overdetermined matrix and a row-sparse matrix. If BSBL-FM is applied to this matrix, the elevation angle estimate is obtained from the coherent processing of all the subcarrier signals, and it would be a better estimate. We will discuss how to apply this method, and the conditions under which this method can be effectively applied are also analyzed.

Estimating the location of targets using MIMO-OFDM signals was also proposed in [17,18]. Both papers exploited the structure of the MIMO-OFDM signal and applied a tensor decomposition method in order to estimate angles, doppler frequencies, and time-delays. In this paper, we will apply a relatively simple algorithm without using tensor decomposition. Although the simplicity is not directly related to fast processing time, our method provides the angle estimates in a single-step process.

The remainder of this document is structured as follows: In Section 2, the multipath environment is discussed, followed by Section 3, where the signal model and the estimation method are introduced. In Section 4, the validity of the proposed method is discussed. Simulation results and conclusions are given in Sections 5 and 6, respectively.

2. Multipath Environment

When a target flies at a very low altitude, the radar signal can return to the radar via the following two paths: the direct path and an indirect path reflected off the ground or sea surface. The indirect path has a different amplitude and phase from the direct one due to the reflection coefficient and the different time delay. On rough surfaces, only diffuse reflections occur, making the effect of the indirect path negligible. However, on smooth surfaces, such as calm water or desert terrain, specular reflection produces a reflected signal strong enough to interfere with the direct signal, hindering the accurate elevation angle estimation of the target.

Figure 1 illustrates the geometry of the multipath environment. The angle θ_0 denotes the known transmission angle, while θ_1 and θ_2 represent the elevation angles for the direct and the reflected paths, respectively. These elevation angles can be determined using geometric equations. The angles θ_1 and θ_2 are given as follows:

$$\theta_1 = \sin^{-1}\left(\frac{h_t - h_r}{R}\right), \tag{1}$$

$$\theta_2 = -\sin^{-1}\left(\frac{h_t + h_r}{R}\right). \tag{2}$$

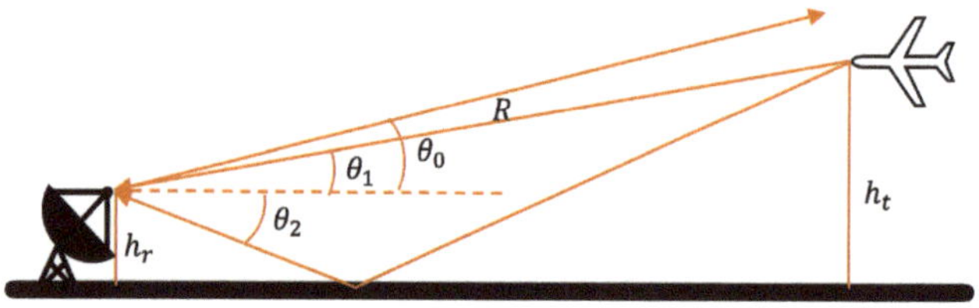

Figure 1. Multipath geometry.

In this paper, it is assumed that the multipath occurs only in the returning signal. So, the total path length of the multipath is given as follows:

$$R_{mul} = R_{dir} + \delta R, \tag{3}$$

where $R_{dir} = 2R$ is the path length of the direct path and δR is the additional path length for the multipath, as follows:

$$\delta R = R\left(\frac{\cos\theta_1}{\cos\theta_2} - 1\right). \tag{4}$$

Note that this is the worst case scenario for angle estimation as both the path length difference and the attenuation due to the reflection coefficient are minimal. This makes it difficult to distinguish the direct signal from the multipath signal.

The reflection coefficient depends on the polarization of the radar signal. For horizontal polarization, the reflection coefficient, Γ_h is as follows:

$$\Gamma_h = \frac{\sin\phi - \sqrt{\epsilon_c - \cos^2\phi}}{\sin\phi + \sqrt{\epsilon_c - \cos^2\phi}}, \tag{5}$$

where ϵ_c is the complex dielectric constant and ϕ is the incident angle, which is the same as $|\theta_2|$ in Figure 1 [19]. The dielectric constant is given as follows:

$$\epsilon_c = \epsilon_r - j60\lambda\sigma_e, \tag{6}$$

where ϵ_r is the relative dielectric constant, σ_e is the conductivity of the surface material, and λ is the wavelength.

When the target distance R is significantly larger than the heights of both the antenna and the target, the elevation angles θ_1 and θ_2 are very close to each other. In such cases, it is not separable using conventional angle estimation algorithms. Furthermore, for small ϕ, the following applies:

$$\Gamma_h \approx \frac{-\sqrt{\epsilon_c - 1}}{\sqrt{\epsilon_c - 1}} = -1. \tag{7}$$

In this case, the reflected signal is as strong as the direct signal with an opposite sign. It can almost cancel out the direct one. This causes significant challenges for monopulse angle estimation methods, as the received signal is a combination of two signals coming from different angles with a nearly identical power. Consequently, this results in considerable estimation errors.

In this paper, a radar with a vertical linear antenna array for both transmission and reception is considered. A single target is approaching the radar at a very low but constant altitude over calm water (sea). The elevation angle of the target, θ_1, is of particular interest since it tells the target's altitude.

Notations: Throughout this paper, bold lowercase letters represent vectors, while uppercase letters denote matrices. The symbols $(\cdot)^T$ and $(\cdot)^H$ indicate transpose and conjugate transpose, respectively. $[\cdot]_{n,n}$ denotes the n-th diagonal element in the matrix and diag$\{\cdot\}$ denotes the diagonal matrix with a diagonal element vector. The symbol $\otimes$ denotes the kronecker product, and $\|\cdot\|_2$ and $\|\cdot\|_F$ represent the $\ell - 2$ norm and the Frobenius norm, respectively.

3. Signal Model and Estimation Method

Suppose that a radar has an M-element vertical uniform linear array (ULA). A MIMO-OFDM signal with N subcarriers at the m-th antenna for a single symbol period is given as follows:

$$x_m(t) = \sum_{n=0}^{N-1} x_{n,m} e^{j2\pi(f_0+n\Delta f)t}, m = 0, 1, \ldots, M-1 \tag{8}$$

where f_0 is the carrier frequency, Δf is the subcarrier frequency interval, and $x_{n,m}$ is the signal transmitted at the m-th antenna for the n-th subcarrier. For each subcarrier, the M-element vector is transmitted from M antennas. Each antenna transmits one element of the vector during a symbol period. The signals transmitted from each antenna do not need to be orthogonal, unlike conventional MIMO radars where the orthogonality plays an important role in increasing spatial resolution. The total bandwidth B of the signal is equal to $(N-1)\Delta f$.

Assuming a one-way multipath reflection from the Earth surface, as illustrated in Figure 1, the received baseband signal vector of the n-th subcarrier and the k-th OFDM block is as follows:

$$\mathbf{y}_{n,k} = b_1 \mathbf{a}(\theta_1)\mathbf{a}^T(\theta_0)\mathbf{x}_{n,k} e^{-j2\pi n\tau_1 \Delta f} e^{j2\pi f_{d,1} k T_s} + b_2 \mathbf{a}(\theta_2)\mathbf{a}^T(\theta_0)\mathbf{x}_{n,k} e^{-j2\pi n\tau_2 \Delta f} e^{j2\pi f_{d,2} k T_s} \tag{9}$$

where b_i and θ_i are the reflection coefficients and the elevation angle of the i-th path, respectively. τ_i and $f_{d,i}$ are the corresponding time delay and the Doppler frequency. The vector $\mathbf{x}_{n,k}$ is an $M \times 1$ signal vector for the n-th subcarrier and the k-th symbol period.

The OFDM symbol period is denoted by T_s and the vector $\mathbf{a}(\theta)$ is an array manifold vector given by the following:

$$\mathbf{a}(\theta) = \left[1, e^{j\pi \sin\theta}, \ldots, e^{j\pi(M-1)\sin\theta}\right]^H. \tag{10}$$

In the above equation, it is assumed that the signal bandwidth B is much smaller than the carrier frequency, f_0, and the antenna spacing is half the wavelength of the carrier frequency. This narrowband assumption, however, may not be suitable for OFDM signals. A more detailed discussion on this will be provided in Section 4. For now, we assume that the narrowband array manifold is applicable.

The reflection coefficient, b_1 is the target's RCS and b_2 is the target's RCS multiplied by the surface reflection coefficient, Γ, as in Equation (5). τ_1 is the two-way travelling time of the signal from the antenna to the target via a direct path and $\tau_2 = \tau_1 + \delta R/C$, where C is the speed of the light. The signal $\mathbf{y}_{n,k}$ can be rewritten in a matrix-vector format as follows:

$$\mathbf{y}_{n,k} = \begin{bmatrix}\mathbf{a}(\theta_1) & \mathbf{a}(\theta_2)\end{bmatrix}\mathbf{C}_n\mathbf{D}_k\begin{bmatrix}\mathbf{a}^T(\theta_0)\\ \mathbf{a}^T(\theta_0)\end{bmatrix}\mathbf{x}_{n,k} \tag{11}$$

where $\mathbf{C}_n$ and $\mathbf{D}_k$ are 2×2 diagonal matrices such as the following:

$$\begin{aligned}\mathbf{C}_n &= \text{diag}\left\{e^{-j2\pi n\tau_1\Delta f}, e^{-j2\pi n\tau_2\Delta f}\right\}\\ \mathbf{D}_k &= \text{diag}\left\{b_1 e^{j2\pi f_{d,1}kT_s}, b_2 e^{j2\pi f_{d,2}kT_s}\right\}.\end{aligned}$$

When the angle difference between the direct path and the multipath is small, it can be assumed that both Doppler frequencies are identical. Thus, $f_d = f_{d,1} = f_{d,2}$ and $\mathbf{D}_k$ can be simplified accordingly.

$$\mathbf{D}_k = \text{diag}\left\{b_1 e^{j2\pi f_d kT_s}, b_2 e^{j2\pi f_d kT_s}\right\}. \tag{12}$$

Define an $N_p \times N_p$ diagonal matrix $\tilde{\mathbf{C}}_n$ whose diagonal elements are as follows:

$$\tilde{\mathbf{C}}_n = \text{diag}\{1, e^{-j2\pi n\alpha}, \cdots, e^{-j2\pi n(N_p-1)\alpha}\}. \tag{13}$$

The parameter α is the resolution or the grid size for the delay estimation and N_p is the total number of grids. Suppose that $\tau_1\Delta f = p\alpha$ and $\tau_2\Delta f = q\alpha$ for $q > p$. Then, Equation (11) can be rewritten in a sparse form as follows:

$$\mathbf{y}_{n,k} = \tilde{\mathbf{A}}(\boldsymbol{\theta})\tilde{\mathbf{C}}_n\tilde{\mathbf{D}}_k\tilde{\mathbf{U}}^T\mathbf{x}_{n,k} \tag{14}$$

The matrix $\tilde{\mathbf{A}}(\boldsymbol{\theta})$ is an $M \times N_p$ matrix such that

$$\tilde{\mathbf{A}}(\boldsymbol{\theta}) = \begin{bmatrix}0 & \cdots & \mathbf{a}(\theta_1) & \cdots & \mathbf{a}(\theta_2) & \cdots & 0\end{bmatrix}, \tag{15}$$

where the p-th column is $\mathbf{a}(\theta_1)$ and the q-th column is $\mathbf{a}(\theta_2)$ and zero elsewhere. $\boldsymbol{\theta}$ is the vector of elevation angles as $\boldsymbol{\theta} = [\theta_1, \theta_2]$. The matrix $\tilde{\mathbf{D}}_k$ is an $N_p \times 2$ matrix, where its elements at the location of the signals' time delays are $e^{j2\pi kf_dT_s}$ and zero elsewhere. That is given as:

$$\tilde{\mathbf{D}}_k = \begin{bmatrix}0 & \cdots & b_1 e^{j2\pi f_d kT_s} & \cdots & \cdots & 0\\ 0 & \cdots & \cdots & b_2 e^{j2\pi f_d kT_s} & \cdots & 0\end{bmatrix}^T. \tag{16}$$

The matrix $\tilde{\mathbf{U}}$ is as follows:

$$\tilde{\mathbf{U}} = [\mathbf{a}(\theta_0), \mathbf{a}(\theta_0)]. \tag{17}$$

The transpose of Equation (14) can be rewritten as follows:

$$\mathbf{y}_{n,k}^T = \mathbf{x}_{n,k}^T\left(\mathbf{c}_n^T \otimes \mathbf{I}_M\right)\mathbf{V}\tilde{\mathbf{A}}^T(\boldsymbol{\theta}). \tag{18}$$

Here, $\mathbf{I}_M$ denotes the $M \times M$ identity matrix and $\mathbf{c}_n$ is the vector of the diagonal element in $\tilde{\mathbf{C}}_n$ as follows:

$$\mathbf{c}_n = \left[1, e^{-j2\pi n\alpha}, \ldots, e^{-j2\pi n(N_p-1)\alpha}\right]^T. \tag{19}$$

The matrix $\mathbf{V}$ is a $MN_P \times N_p$ block sparse matrix. The p-th column of the p-th block matrix is as follows:

$$(\mathbf{V}_p)_p = b_1 \exp\{j2\pi f_d k T_s\}\mathbf{a}(\theta_0) \tag{20}$$

Likely, the q-th column of the q-th block matrix is as follows:

$$(\mathbf{V}_q)_q = b_2 \exp\{j2\pi f_d k T_s\}\mathbf{a}(\theta_0). \tag{21}$$

All other columns are zeros. Then, $\mathbf{V}\tilde{\mathbf{A}}^T(\boldsymbol{\theta})$ is as follows:

$$\mathbf{V}\tilde{\mathbf{A}}^T(\boldsymbol{\theta}) = \begin{bmatrix} \mathbf{0} \\ \vdots \\ b_1 \exp\{j2\pi f_d k T_s\}\mathbf{a}(\theta_0)\mathbf{a}^T(\theta_1) \\ \vdots \\ b_2 \exp\{j2\pi f_d k T_s\}\mathbf{a}(\theta_0)\mathbf{a}^T(\theta_2) \\ \vdots \\ \mathbf{0} \end{bmatrix} \tag{22}$$

Since the above sparse block matrix does not depend on the subcarrier index n, a data matrix $\mathbf{Y}_k$ can be defined as follows:

$$\begin{aligned} \mathbf{Y}_k &= \begin{bmatrix} \mathbf{y}_{1,k} & \cdots & \mathbf{y}_{N,k} \end{bmatrix}^T \\ &= \mathbf{W}\mathbf{V}\tilde{\mathbf{A}}^T(\boldsymbol{\theta}). \end{aligned} \tag{23}$$

In the above equation, $\mathbf{W}$ is a $N \times MN_p$ sensing matrix such as the following:

$$\mathbf{W} = \begin{bmatrix} \mathbf{x}_{1,k}^T(\mathbf{c}_1^T \otimes \mathbf{I}_M) \\ \vdots \\ \mathbf{x}_{N,k}^T(\mathbf{c}_N^T \otimes \mathbf{I}_M) \end{bmatrix}. \tag{24}$$

The matrix $\mathbf{Y}_k$ is a product of a sensing matrix and a sparse matrix. The problem now is finding a block sparse matrix $\mathbf{G}$ such that $\mathbf{Y} = \mathbf{W}\mathbf{G}$. We can apply block sparse Bayesian learning-fast marginalized (BSBL-FM) algorithm [14,16] to find $\mathbf{G}$ by solving the following:

$$\min \|\mathbf{G}\|_0 \text{ s.t. } \|\mathbf{Y} - \mathbf{W}\mathbf{G}\|_F^2 < \eta. \tag{25}$$

The parameter η is a value to be chosen according to the noise level. The BSBL-FM algorithm reconstructs the sparse matrix by reformulating the above equation in Bayesian terms, transforming it into a Gaussian likelihood function [14]. Then Equation (25) is transformed as follows:

$$\begin{aligned} \mathbf{G}_0(\zeta) &= \arg\max_{\mathbf{G}} p(\mathbf{G}|\mathbf{Y}) \\ &= \arg\max_{\mathbf{G}} p(\mathbf{Y}|\mathbf{G})p(\mathbf{G}) \end{aligned} \tag{26}$$

where ζ is a parameter which balances the estimation quality and the sparsity. $p(\mathbf{Y}|\mathbf{G})$ is a Gaussian probability density function obtained from the following:

$$p(\mathbf{Y}_i|\mathbf{G}_i) = (2\pi\sigma^2)^{-N/2} \exp\left(-\frac{1}{2\sigma^2}\|\mathbf{Y}_i - \mathbf{W}\mathbf{G}_i\|_2^2\right),\ i = 1, \ldots, M \tag{27}$$

where $\mathbf{Y}_i$ and $\mathbf{G}_i$ are the i-th column vector of $\mathbf{Y}$ and $\mathbf{G}$, respectively. The algorithm to obtain $\mathbf{G}_0$ can be found in [14]. Since $\mathbf{G} = \mathbf{V}\tilde{\mathbf{A}}^T(\boldsymbol{\theta})$, there will be two nonzero block matrices in

$\mathbf{G}_0$. Let $\mathbf{B}$ be the nonzero block matrix in $\mathbf{G}_0$. Then, the elevation angle can be calculated as follows:

$$\sin(\theta) \approx \frac{1}{\pi}\angle\left(\sum_{r=1}^{M-1} \mathbf{B}_r^* \mathbf{B}_{r+1}\right), \tag{28}$$

where $\mathbf{B}_r$ is the r-th column vector of $\mathbf{B}$.

Figure 2 summarizes the process of the proposed method.

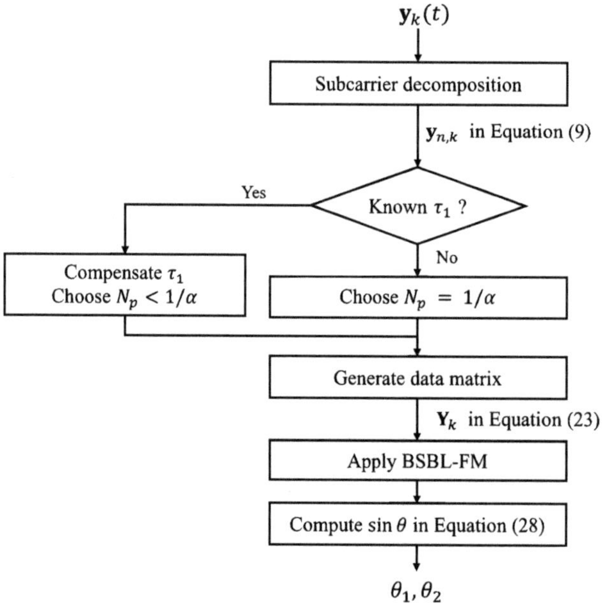

Figure 2. Flowchart for the proposed elevation angle estimation method.

4. Validity of the Signal Model

In the previous section, it was implied that the signal could be approximated as a narrowband signal. This section explores specific conditions under which this assumption is valid.

The OFDM-MIMO signal transmitted from the m-th antenna is $x_m(t)$ for $m = 0, 1, \ldots,$ $M-1$ is given as follows:

$$x_m(t) = \sum_{n=0}^{N-1} x_{m,n} e^{j2\pi(f_0+n\Delta f)t}. \tag{29}$$

Then, the signal at the target location is as follows:

$$s(t) = \sum_{m=0}^{M-1} x_m\left(t - \tau - \frac{md\sin\theta_0}{C}\right), \tag{30}$$

where τ is the signal traveling time from the first antenna to the target and d is the antenna displacement. The received signal at the l-th antenna is given as follows:

$$\begin{aligned} y_l(t) &= s\left(t - \tau - \frac{ld\sin\theta_1}{C}\right) \\ &= \sum_{m=0}^{M-1} x_m\left(t - 2\tau - \frac{md\sin\theta_0}{C} - \frac{ld\sin\theta_1}{C}\right). \end{aligned} \tag{31}$$

For simplicity, assume that only the m-th antenna transmitted the signal and $\theta_1 \approx \theta_0$. Then,

$$\begin{aligned} y_l(t) &= x_m\left(t - 2\tau - \frac{(l+m)d\sin\theta_1}{C}\right) \\ &= \sum_{n=0}^{N-1} x_{m,n} e^{j2\pi(f_0+n\Delta f)(t-2\tau-\frac{(l+m)d\sin\theta_1}{C})}. \end{aligned} \tag{32}$$

The down-converted signal is

$$\begin{aligned} \bar{y}_l(t) &= y_l(t)e^{-j2\pi f_0 t} \\ &= e^{-j2\pi f_0(2\tau+(l+m)\beta)} \sum_{n=0}^{N-1} x_{m,n} e^{j2\pi n\Delta f(t-2\tau-(l+m)\beta)}, \end{aligned} \tag{33}$$

where $\beta = d\sin\theta_1/C$. After sampling with the sampling interval T_f,

$$\begin{aligned} \bar{y}_l[k] &= \bar{y}_l(t)|_{t=kT_f} \\ &= e^{-j2\pi f_0(2\tau+(l+m)\beta)} \sum_{n=0}^{N-1} x_{m,n} e^{j2\pi n\Delta f(kT_f-2\tau-(l+m)\beta)}. \end{aligned} \tag{34}$$

Here, the number of samples is the same as the number of the subcarrier, N. The s-th subcarrier element of $\bar{y}_l[k]$ is obtained as follows:

$$\begin{aligned} Y_{l,s} &= \sum_{k=0}^{N-1} \bar{y}_l[k]e^{-j2\pi\frac{sk}{N}} \\ &= e^{-j2\pi f_0(2\tau+(l+m)\beta)} \sum_{k=0}^{N-1}\sum_{n=0}^{N-1} x_{m,n} e^{j2\pi n\Delta f(kT_f-2\tau-(l+m)\beta)} e^{-j2\pi\frac{sk}{N}}. \end{aligned} \tag{35}$$

For $T_f = 1/(N\Delta f)$,

$$Y_{l,s} = e^{-j2\pi f_0(2\tau+(l+m)\beta)} \sum_{k=0}^{N-1}\sum_{n=0}^{N-1} x_{m,n} e^{j2\pi\frac{k(n-s)}{N}} e^{-j2\pi n\Delta f(2\tau+(l+m)\beta)} \tag{36}$$

The summation is nonzero only for $n = s$. Then,

$$\begin{aligned} Y_{l,s} &= e^{-j2\pi f_0(2\tau+(l+m)\beta)} \sum_{k=0}^{N-1} x_{m,s} e^{-j2\pi s\Delta f(2\tau+(l+m)\beta)} \\ &= e^{-j2\pi f_0(2\tau+(l+m)\beta)} e^{-j2\pi s\Delta f(2\tau+(l+m)\beta)} N x_{m,s} \end{aligned} \tag{37}$$

In order to apply the narrowband assumption, the phase term that depends on the subcarrier number s and the array element number l must vanish. In other words, the second phase term in the second exponential of the above equation should be negligible, as follows:

$$s\Delta f(l+m)\beta = s\Delta f(l+m)\frac{d\sin\theta_1}{C} \ll 1. \tag{38}$$

Since $d = \lambda/2 = C/2f_0$,

$$s\Delta f(l+m)\beta = s\Delta f(l+m)\frac{\sin\theta_1}{2f_0} \tag{39}$$

The maximum value of the above phase term is for $\sin\theta_1 = 1, s = N-1$, and $l = m = N-1$.

$$s\Delta f(l+m)\beta \leq (N-1)\frac{B}{f_0} \ll 1 \tag{40}$$

Note that $B = (N-1)\Delta f$ is the total bandwidth of the signal. Therefore, when the signal bandwidth to the center frequency ratio is very small, the above phase term can be ignored. Then,

$$Y_{l,s} = e^{-j2\pi f_0(2\tau+(l+m)\beta)} e^{-j2\pi s\Delta f 2\tau} N x_{m,s}. \quad (41)$$

Typically, the narrowband condition is defined by the ratio of the signal bandwidth to the center frequency. In this case, the ratio is multiplied by $N-1$, necessitating an even narrower signal for the proposed method to be applicable. However, for low-altitude targets, $\sin\theta_1 \ll 1$ due to small θ_1. For example, when the antenna height is 20 m and the target height is 90 m and 20 km away, the elevation angle is 0.13 degree ($\sin\theta_1 \approx 0.0023$). For $f_0 = 3$ GHz, $\Delta f = 100$ MHz, and $N = 20$, the phase term is as small as

$$(N-1)\frac{B \cdot (0.0023)}{f_0} \approx 2.7 \times 10^{-2}. \quad (42)$$

Therefore, the assumption in Equation (40) for the narrowband array response is valid.

The parameters α and N_p in Equation (13) should also be properly chosen for the algorithm to be valid. The value of α represents the grid size for the time-delay estimation. It should be small to achieve a higher resolution and, at the same time, reduce off-grid error [20]. The product $N_p\alpha$ is the maximum value of $\tau_i\Delta f$ that can be estimated. Since the phase of the exponential to be estimated lies between 0 and 2π, $N_p\alpha$ should be 1 to cover the full range from 0 to 1. However, when the upper limit of $\tau_i\Delta f$ is available, N_p can be chosen to be less than $1/\alpha$ to reduce the size of the matrix $\tilde{\mathbf{C}}_n$ and the processing time. The value of τ_1, which is the two-way traveling time of the signal between radar and target, is not zero. However, once it is estimated by preprocessing or from the previous dwell, it can be compensated accordingly. Then, τ_1 becomes zero and $\tau_2 = \delta R/C$.

$$\tau_2\Delta f = \frac{\delta R}{C}\Delta f = \frac{\delta R}{\lambda_{\Delta f}}, \quad (43)$$

where $\lambda_{\Delta f}$ is the wavelength of the subcarrier frequency interval Δf. If δR is less than half the wavelength of Δf, the following applies:

$$\frac{\delta R}{\lambda_{\Delta f}} < 0.5 = N_p\alpha. \quad (44)$$

Then, $N_p = 1/2\alpha$ can be used, and it causes the size of $\tilde{\mathbf{C}}_n$ to be halved. If τ_1 is not compensated, there is no guarantee that $\delta R/\lambda_{\Delta f}$ will be less than 0.5. In this case, $N_p = 1/\alpha$ should be used. Although the estimated value of τ_1 has errors and might not be fully compensated, if the error is sufficiently small, it is still possible to use $N_p < 1/\alpha$. On the other hand, given a fixed α, a larger N_p is preferable, as it reduces the coherency between the column vectors of matrix $\mathbf{W}$ in Equation (24), thereby decreasing the estimation error [21]. The value of N_p should be determined to balance estimation accuracy and processing time.

5. Simulations

A computer simulation was conducted to show the performance of the proposed method. Since the signal used in this method differs from those in previous approaches, direct comparisons are not feasible. Instead, we compared the performance of our method with that of the approach in [7] to show the advantage of the MIMO-OFDM signal and BSBL-FM algorithm over CS with narrowband signals. In [7], Liu, et al. also used compressive sensing algorithm for a MIMO radar but with narrowband signals.

In this simulation, very calm sea surface is considered using the appropriate reflection coefficients. A single target at an altitude of 90 m approaches the radar, moving from a distance of 20 km to 8 km at a speed of 20 m/s (72 km/h). Although the target's speed does not directly impact the elevation angle estimation, the scenario assumes a drone as the target, and the speed is therefore set to typical drone speed. A vertical uniform linear array (ULA) with 20 antenna elements is used for both transmission and reception, with the highest antenna element positioned 20 m above the sea surface, simulating a radar on a frigate ship. Therefore, the height difference between the target and the antenna is 70 m. The carrier frequency f_0 is 3 GHz, and the antenna element spacing is 0.05 m, corresponding to half the wavelength of f_0. Figure 3 shows the elevation angles of both the direct and the reflected signals as a function of the distance from the antenna. The elevation angle in this simulation scenario remains below 1 degree at all times, with the maximum difference between the two signals being just 1.2 degrees. These values are much less than those of other simulation scenarios, where the minimum value of θ_1 is equal to 1.19 degrees [4,7,8] .

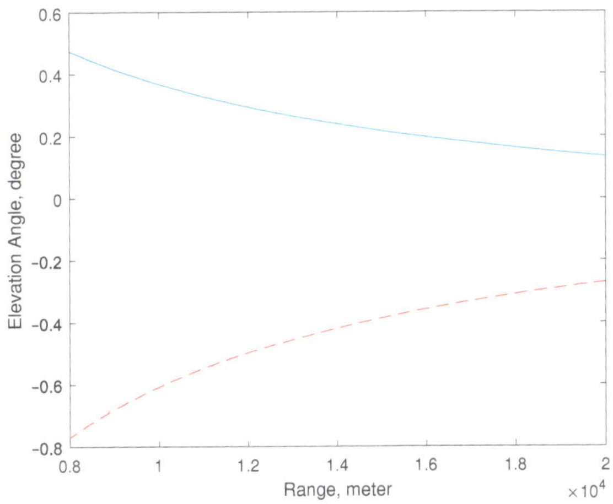

Figure 3. Elevation angles of the target and the reflected signal according to the ground range from the radar. The elevation angle is less than 0.5 degree and the difference between the two angles is very small.

Figure 4 depicts the power gain of the single-frequency signal and OFDM signal over the range in the same scenario. It represents the average power over antenna elements. Due to the reflection coefficients, the amplitude of the single-frequency signal can double at certain ranges, resulting in a power gain up to 6 dB. Conversely, at specific ranges, such as around 16km, the reflected signal can nearly cancel out the direct signal. As such, these multipath effects not only hinder accurate angle estimation, but can also obstruct the detection of the signal. However, an OFDM signal does not exhibit this power fluctuation, as the locations of high and low gain differ across frequencies. Consequently, OFDM, being the sum of multiple frequency signals, maintains a more stable power gain. This stability is another benefit of using multiple frequencies.

First, we applied Liu's algorithm in [7] with a single-frequency signal ($N = 1$). This method estimates the hybrid-oriented vector, which is the sum of the direct signal and reflected signals. Then, it applies the CS method to estimate the elevation angles. In the simulation, it is assumed that the range of elevation angle is defined as spanning from -3.0 degrees to 3.0 degrees. The dictionary matrix for the CS method consists of array response vectors from -3.0 degrees to 3.0 degrees with a 0.002-degree step, resulting in a 20 by 3001 matrix.

Figure 5 shows the result of Liu's algorithm. In Figure 5a, blue dots and red dots indicate the elevation angles of the target signal and the reflected signal, respectively. At each point, the estimator estimated two angles. The cross and triangles indicate the estimated values. Where Liu's algorithms could not resolve two angles, no mark is plotted. Figure 5b represents the plot of the probability of resolution versus range out of 100 trials. It shows that the probability is always less than one and, at some ranges, even zero. Due to the very close proximity of the elevation angles, this algorithm seems unable to accurately estimate the elevation angles.

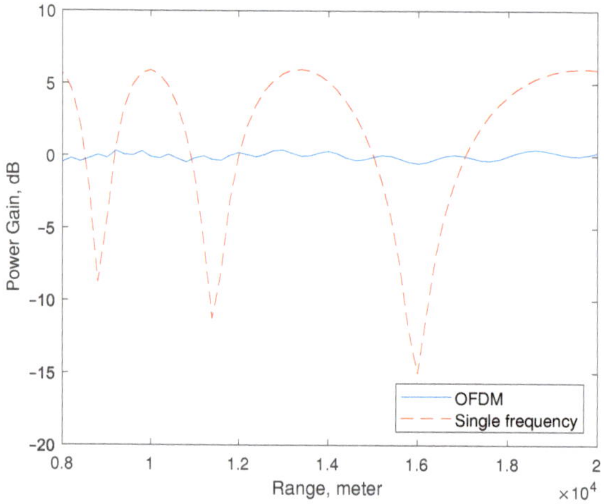

Figure 4. Power gain of the received signal. For a single-frequency signal, when the direct signal and the reflected signal are in phase, its magnitude becomes almost two, resulting in a 6 dB power gain. However, at some range, it becomes out of phase. Then, its power is significantly lower. On the other hand, the power of the OFDM signal is relatively stable. This is another advantage of using multi-frequency signals.

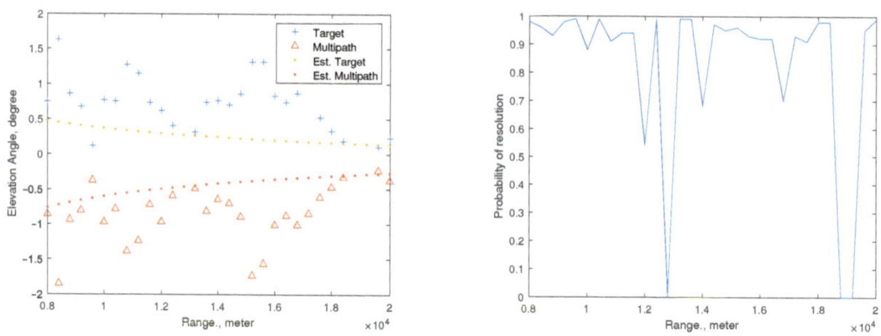

(a) (b)

Figure 5. Estimation results of Liu's algorithm [7] for SNR = 10 dB and 100 trials. (**a**) Estimation results; Dots indicate the target signal and the reflected signal. Triangles and crosses are estimated values. Where it could not resolve two values, no mark is plotted. (**b**) Probability resolution; It never reached 100% and at some range, it was 0%.

In order to resolve this problem, a MIMO-OFDM signal, which is a multi-frequency signal, is generated according to Equation (8). For the signal $x_{n,k}$, a set of complex Gaussian random numbers is used. To ensure consistent signal energy across all antennas, the signal power for each antenna element is normalized. White Gaussian noise was added to the received signal. For a fair comparison, the same SNR at a single antenna element is maintained for both MIMO-OFDM and single-frequency signals. As a result, the noise level for MIMO-OFDM signals is higher compared to single-frequency signals. The subcarrier frequency interval, Δf, is 100 MHz and the number of frequency bins, N, is 20. Therefore, the total signal bandwidth, ($B \approx (N-1)\Delta f$), is 1800 MHz. The bandwidth to the center frequency ratio is $1.8/4 = 0.45$. making it a wideband signal. However, since the elevation angle is very low, it is still valid to approximate it as a narrowband signal in terms of the array response vector. We used $N_p = 120$ and $\alpha = 1/240$ assuming that δR is less than half the wavelength of Δf. Figure 6 shows the value of $\tau_2 \Delta f$ for various height differences between the antenna and the target. Note that this value is less than 0.5, even at a distance as close as 5 km away, with a 160 m height difference. In the simulation scenario, its value is less than 0.25. If the value is over 0.5, it means that the target is much higher than the antenna and/or very close to the antenna. In those cases, the elevation angle is not that small and this proposed algorithm is no longer necessary. Table 1 summarizes the simulation scenario.

Table 1. Simulation Parameters.

Parameter	Symbol	Value
Radar height	h_r	20 m
Target height	h_t	90 m
Target range	R	8–20 km
Number of antennas	M	20
Number of subcarriers	N	20
Center frequency	f_0	3 GHz
Subcarrier frequency interval	Δf	100 MHz
Time-delay grid size	α	$1/240 \approx 0.0042$
Number of grid	N_p	120

The estimation results of BSBL-FM are presented in Figure 7. Unlike the single-frequency case, it shows that the estimates follow both elevation angles well. The root mean square error (RMSE) of the estimates versus the target range is shown in Figure 8. The solid lines represent the proposed algorithm, while the dotted lines represent Liu's algorithm. Across all SNR levels, the BSBL-FM algorithm outperforms Liu's algorithm except for the no-noise cases. However, note that the probability of resolution for Liu's algorithm is less than one, meaning that RMSEs are calculated only for the cases where successful resolution occurred. In contrast, the probability of resolution for BSBL-FM is always one. Even at a low SNR of 0 dB, the RMSE value of the BSBL-FM is as low as approximately 1 degree. When SNR exceeds 20 dB, the RMSE drops to less than 0.1 degrees. The fluctuations observed in the figure are due to the proximity of the true ground values to the grid points over the range. A smaller grid size α would reduce this error further. However, this would increase processing time. Therefore, it is essential to balance α and N_p, taking into account the RMSE value and the processing time. The computational complexity of BSBL-FM is $O(N^2 M N_p)$ per iteration [14]. The iteration continues until the computation result converges. The computational cost for Liu's algorithm is not analyzed in [7]. Since it consists of single eigenvalue decomposition and ℓ_1 optimization, its computational complexity would be $O(M^3) + 2KNL + 2K^2N + 2K(N+L) + K^3 = O(M^3) + O(KNL)$, where N is the number of observations, L is the number of elements in the dictionary, and $K(\ll L)$ is the sparsity [22]. If the parameters used in the simulation are applied, the computational complexity of BSBL-FM is $O(20^2 \times 2400)$, and that of Liu's algorithm is $O(20^3) + O(2^3 \cdot 20 \cdot 3001)$. Although the number of iterations cannot be determined, it shows that BSBL-FM is a more expensive processing method than Liu's algorithm since its processing cost per single iteration is already larger than that of Liu's algorithm. This is the cost to be paid for estimating very low flying targets.

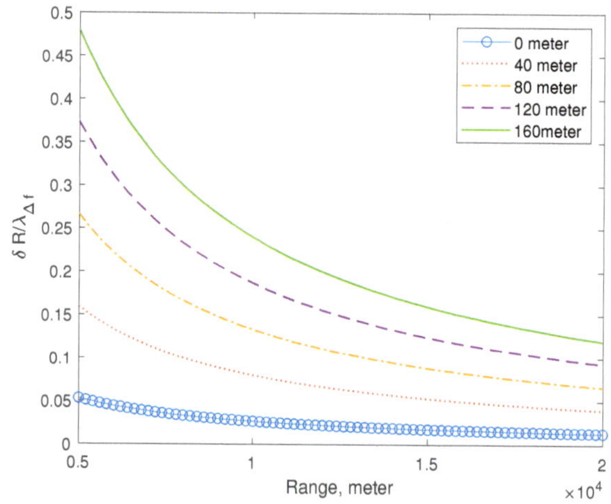

Figure 6. $\delta R/\lambda_{\Delta f}$ value for various height differences between the target and the antenna. When the difference is large and the target is close, the value is large. However, for the cases of low altitude and a far-range target, where the elevation angle is small, its value is much less than 0.5.

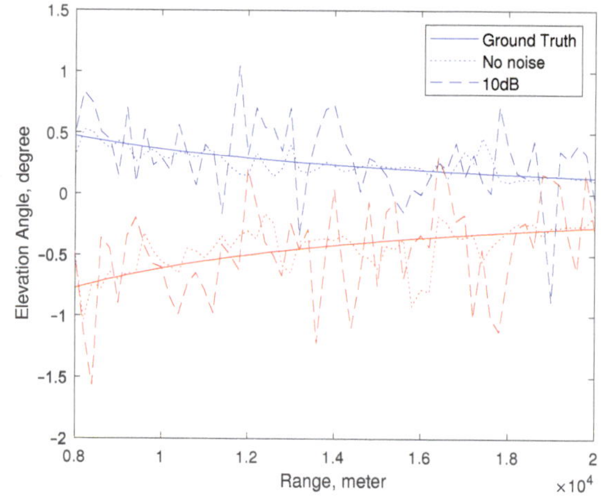

Figure 7. Elevation angles and their estimates obtained using the BSBL-FM algorithm with MIMO-OFDM signals. Solid lines represent the ground truth, dotted lines correspond to the no-noise case, and dashed lines indicate the 10 dB SNR case. Blue lines are for the target and red lines are for the reflected signal.

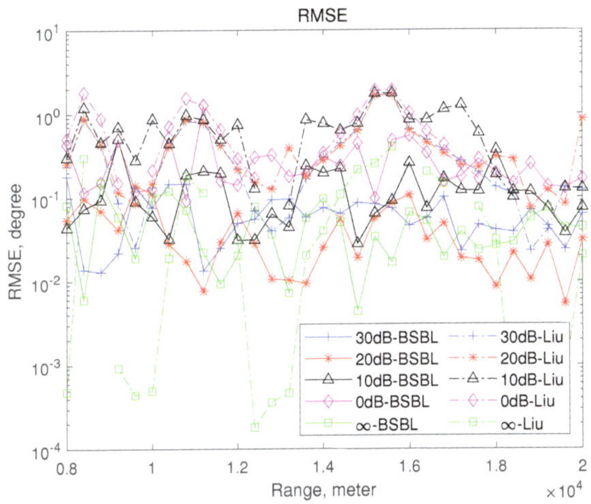

Figure 8. RMSE of elevation angle of the target for various SNRs (200 simulations per each SNR). The solid lines represent the proposed method and the dotted lines represent Liu's method.

6. Conclusions

In this paper, we introduced MIMO-OFDM signals with the BSBL-FM method to mitigate the challenges posed by multipath effects on low-altitude target signals, and discussed the validity of the proposed method in this context. Unlike previous methods, which struggle to estimate the elevation angles of both direct and reflected signals for targets at extremely low altitudes, the proposed approach effectively estimates both angles. Compared with single-frequency signals, MIMO-OFMD signals require more complex hardware, have a shorter detection range, and need a longer processing time. Although the processing time could be reduced by using smaller values for N_p and α, this would compromise the estimation performance. Despite these limitations, which may limit its widespread adoption, this method offers a viable solution for situations where slow-moving targets, such as drones, fly very low.

Funding: This work was supported by the faculty research fund of Sejong University in 2023 and grant-in-aid of Hanwha Systems Company.

Data Availability Statement: The raw data supporting the conclusions of this article will be made available by the authors upon request.

Conflicts of Interest: The author declares no conflicts of interest.

Abbreviations

The following abbreviations are used in this manuscript:

MIMO	Multi-input multi-output
OFDM	Orthogonal frequency division multiplexing
CS	Compressive sensing
BSBL-FM	Block sparse Bayesian learning fast marginalized

References

1. Schmidt, R. Multiple emitter location and signal parameter estimation. *IEEE Trans. Antennas Propag.* **1986**, *34*, 276–280. [CrossRef]
2. Guo, Z.; Wang, X.; Heng, W. Millimeter-wave channel estimation based on 2-D beamspace MUSIC method. *IEEE Trans. Wirel. Commun.* **2017**, *16*, 5384–5394. [CrossRef]
3. Pillai, S.; Kwon, B. Forward/backward spatial smoothing techniques for coherent signal identification. *IEEE Trans. Acoust. Speech Signal Process.* **1989**, *37*, 8–15. [CrossRef]

4. Wu, J.; Zhu, W.; Chen, B. Compressed sensing techniques for altitude estimation in multipath conditions. *IEEE Trans. Aerosp. Electron. Syst.* **2015**, *51*, 1891–1900. [CrossRef]
5. Donoho, D. Compressed Sensing. *IEEE Trans. Inf. Theory* **2006**, *52*, 1289–1306. [CrossRef]
6. Yoon, Y.S.; Amin, M. Compressed sensing technique for high-resolution radar imaging. In Proceedings of the SPIE Signal Processing, Sensor Fusion, and Target Recognition XVII, Orlando, FL, USA, 16–20 March 2008; Volume 6958.
7. Liu, Y.; Jiu, B.; Xia, X.G.; Liu, H.; Zhang, L. Height measurement of low-angle target using MIMO radar under multipath interference. *IEEE Trans. Aerosp. Electron. Syst.* **2017**, *54*, 808–818. [CrossRef]
8. Chen, S.; Zhao, Y.; Hu, Y. Beamspace phase solving algorithm for elevation angle estimation. *IEEE Signal Process. Lett.* **2022**, *29*, 742–746. [CrossRef]
9. Shi, J.; Hu, G.; Zong, B.; Chen, M. DOA estimation using multiple echo power for MIMO radar in low-grazing angle. *IEEE Sens. J.* **2016**, *16*, 6087–6094. [CrossRef]
10. Zhang, H.; Liu, W.; Fei, T.; Zhou, H.; Xie, J. An efficient radar-target assignment and power allocation strategy for low-angle tracking in the MIMO-multisite radar system. *Signal Process.* **2022**, *200*, 108645. [CrossRef]
11. Zhu, Y.; Zhao, Y.; Shui, P. Low-angle target tracking using frequency-agile refined maximum likelihood algorithm. *IET Radar Sonar Navig.* **2017**, *11*, 491–497. [CrossRef]
12. Wu, X.; Kishk, A.; Glisson, A. MIMO-OFDM radar for direction estimation. *IET Radar Sonar Navig.* **2010**, *4*, 28–36. [CrossRef]
13. Sen, S.; Nehorai, A. Adaptive OFDM radar for target detection in multipath scenarios. *IEEE Trans. Signal Process.* **2011**, *59*, 78–90. [CrossRef]
14. Wipf, D.; Rao, B. An empirical Bayesian strategy for solving the simultaneous sparse approximation problem. *IEEE Trans. Signal Process.* **2007**, *55*, 3704–3716. [CrossRef]
15. Liu, B.; Zhang, Z.; Xu, G.; Fan, H.; Fu, Q. Energy efficient telemonitoring of physiological signals via compressed sensing: A fast algorithm and power consumptionevaluation. *Biomed. Signal Process. Control.* **2014**, *11*, 80–88. [CrossRef]
16. Rahman, M.L.; Zhang, J.A.; Huang, X.; Guo, Y.J.; Heath, R.W. Framework for a perceptive mobile network using joint communication and radar sensing. *IEEE Trans. Aerosp. Electron. Syst.* **2020**, *56*, 1926–1941. [CrossRef]
17. Li, Y.; Wang, X.; Ding, Z. Multidimensional spectral super-resolution with prior knowledge with application to high mobility channel estimation. *IEEE J. Sel. Areas Commun.* **2020**, *38*, 2836–2852. [CrossRef]
18. Zhang, R.; Wu, X.; Lou, Y.; Yan, F.G.; Zhou, Z.; Wu, W.; Yuen, C. Channel training-aided target sensing for terahertz integrated sensing and massive MIMO communications. *IEEE Internet Things J.* 2024, *early access*. [CrossRef]
19. Barton, D. *Radar System Analysis and Modeling*; Artech House: Boston, MA, USA, 2005.
20. Fannjiang, A.; Liao, W. Coherence pattern–guided compressive sensing with unresolved grids. *SIAM J. Imaging Sci.* **2012**, *5*, 179–202. [CrossRef]
21. Candes, E.J.; Eldar, Y.C.; Needell, D.; Randall, P. Compressed sensing with coherent and redundant dictionaries. *Appl. Comput. Harmon. Anal.* **2011**, *31*, 59–73. [CrossRef]
22. Rubinstein, R.; Zibulevsky, M.; Elad, M. Efficient implementation of the K-SVD algorithm using batch orthogonal matching pursuit. *Cs Tech.* **2008**, *40*, 1–15.

MDPI AG
Grosspeteranlage 5
4052 Basel
Switzerland
Tel.: +41 61 683 77 34

Remote Sensing Editorial Office
E-mail: remotesensing@mdpi.com
www.mdpi.com/journal/remotesensing

www.ingramcontent.com/pod-product-compliance
Lightning Source LLC
Chambersburg PA
CBHW041418210926
51850CB00013B/3302
* 9 7 8 3 7 2 5 8 4 6 0 5 4 *